Prime numbers

William and Fern Ellison

Prime numbers

A Wiley-Interscience Publication

JOHN WILEY & SONS
New York . London . Sydney . Toronto

HERMANN
Publishers in arts and science Paris

Published by Hermann and in the USA by Wiley-Interscience,
a division of John Wiley & Sons, Inc., New York.

Les nombres premiers was originally published in French by Hermann in 1975.

ISBN 2 7056 5968 4 (Hermann)

ISBN 0 471 82653 7 (Wiley)

10 9 8 7 6 5 4 3 2 1

To the memory of Albert and Ari

Contents

Notations

We give here a list of notations and conventions which are not defined explicitly within the text.

1. Whenever we write a real number in decimal notation we shall always separate the integral and fractional parts by a comma.

2. If $s = \sigma + it$ where σ and t are real, then

$$\mathfrak{R}(s) = \sigma \quad \text{and} \quad \mathfrak{Im}(s) = t\,.$$

3. The function log will always denote logarithms to base e.

4. If $g(x)$ is defined and positive for all $x \geqslant 0$ and $f(x)$ is defined for all $x > x_0$, then

$$f(x) = \mathrm{O}(g(x)) \quad \text{means} \quad \limsup_{x \to \infty} \frac{|f(x)|}{g(x)} < \infty\,,$$

$$f(x) = \mathrm{o}(g(x)) \quad \text{means} \quad \lim_{x \to \infty} \frac{|f(x)|}{g(x)} = 0\,,$$

$$f(x) \sim g(x) \quad \text{means} \quad \lim_{x \to \infty} \frac{f(x)}{g(x)} = 1\,.$$

The above notation will also be used with the limit operation $x \to \infty$ replaced by $x \to c$. It will always be obvious which limit operation is involved.

5. An arithmetic function is a real or complex valued function defined on the positive integers.

6. We will always use p to denote a prime and n to denote an integer.

7. The symbols $\sum_{p \leqslant x}$ and $\prod_{p \leqslant x}$ denote a sum and a product taken over all primes satisfying $2 \leqslant p \leqslant x$ and the symbols $\sum_p$ and $\prod_p$ denote $\lim_{x \to \infty} \sum_{p \leqslant x}$ and $\lim_{x \to \infty} \sum_{p \leqslant x}$ respectively.

8. If m and n are positive integers, then

$m \mid n$ means m divides n,

$m \nmid n$ means m does not divide n,

$p^\alpha \parallel n$ means $p^\alpha \mid n$ and $p^{\alpha+1} \nmid n$.

9. If x is a real number, then

$[x]$ is the unique integer defined by $x - 1 < [x] \leqslant x$,

$\| x \|$ denotes the distance to the nearest integer, that is

$$\| x \| = \min_n | x - n | ,$$

$\{ x \} = x - [x]$, the fractional part of x.

CHAPTER 1

Tchebycheff 's theorems

§1 INTRODUCTION

1.1 A prime number p is an integer, strictly greater than 1, whose only divisors are 1 and p. The sequence of primes begins thus :

$$2, 3, 5, 7, 11, 13, 17, 19, \ldots, 2^{19937} - 1, \ldots, 2^{44497}$$

It has been known since the time of Euclid that the sequence of primes is infinite. The proof is simple. Suppose that the number of primes is finite and that P is the largest prime; consider the integer $1 + P\,!$, this integer has a prime factor $q > P$ and so we have a contradiction.

It is natural to wonder how the prime numbers are distributed among the integers. A reasonable question to ask is : " How many primes are there in the interval $[1, x]$? " To facilitate the discussion of this question we define, for $x \geqslant 0$, the function π by

$$\pi(x) = \sum_{p \leqslant x} 1 .$$

It is trivial that $\pi(x) = 0$ for $0 \leqslant x < 2$ and a consequence of Euclid's theorem that $\pi(x)$ tends to infinity with x. A considerable portion of this book will be devoted to the study of $\pi(x)$ and related functions to be defined later.

The purpose of this chapter is twofold. Firstly, we wish to introduce notation and basic techniques which are fundamental to the study of prime numbers. Secondly, we shall begin our serious study of $\pi(x)$ by finding its exact order of magnitude as x tends to infinity.

1.2 Towards the end of the 18th century Legendre, Gauss and others conjectured, in various ways, that $\pi(x) \sim x/\operatorname{Log} x$ as x tends to infinity. The conjecture was suggested by information obtained from tables of prime numbers. We now know that the conjecture is true; the result is known as the Prime Number Theorem. This theorem was proved independently by Hadamard and de la Vallée Poussin in 1896 and was the culmination of one hundred years effort on the part of numerous mathematicians. The search for a proof of the Prime Number Theorem provided the incentive for a considerable body of complex function theory. We shall study various aspects of the Prime Number Theorem in chapters 2, 3, 4, 6 and 11.

Tchebycheff was the first to *prove* a non-trivial result about $\pi(x)$ in 1849. He showed that *if* $\lim \pi(x) \log x/x$ exists, then its value must be 1. But he could not prove that the limit existed. In 1851 Tchebycheff found the true order of magnitude of $\pi(x)$ by proving that there exist positive real numbers A and B such that for all $x \geqslant 2$ we have

$$\frac{Ax}{\operatorname{Log} x} < \pi(x) < \frac{Bx}{\operatorname{Log} x}.$$

Tchebycheff's theorems, which represented an advance of great importance in the subject, were motivated by his desire to prove " Bertrand's Postulate ", which asserts that for all integers $n \geqslant 2$ there is at least one prime in the interval $[n, 2n]$. The proof of this " postulate " which was given by Tchebycheff depended on finding " good " explicit values for the numbers A and B. For more information on this topic see the *Notes* at the end of the chapter.

1.3 The contents of this chapter are necessarily a *pot-pourri* of ideas and their connection with prime number theory may not be immediately apparent. For the readers convenience we shall a fairly detailed outline of the structure of the chapter.

In § 2 we shall discuss an identity, due essentially to Euler, which relates a product over primes to a sum over the positive integers. This identity is the starting point of modern prime number theory. We define in § 3 the Riemann zeta function $\zeta(s)$, the study of which is essential to a deeper knowledge of the distribution of prime numbers. Associated with $\zeta(s)$ are certain arithmetic functions which are intimately related to $\pi(x)$. They are introduced in § 4 where we show how their behaviour at infinity gives us information about $\pi(x)$. Before proving the theorems of Tchebycheff in § 7 we study in § 5 a summation formula due to Abel and in § 6 the Möbius function. The Abel summation formula is a fundamental tool which will be used throughout the book and the topics discussed in § 6 will be used constantlv in chapter 3.

§ 2 EULER PRODUCTS

2.1 Multiplicative functions. Before we prove Euler's identity we need to define an important class of arithmetic functions. Let f be defined on $\mathbb{N}$ and taking values in $\mathbb{C}$. (Sometimes it will be convenient to consider f as a sequence of complex numbers $\{ f(n) \}$.) The function f is *multiplicative* if $(m, n) = 1$ implies that

$$f(mn) = f(m)\,f(n) \tag{1.1}$$

The equality $f(n) = f(1 . n) = f(1)\, f(n)$ shows that there are only two possible values for $f(1)$, namely $f(1) = 1$ or $f(1) = 0$. Il $f(1) = 0$, then f is identically zero; otherwise we have $f(1) = 1$ and

$$f(n) = \prod_{p^\alpha \| n} f(p^\alpha) . \tag{1.2}$$

In order to avoid trivial exceptional cases to theorems it is convenient to adopt the convention that a multiplicative function is not identically zero. Equation (1.2) shows that a multiplicative function is determined by its values at prime powers.

A function f is *completely multiplicative* if equation (1.1) is true for all pairs of positive integers (m, n). For completely multiplicative functions we have

$$f(n) = \prod_{p^\alpha \| n} (f(p))^\alpha ;$$

Thus a completely multiplicative function is determined by its values on the primes.

2.2 Euler's identity. The theorem which we are now going to prove is of fundamental importance in the theory of prime numbers.

Theorem 1.1 *Let f be a multiplicative function satisfying one of the following conditions :*

(i) $\sum_{n=1}^{\infty} | f(n) | < \infty$,

(ii) $\prod_p \left(1 + \sum_{\nu=1}^{\infty} | f(p^\nu) |\right) < \infty$.

Then the following equality holds :

$$\sum_{n=1}^{\infty} f(n) = \prod_p (1 + f(p) + f(p^2) + \cdots) .$$

Proof. Define the quantities S^* and P^* by

$$S^* = \sum_{n=1}^{\infty} | f(n) | \quad \text{and} \quad P^* = \prod_p \sum_{\nu=0}^{\infty} | f(p^\nu) | .$$

Thus we obviously have

$$1 \leqslant S^* \leqslant \infty \quad \text{and} \quad 1 \leqslant P^* \leqslant \infty .$$

Suppose first that condition (i) is satisfied, i.e. $S^* < \infty$. The series $\sum f(n)$ is absolutely convergent; denote its sum by S. For each prime p the series $\sum_{\nu=0}^{\infty} f(p^\nu)$

is convergent, its sum being majorised by S*. Hence, for each prime p the series $\sum_{\nu} f(p^{\nu})$ is absolutely convergent. Consequently the following two products are defined :

$$\mathrm{P}(x) = \prod_{p \leqslant x} \sum_{\nu=0}^{\infty} f(p^{\nu}), \qquad \mathrm{P}^*(x) = \prod_{p \leqslant x} \sum_{\nu=0}^{\infty} |f(p^{\nu})|$$

By virtue of the absolute convergence of the series in each of the products we can formally multiply and rearrange the terms of the series. In particular we can write P(x) as a sum in the following two ways :

$$\mathrm{P}(x) = \sum{}' f(n) = \mathrm{S} - \sum{}'' f(n)$$

where $\sum'$ denotes a summation over all integers n which have *no* prime factors greater than x and $\sum''$ denotes a summation extended over all integers which have *at least one* prime factor greater than x. It now follows that

$$|\mathrm{P}(x) - \mathrm{S}| \leqslant \sum{}'' |f(n)| \leqslant \sum_{n \geqslant x} |f(n)|.$$

from which we conclude that P(x) tends to S as x tends to infinity. The same reasoning also proves that P*(x) tends to S* as x tends to infinity.

Now suppose that condition (ii) is satisfied, namely $\mathrm{P}^* < \infty$. It follows that

$$\sum_{n \leqslant x} |f(n)| \leqslant \sum{}' |f(n)| = \mathrm{P}^*(x) \leqslant \mathrm{P}^* < \infty .$$

from which we conclude that $\mathrm{S}^* < \infty$, i.e. condition (i) is satisfied and the theorem follows. ■

It is easily seen that the two conditions in theorem 1.1 are equivalent. Our reason for including both conditions in the ennunciation of the theorem is as follows. When one applies the theorem to a specific function f one often encounters situations when it is trivially obvious that one of the two conditions is satisfied and it is not at all clear that the other condition is satisfied. By having the two conditions explicitly stated in the theorem it will enable us to refer to the obvious condition as our justification for assuming the equality.

The next two theorems are easy consequences of theorem 1.1 and represent important special cases. They will be used often in later chapters.

Theorem 1.2 *Let f be a completely multiplicative function such that for all primes p*

$$|f(p)| < 1 \quad \text{and} \quad \sum_{p} |f(p)| < \infty .$$

We then have

$$\sum_{n=1}^{\infty} f(n) = \prod_p (1 - f(p))^{-1} .$$

Proof. Because $| f(p) | < 1$ and f is completely multiplicative we see that

$$\sum_{\nu=0}^{\infty} f(p^\nu) = \sum_{\nu=0}^{\infty} f(p)^\nu = (1 - f(p))^{-1} .$$

The convergence of the series $\sum_p | f(p) |$ implies the convergence of the following series and products :

$$\prod_p (1 - | f(p) |), \quad \prod_p (1 - | f(p) |)^{-1}, \quad \prod_p \left(\sum_{\nu=0}^{\infty} | f(p^\nu) | \right) .$$

Thus, condition (ii) of theorem 1.1 is satisfied, so the theorem is proved. ■

Theorem 1.3 *If f is a real, non-negative multiplicative function, then*

$$\sum_{n=1}^{\infty} f(n) = \prod_p \left(\sum_{\nu=0}^{\infty} f(p^\nu) \right)$$

is always true, provided that we allow both sides to be infinite.

Proof. If one of the two numbers

$$\sum_{n=1}^{\infty} f(n) \text{ or } \prod_p \left(\sum_{\nu=0}^{\infty} f(p^\nu) \right)$$

is finite, then the result is an immediate consequence of theorem 1.1. On the other hand if one of these numbers is infinite, then so is the other and we again have equality. ■

§ 3 THE RIEMANN ZETA FUNCTION

3.1 As we shall soon see, there is an intimate connection between the distribution of prime numbers and the behaviour of the series $\sum_{n=1}^{\infty} n^{-s} = \zeta(s)$. In this section we shall formally define $\zeta(s)$ and discuss some of its more elementary properties.

Let $s = \sigma + it$ be a complex number satisfying $\Re(s) > 1$. The series

$$\sum_{n=1}^{\infty} n^{-s} \tag{1.3}$$

is absolutely convergent; we represent its sum by $\zeta(s)$. The function thus defined in the half plane $\mathfrak{R}(s) > 1$ is called the *Riemann zeta function*. Theorem 1.2 allows us to give an equivalent representation of $\zeta(s)$:

$$\zeta(s) = \prod_p (1 - p^{-s})^{-1} . \tag{1.4}$$

If we put $s = 1$ in (1.3) and use theorem 1.3, then

$$\prod_p \left(1 - \frac{1}{p}\right)^{-1} = + \infty .$$

and so we have proved the following result.

Theorem 1.4 *The product* $\prod_p (1 - p^{-1})$ *diverges to zero and the series* $\sum_p p^{-1}$ *diverges to infinity.*

The above theorem gives an alternative proof of the fact that there are an infinity of prime numbers. The representation (1.4), due essentially to Euler, represents the starting point of analytic number theory. All the modern developments stem from the detailed analysis of equation (1.4) and its consequences.

3.2 All the series which we are going to consider in this chapter will be absolutely convergent in the half plane $\mathfrak{R}(s) > 1$ and uniformly convergent in the half plane $\mathfrak{R}(s) \geqslant 1 + \delta$, for any fixed $\delta > 0$. Consequently all formal analytic operations can be easily justified and we shall omit such justifications from our proofs. By the symbol " Log " we shall always mean the principal value of the logarithmic function, namely the branch of the function which is real on the real axis.

3.3 Equation (1.4) shows that $\zeta(s)$ has no zeros in the half plane $\mathfrak{R}(s) > 1$. Thus, in this half plane we can write

$$\operatorname{Log} \zeta(s) = - \sum_p \operatorname{Log} (1 - p^{-s}) .$$

By differentiation and rearràngement of terms we obtain

$$\begin{aligned} \frac{\zeta'(s)}{\zeta(s)} &= - \sum_p p^{-s} . \operatorname{Log} p . (1 - p^{-s})^{-1} \\ &= - \sum_p p^{-s} \operatorname{Log} p \sum_{v=0}^{\infty} p^{-vs} . \end{aligned}$$

If we define the function $\Lambda(n)$ by

$$\Lambda(n) = \begin{cases} \operatorname{Log} p & \text{if } n = p^v \\ 0 & \text{if } n \neq p^v , \end{cases}$$

and the function $Z(s)$ by

$$Z(s) = \frac{-\zeta'(s)}{\zeta(s)},$$

then we can write the last double sum as

$$Z(s) = \sum_{n=1}^{\infty} \frac{\Lambda(n)}{n^s}.$$

As we shall see later, the function $Z(s)$ will be very important to our study of prime numbers.

§ 4 THE FUNCTIONS π, Π, θ, ψ

4.1 The function $\pi(x)$ has already been introduced in § 1. For analytic reasons it is necessary to introduce the following functions, which are closely related to $\pi(x)$. First of all we define $\Pi(x)$ by

$$\Pi(x) = \sum_{p^\nu \leqslant x} \frac{1}{\nu},$$

the summation being over *all* prime powers less than or equal to x. It is clear that

$$\Pi(x) = \pi(x) + \frac{1}{2}\pi(x^{1/2}) + \frac{1}{3}\pi(x^{1/3}) + \cdots. \tag{1.5}$$

and it follows from the definition of $\Lambda(n)$ that

$$\Pi(x) = \sum_{n \leqslant x} \frac{\Lambda(n)}{\operatorname{Log} n}.$$

The function $\theta(x)$ is defined by

$$\theta(x) = \sum_{p \leqslant x} \operatorname{Log} p$$

and finally the function $\psi(x)$ is

$$\psi(x) = \sum_{n \leqslant x} \Lambda(n).$$

The following relations are easily verified :

$$\begin{aligned} \psi(x) &= \sum_{p^\nu \leqslant x} \operatorname{Log} p \\ &= \sum_{p \leqslant x} \left[\frac{\operatorname{Log} x}{\operatorname{Log} p}\right] \operatorname{Log} p && (1.4) \\ &= \theta(x) + \theta(x^{1/2}) + \theta(x^{1/3}) + \cdots. && (1.5) \end{aligned}$$

Readers should resign themselves to the fact that, from the analytic point of view, the fundamental functions to be studied are $\psi(x)$ and $\Pi(x)$ rather than the superficially more natural function $\pi(x)$. The reason why this is so can be appreciated once one knows that the behaviour of the function

$$f(s) = \sum_{n=1}^{\infty} \frac{a_n}{n^s}$$

in the complex plane and the asymptotic behaviour of the summatory function

$$A(x) = \sum_{n \leqslant x} a_n .$$

are very closely related. In chapter 2 we shall see that the functions associated with $\psi(x)$ and $\Pi(x)$ are " well behaved " analytic functions. However the function associated with $\pi(x)$ is

$$\mathrm{P}(s) = \sum_p p^{-s}$$

and it can be shown that this function has the line $\mathfrak{R}(s) = 0$ as a natural boundary. Thus $\mathrm{P}(s)$ is a complicated function and the discussion of relationships between $\pi(x)$ and $\mathrm{P}(s)$ tends to be very delicate.

4.2 We now prove an identity which will be used several times in this and later chapters. The identity in question is

$$\sum_{n \leqslant x} \psi\left(\frac{x}{n}\right) = \operatorname{Log}([x]\,!) .$$

One can easily verify the following equations :

$$\begin{aligned} \sum_{n \leqslant x} \psi\left(\frac{x}{n}\right) &= \sum_{n \leqslant x} \sum_{m \leqslant x/n} \Lambda(m) \\ &= \sum_{m \leqslant x} \Lambda(m) \sum_{n \leqslant x/m} 1 \\ &= \sum_{m \leqslant x} \left[\frac{x}{m}\right] \Lambda(m) \\ &= \sum_{p^\nu \leqslant x} \left[\frac{x}{p^\nu}\right] \operatorname{Log} p . \end{aligned}$$

If we now write

$$i_p(x) = \left[\frac{x}{p}\right] + \left[\frac{x}{p^2}\right] + \cdots,$$

then we have

$$\sum_{n \leqslant x} \psi\left(\frac{x}{n}\right) = \sum_{p \leqslant x} i_p(x) \operatorname{Log} p .$$

The identity (1.6) now follows upon recalling the elementary fact that

$$[x]! = \prod_{p \leqslant x} p^{i_p(x)} .$$

4.3 The next theorem relates the asymptotic behaviour of the functions $\pi(x)$, $\psi(x)$, $\theta(x)$ and $\Pi(x)$. It is the simplest theorem of its type, more precise results will be given in chapters 2 and 4.

Theorem 1.5 *The following relations hold :*

$$\liminf_{x\to\infty} \frac{\pi(x) \operatorname{Log} x}{x} = \liminf_{x\to\infty} \frac{\Pi(x) \operatorname{Log} x}{x} = \liminf_{x\to\infty} \frac{\theta(x)}{x} = \liminf_{x\to\infty} \frac{\psi(x)}{x}$$

and

$$\limsup_{x\to\infty} \frac{\pi(x) \operatorname{Log} x}{x} = \limsup_{x\to\infty} \frac{\Pi(x) \operatorname{Log} x}{x} = \limsup_{x\to\infty} \frac{\theta(x)}{x} = \limsup_{x\to\infty} \frac{\psi(x)}{x} .$$

Proof. Let G_1, G_2, G_3, G_4 denote respectively the four quantities

$$\limsup_{x\to\infty} \frac{\pi(x) \operatorname{Log} x}{x}, \quad \limsup_{x\to\infty} \frac{\Pi(x) \operatorname{Log} x}{x}, \quad \limsup_{x\to\infty} \frac{\theta(x)}{x}, \quad \limsup_{x\to\infty} \frac{\psi(x)}{x},$$

and g_1, g_2, g_3, g_4 the corresponding lower limits. From (1.5) it follows that

$$0 \leqslant \Pi(x) - \pi(x) \leqslant \frac{1}{2}\pi(x^{1/2}) \left[\frac{\operatorname{Log} x}{\operatorname{Log} 2}\right]$$

and hence

$$0 \leqslant \Pi(x) - \pi(x) = O(x^{1/2} \operatorname{Log} x) .$$

Consequently there exists a constant $A > 0$ such that

$$0 \leqslant \frac{\Pi(x) \operatorname{Log} x}{x} - \frac{\Pi(x) \operatorname{Log} x}{x} \leqslant \frac{A \operatorname{Log}^2 x}{\sqrt{x}}$$

and we can conclude that $G_1 = G_2$, $g_1 = g_2$.

From the definition of the functions θ, ψ and π it is clear that for all x

$$\theta(x) \leqslant \psi(x) \leqslant \pi(x) \operatorname{Log} x$$

which immediately implies

$$G_3 \leqslant G_4 \leqslant G_1 \qquad \text{and} \qquad g_3 \leqslant g_4 \leqslant g_1 .$$

We are now going to prove that $G_3 \geqslant G_2$ and $g_3 \geqslant g_2$. These inequalities, when combined with our preliminary observations, prove the theorem.

Let $\xi = \xi(x)$ be a function of x, to be chosen explicitly later, satisfying $1 < \xi < x$. The following inequalities are then trivially satisfied :

$$\theta(x) \geqslant \sum_{\xi < p \leqslant x} \operatorname{Log} p \geqslant \{ \pi(x) - \pi(\xi) \} \operatorname{Log} \xi$$

Hence we certainly have

$$\frac{\theta(x)}{x} \geqslant \frac{\pi(x) \operatorname{Log} \xi}{x} - \frac{\pi(\xi) \operatorname{Log} \xi}{x}$$
$$\geqslant \frac{\pi(x) \operatorname{Log} \xi}{x} - \frac{\xi \operatorname{Log} \xi}{x} .$$

If ξ is chosen so that when x tends to infinity

$$\frac{\operatorname{Log} \xi}{\operatorname{Log} x} \to 1 \qquad \text{et} \qquad \frac{\xi \operatorname{Log} \xi}{x} \to 0 ,$$

then we can deduce that $G_3 \geqslant G_2$ and $g_3 \geqslant g_2$. Such a choice of ξ is possible, for example

$$\xi = \xi(x) = x(\operatorname{Log} x)^{-2} .$$

■

4.4 If we denote the common value of G_1, G_2, G_3, G_4 by G and that of g_1, g_2, g_3, g_4 by g, then the Prime Number Theorem, which asserts that $\pi(x) \sim x/\operatorname{Log} x$, is equivalent to the assertion that $g = G = 1$. This result will be proved in chapter 2. Our ambition in this chapter is more modest. We shall prove in § 7 two theorems due to Tchebycheff which imply that

$$0 < g \leqslant 1 \leqslant G < \infty .$$

§ 5 THE ABEL SUMMATION FORMULA

5.1 The following theorem is a very frequently used tool in analytic number theory. It allows one to replace a sum by an integral and integrals are often much more manageable than sums.

Theorem 1.6 *Let $\{ a_n \}$ be a sequence of complex numbers and let $\mathcal{C}$ be defined for $x \geqslant 1$ and possess a continuous derivative for $x > 1$. If we write*

$$A(x) = \sum_{n \leqslant x} a_n .$$

then for all $x > 1$ we have

$$\sum_{n \leqslant x} a_n \varphi(n) = A(x) \varphi(x) - \int_1^x A(u) \varphi'(u) \, du .$$

Proof. One easily verifies the following equalities :

$$\sum_{n \leqslant x} a_n \{ \varphi(x) - \varphi(n) \} = \sum_{n \leqslant x} a_n \int_n^x \varphi'(u) \, du$$

$$= \int_1^x \sum_{n \leqslant u} a_n \varphi'(u) \, du$$

$$= \int_1^x A(u) \varphi'(u) \, du ,$$

thus

$$\sum_{n \leqslant x} a_n \varphi(n) = A(x) \varphi(x) - \int_1^x A(u) \varphi'(u) \, du . \qquad \blacksquare$$

An immediate and important consequence of theorem 1.6 is the following result.

Theorem 1.7 *With the notation of theorem 1.6 and the additional hypothesis that $A(x) \varphi(x)$ tends to zero as x tends to infinity we have*

$$\sum_{n=1}^{\infty} a_n \varphi(n) = - \int_1^{\infty} A(u) \varphi'(u) \, du$$

provided at least one of the two sides exists.

The basic technique behind the proof of theorem 1.6 has many applications in analysis and number theory. For example, the same idea is used in the proof of Abel's test for the convergence of infinite series. We shall now give a slight variation of the technique and prove a theorem which will be used often in later chapters.

Theorem 1.8 *If $\{ a_n \}$ is a monotone sequence of positive real numbers and $\{ b_n \}$ a sequence of complex numbers, then for any integers $1 \leqslant M < N$*

$$\left| \sum_{n=M+1}^{N} a_n b_n \right| \leqslant 2 \max \{ a_{M+1}, a_N \} . \max_{M \leqslant n \leqslant N} \left| \sum_{m=M}^{n} b_m \right| .$$

Proof. Write $s_n = \sum_{m=M}^{n} b_m$, then $b_n = s_n - s_{n-1}$. Thus we have

$$\sum_{n=M+1}^{N} a_n b_n = \sum_{n=M+1}^{N} a_n (s_n - s_{n-1})$$

$$= - a_{M+1} s_M + \sum_{n=M+1}^{N-1} (a_n - a_{n+1}) s_n + a_N s_N ,$$

and consequently

$$\left|\sum_{n=M+1}^{N} a_n b_n\right| \leqslant \left\{|a_{M+1}| + \sum_{n=M+1}^{N-1} |a_n - a_{n+1}| + |a_N|\right\} \max_{M \leqslant n \leqslant N} |s_n| .$$

since the sequence $\{a_n\}$ is monotonic and positive the above expression simplifies and is equal to

$$2 \max\{a_{M+1}, a_N\} \max_{M \leqslant n \leqslant N} |s_n| .$$ ■

5.2 Some applications. We now illustrate the use of the Abel summation formula with three examples. Each of the results which we obtain will be used repeatedly in later chapters.

(*a*) Take $a_n = 1$ and $\varphi(n) = \operatorname{Log} n$ in theorem 1.6 and obtain

$$\begin{aligned}\operatorname{Log}([x]!) &= \sum_{n \leqslant x} \operatorname{Log} n \\ &= [x] \operatorname{Log} x - \int_1^x \frac{[u]}{u} du .\end{aligned}$$

It is trivial that

$$\int_1^x \frac{[u]}{u} du = \int_1^x \frac{u - \{u\}}{u} du = x + \mathrm{O}(\operatorname{Log} x) ,$$

and now we have proved that

$$\operatorname{Log}([x]!) = x \operatorname{Log} x - x + \mathrm{O}(\operatorname{Log} x) .$$

(*b*) For $x \geqslant 1$ we define $\mathrm{U}(x)$ by

$$\mathrm{U}(x) = \sum_{n \leqslant x} \frac{1}{n} .$$

Upon choosing $a_n = 1$ and $\varphi(x) = 1/x$ in theorem 1.6 we have

$$\begin{aligned}\mathrm{U}(x) &= \frac{[x]}{x} + \int_1^x \frac{[u]}{u^2} du \\ &= 1 - \frac{\{x\}}{x} + \operatorname{Log} x - \int_1^x \frac{\{u\}}{u^2} du .\end{aligned}$$

If we define the real number γ by

$$\gamma = 1 - \int_1^\infty \frac{\{u\}}{u^2} du = 0.57721\ldots ,$$

then

$$\mathrm{U}(x) = \operatorname{Log} x + \gamma - \frac{\{x\}}{x} + \int_x^\infty \frac{\{u\}}{u^2}\, du$$

$$= \operatorname{Log} x + \gamma + \mathrm{O}\left(\frac{1}{x}\right).$$

The number γ is known as Euler's constant, it is conjectured to be a transcendental number, but one does not even know whether or not it is rational !

(*c*) For $x \geqslant 1$ we write $V(x)$ for

$$V(x) = \sum_{n \leqslant x} \frac{1}{n} \operatorname{Log} \frac{x}{n}.$$

By taking $a_n = n^{-1}$ and $\varphi(u) = \operatorname{Log}(x/u)$ in theorem 1.6 one easily finds that

$$V(x) = \frac{1}{2}(\operatorname{Log} x)^2 + \gamma \operatorname{Log} x + \delta + \mathrm{O}\left(\frac{1}{x}\right),$$

where γ is Euler's constant and δ is the real number defined by

$$\delta = 1 - \gamma + \int_1^\infty \frac{du}{u} \int_u^\infty \frac{\{v\}}{v^2}\, dv = 2.723 \ldots$$

§ 6 MÖBIUS PAIRS

6.1 A situation which arises frequently in prime number theory is as follows. Two real valued functions F and f are related by

$$F(x) = \sum_{n \leqslant x} f\left(\frac{x}{n}\right). \tag{1.13}$$

For example, we have already encountered the special cases :

(i) $f(x) \equiv 1$, $\quad F(x) \equiv [x]$

(ii) $f(x) \equiv x$, $\quad F(x) \equiv x\mathrm{U}(x)$

(iii) $f(x) = \psi(x)$, $\quad F(x) \equiv \operatorname{Log}([x]\,!)$.

Other examples will occur in later sections of the book.

In such situations one usually possesses information about the asymptotic behaviour of F and one would like to deduce information about $f(x)$. For example, in (iii) we showed that $F(x) = x \operatorname{Log} x - x + \mathrm{O}(\operatorname{Log} x)$ and we would like to be able to conclude that $\psi(x) \sim x$ or at the very least $\psi(x) = \mathrm{O}(x)$. The main result of this section, theorem 1.10 will enable us to deduce some information about $f(x)$ from knowledge about the rate of growth of $F(x)$.

To prove theorem 1.10 we shall "invert" the relationship between f and F and express f as a function of F. To do this we need to introduce an important arithmetic function, known as the Möbius μ function, and to discuss several of its formal properties.

6.2 The Möbius μ function is defined on $\mathbb{N}$ as follows :

$$\mu(1) = 1\,,$$
$$\mu(n) = (-1)^r \quad \text{if } n \text{ is product of } r \text{ distinct prime numbers, otherwise}\,.$$
$$\mu(n) = 0$$

It is quite easy to verify that μ is a multiplicative function, but not completely multiplicative. The Möbius function arises quite naturally from the Dirichlet series representation for $\zeta^{-1}(s)$. For if $\mathcal{R}(s) > 1$, then from (1.4), we have

$$\zeta^{-1}(s) = \prod_p (1 - p^{-s}) = \sum_{n=1}^{\infty} \frac{\mu(n)}{n^s}\,.$$

We now prove the formal properties of the Möbius function which will be needed for the proof of theorem 1.10 and in chapter 3.

Theorem 1.9 (a) *For all integers $n \geqslant 1$ we have*

$$\sum_{k|n} \mu(k) = \begin{cases} 1 & \text{if} \quad n = 1 \\ 0 & \text{if} \quad n > 1\,. \end{cases}$$

(b) *If the functions f and F are related by*

$$F(x) = \sum_{n \leqslant x} f\left(\frac{x}{n}\right),$$

then

$$f(x) = \sum_{n \leqslant x} \mu(n)\, F\left(\frac{x}{n}\right).$$

Proof. (*a*) The result is trivial when $n = 1$, so we shall suppose that $n > 1$. Let $m(n) = \prod_{p|n} p$ and observe that

$$\sum_{k\,|\,n} \mu(k) = \sum_{k\,|\,m(n)} \mu(k)\,.$$

If r denotes the number of prime factors of $m(n)$, then trivially

$$\sum_{k\,|\,m} \mu(k) = 1 + C_r^1(-1) + C_r^2(-1)^2 + \cdots + C_r^r(-1)^r$$
$$= (1-1)^r = 0\,.$$

and so (*a*) is proved.

(*b*) Let us write $\delta(n) = \sum_{k \mid n} \mu(k)$, then from (*a*) we know that $\delta(1) = 1$ and $\delta(n) = 0$ if $n > 1$. Thus we have

$$\begin{aligned} f(x) &= \sum_{n \leqslant x} \delta(n) f\left(\frac{x}{n}\right) = \sum_{n \leqslant x} f\left(\frac{x}{n}\right) \sum_{k \mid n} \mu(k) \\ &= \sum_{kl \leqslant x} \mu(k) f\left(\frac{x}{kl}\right) = \sum_{k \leqslant x} \mu(k) \sum_{l \leqslant x/k} f\left(\frac{x}{kl}\right) \\ &= \sum_{k \leqslant x} \mu(k) F\left(\frac{x}{k}\right). \end{aligned}$$ ∎

When $f(x) = 0$ for non-integral values of x, then part (*b*) can be formulated as :

If

$$F(n) = \sum_{m|n} f(m)$$

then

$$f(n) = \sum_{m|n} \mu(m) F\left(\frac{n}{m}\right)$$

This statement is usually called the *Möbius inversion formula.*

6.3 We are now in a position to prove the main result of this section. First we shall make a definition. If a pair of functions f and F are related by

$$F(x) = \sum_{n \leqslant x} f\left(\frac{x}{n}\right),$$

then we say that (f, F) is a *Möbius pair of functions.*

Theorem 1.10 *Let (f, F) be a Möbius pair of functions. Suppose that as x tends to infinity*

$$F(x) = Ax(\operatorname{Log} x)^2 + Bx \operatorname{Log} x + Cx + O(x^{\beta})$$

where A, B, C, β are real constants and $0 \leqslant \beta < 1$. Then, as x tends to infinity

$$f(x) = 2\,Ax \operatorname{Log} x + O(x)\,.$$

Proof. First we shall consider the special case $A = B = C = 0$. Under these conditions there exists $H > 0$ such that for all $x \geqslant 1$ we have

$$|\,F(x)\,| \leqslant Hx^{\beta}\,.$$

Consequently it trivially follows that

$$\begin{aligned} | f(x) | &= \left| \sum_{n \leqslant x} \mu(n) F\left(\frac{x}{n}\right) \right| \\ &\leqslant \sum_{n \leqslant x} H\left(\frac{x}{n}\right)^{\beta} \\ &= Hx^{\beta} \sum_{n \leqslant x} n^{-\beta} \\ &\leqslant Hx^{\beta}\left(1 + \int_1^x u^{-\beta}\, du\right) \\ &= \frac{Hx}{1-\beta} - \frac{H\beta x^{\beta}}{1-\beta} = \mathrm{O}(x)\,. \end{aligned}$$

We now reduce the general case to the special case considered above. Define the function f_0 by

$$f_0(x) = f(x) - ax \operatorname{Log} x - bx - c\,,$$

where a, b, c are constants to be chosen explicitly later. If F_0 is the other half of the Möbius pair (f_0, F_0), then

$$F_0(x) = F(x) - a \sum_{n \leqslant x} \frac{x}{n} \operatorname{Log} \frac{x}{n} - b \sum_{n \leqslant x} \frac{x}{n} - c \sum_{n \leqslant x} 1\,,$$

From the hypothesis on $F(x)$ and the formulae proved in § 5.2 we have

$$F_0(x) = \left(A - \frac{1}{2}a\right) x(\operatorname{Log} x)^2 + \\ + (B - b - a\gamma)\, x \operatorname{Log} x + (C - c - a\delta - b\gamma)\, x + \mathrm{O}(x^{\beta})\,.$$

Thus, if we choose

$$a = 2A, \qquad b = B - a\gamma, \qquad c = C - a\delta - b\gamma$$

then $F_0(x) = \mathrm{O}(x^{\beta})$ and by the special case we conclude that $f_0(x) = \mathrm{O}(x)$, i.e.

$$f(x) = 2Ax \operatorname{Log} x + \mathrm{O}(x)\,.$$

■

It is possible to deduce a stronger conclusion from the same hypotheses, but to do this one needs to know that $\sum_{n \leqslant x} \mu(n) = \mathrm{o}(x)$ as x tends to infinity. This will be proved in chapter 3 as will the refinement of theorem 1.10. If the reader so desires it is perfectly possible to proceed directly to chapter 3 and to study the stronger version of theorem 1.10, from which the Prime Number Theorem is a trivial consequence.

§ 7 TCHEBYCHEFF'S THEOREMS

7.1 Recall the notation of § 4.4. We are going to prove two theorems of Tchebycheff, the first of which asserts that $g \leqslant 1 \leqslant G$ and the second that g^{-1} and G are both finite. Our proof of the first fact will be a simple consequence of theorem 1.11, which is a general result relating a weighted mean of a real function $t(x)$ to the asymptotic behaviour of $t(x)$. The second Tchebycheff's theorem will be a consequence of theorem 1.10.

Many different proofs of Tchebycheff's theorems are known. For example a study of $\zeta(s)$ near $s=1$ quickly leads to a proof that $g \leqslant 1 \leqslant G$ (see exercises 1.18 and 1.19), and an analysis of binomial coefficients can be used to obtain explicit upper bounds for g^{-1} and G (see exercises 1.13 and 1.14). The reason for our particular approach is explained in the *Notes* to this chapter, when we discuss "generalised prime numbers". In these more general systems the analogues of Tchebycheff's theorems are true, but proofs, *via* binomial coefficients are no longer possible. However, our proofs will only require trivial modifications to be still valid.

7.2 Theorem 1.11 will be stated in terms of the following notation. Let t be a real function defined for $x \geqslant 1$ and $\{ a_n \}$ a sequence of non-negative real numbers, not all of which are zero. Suppose that the function A defined by

$$A(x) = \sum_{n \leqslant x} a_n$$

is "slowly increasing", that is, for each fixed $\theta > 0$

$$\lim_{x \to \infty} \frac{A(\theta x)}{A(x)} = 1 .$$

Theorem 1.11 *With the above notation we have*

$$\liminf_{x \to \infty} t(x) \leqslant \liminf_{x \to \infty} \frac{1}{A(x)} \sum_{n \leqslant x} a_n \, t\left(\frac{x}{n}\right) \leqslant$$

$$\leqslant \limsup_{x \to \infty} \frac{1}{A(x)} \sum_{n \leqslant x} a_n \, t\left(\frac{x}{n}\right) \leqslant \limsup_{x \to \infty} t(x) .$$

Proof. We first prove the inequalities relating the lim sup. If lim sup $t(x) = \infty$ there is nothing more to prove, so from now on we suppose that it is finite. The proof will be by contradiction. If

$$\limsup_{x \to \infty} t(x) < \limsup_{x \to \infty} \frac{1}{A(x)} \sum_{n \leqslant x} a_n \, t\left(\frac{x}{n}\right),$$

then there exists a real number B satisfying

$$\limsup_{x\to\infty} t(x) < B < \limsup_{x\to\infty} \frac{1}{A(x)} \sum_{n\leqslant x} a_n\, t\left(\frac{x}{n}\right).$$

Thus there is a number $\xi = \xi(B) > 1$ such that

$$u \geqslant \xi \quad \text{implies} \quad t(u) < B\,.$$

It $H = H(\xi)$ is an upper bound for t in the interval $[1, \xi]$, then we have

$$\begin{aligned}\sum_{n\leqslant x} a_n\, t\left(\frac{x}{n}\right) &= \sum_{n\leqslant x/\xi} a_n\, t\left(\frac{x}{n}\right) + \sum_{x/\xi<n\leqslant x} a_n\, t\left(\frac{x}{n}\right)\\ &\leqslant B \sum_{n\leqslant x/\xi} a_n + H \sum_{x/\xi<n\leqslant x} a_n\\ &= BA\left(\frac{x}{\xi}\right) + H\left(A(x) - A\left(\frac{x}{\xi}\right)\right).\end{aligned}$$

Hence for $x > \xi$ we have

$$\frac{1}{A(x)} \sum_{n\leqslant x} a_n\, t\left(\frac{x}{n}\right) \leqslant \frac{BA(x\xi^{-1})}{A(x)} + H\left(1 - \frac{A(x\xi^{-1})}{A(x)}\right)$$

and using the fact that A is "*slowly increasing*" *we conclude that*

$$\limsup_{x\to\infty} \frac{1}{A(x)} \sum_{n\leqslant x} a_n\, t\left(\frac{x}{n}\right) \leqslant B\,,$$

which is a contradiction. This proves the inequalities between the lim sups. Exactly the same reasoning applied to $-t$ proves the remaining inequalities. ∎

The first of Tchebycheff's theorems is, as we have already mentioned, a simple corollary of the preceding result.

Theorem 1.12 *The following inequalities hold* :

$$g = \liminf_{x\to\infty} \frac{\psi(x)}{x} \leqslant 1 \leqslant \limsup_{x\to\infty} \frac{\psi(x)}{x} = G\,.$$

Proof. We apply theorem 1.11 with $a_n = 1/n$ and

$$t(x) = \frac{\psi(x)}{x} = \frac{1}{x} \sum_{n\leqslant x} \Lambda(n)\,.$$

It is clear that $A(x) = \sum_{n \leqslant x} n^{-1}$ satisfies the condition of being slowly increasing, since $A(x) \sim \operatorname{Log} x$. From the definition of t and (1.6)

$$\sum_{n \leqslant x} \frac{1}{n} t\left(\frac{x}{n}\right) = \frac{1}{x} \sum_{n \leqslant x} \psi\left(\frac{x}{n}\right) = \frac{1}{x} \operatorname{Log}([x]\,!)$$

and the asymptotic formula for Log ($[x]$!) proved in § 5.2 implies that

$$\sum_{n \leqslant x} \frac{1}{n} t\left(\frac{x}{n}\right) \sim \operatorname{Log} x,$$

hence we have

$$\lim_{x \to \infty} \frac{1}{A(x)} \sum_{n \leqslant x} \frac{1}{n} t\left(\frac{x}{n}\right) = 1$$

and the theorem is proved. ■

Theorem 1.13 *The following inequalities hold* :

$$0 < \liminf_{x \to \infty} \frac{\psi(x)}{x} \leqslant \limsup_{x \to \infty} \frac{\psi(x)}{x} < \infty .$$

Proof. From § 4.2 we see that ($\psi(x)$, Log ($[x]$!)) is a Möbius pair of functions and recalling that

$$\begin{aligned}\operatorname{Log}([x]\,!) &= x \operatorname{Log} x - x + O(\operatorname{Log} x) \\ &= x \operatorname{Log} x - x + O(x^{\beta}),\end{aligned}$$

for any fixed $\beta > 0$ we appeal to theorem 1.10 to conclude that $\psi(x) = O(x)$. Thus, lim sup $\psi(x)/x$ is finite.

In order to prove that lim inf $\psi(x)/x$ is non-zero we need the following asymptotic formula :

$$\sum_{n \leqslant x} \frac{\Lambda(n)}{n} = \operatorname{Log} x + O(1) \tag{1.7}$$

The proof is simple, we write Log ($[x]$!) as

$$\begin{aligned}\operatorname{Log}([x]\,!) &= \sum_{n \leqslant x} \Lambda(n)\left[\frac{x}{n}\right] \\ &= x \sum_{n \leqslant x} \frac{\Lambda(n)}{n} + O\left(\sum_{n \leqslant x} \Lambda(n)\right) \\ &= x \sum_{n \leqslant x} \frac{\Lambda(n)}{n} + O(\psi(x)) .\end{aligned}$$

whence the result, since $\psi(x) = O(x)$ and Log ($[x]$!) $\sim x \operatorname{Log} x$.

Let $\xi = \xi(x)$ be a function of x satisfying $1 \leqslant \xi \leqslant x$ and which will be chosen explicitly later. From (1.7) we have

$$\sum_{\xi < n \leqslant x} \frac{\Lambda(n)}{n} = \operatorname{Log} x - \operatorname{Log} \xi + O(1) .$$

and trivially

$$\sum_{\xi < n \leqslant x} \frac{\Lambda(n)}{n} \leqslant \frac{\psi(x)}{\xi},$$

which implies that

$$\frac{\psi(x)}{\xi} \geqslant \operatorname{Log} \frac{x}{\xi} + O(1) .$$

Thus, there exists a constant $K \geqslant 0$ such that for all $x \geqslant 1$

$$\frac{\psi(x)}{\xi} \geqslant \operatorname{Log} \frac{x}{\xi} - K .$$

If we write $H = e^{1+K}$ and choose $\xi(x) = xH^{-1}$ if $x \geqslant H$ and $\xi(x) = 1$ if $1 \leqslant x \leqslant H$, then for all $x \geqslant H$

$$\frac{\psi(x)}{x} > \frac{1}{H}(\operatorname{Log} H - K) = \frac{1}{H} > 0$$

Thus $\liminf_{x \to \infty} \psi(x)/x > 0$ and the proof is complete. ■

Notes to chapter 1

TCHEBYCHEFF'S EFFECTIVE THEOREMS

In our proof of Tchebycheff's theorems we were content to show the existence of x_0, $A(x_0)$, $B(x_0)$ such that

$$0 < A \leqslant 1 \leqslant B < \infty$$

and for all $x > x_0$

$$A \frac{x}{\operatorname{Log} x} \leqslant \pi(x) \leqslant B \frac{x}{\operatorname{Log} x}.$$

with analogous results for $\theta(x)$, $\psi(x)$ and $\Pi(x)$. It is possible to follow our method of proof and give explicit numerical values to x_0, $A(x_0)$, $B(x_0)$, but we do not trouble to do so. The original proofs of Tchebycheff, [1] and [2], showed that for $x_0 = 30$ one could take

$$A = \operatorname{Log} \frac{2^{1/2}\, 3^{1/3}\, 4^{1/4}}{30^{1/30}} = 0.921\ldots \text{ and } B = \frac{6}{5} A = 1.105\ldots$$

His methods can be refined somewhat to give slightly better values for A and B. For more information see Sylvester [1], [2]; Schur [1] and Breusch [1]. It seems highly unlikely that Tchebycheff's technique can be improved to the extent of showing that $\lim \pi(x) \operatorname{Log} x/x$ exists. Erdös, in 1936, proved that given any $\varepsilon > 0$ one can refine Tchebycheff's technique to prove that $A > 1 - \varepsilon$ and $B < 1 + \varepsilon$. However, to prove that this refinement is possible Erdös had to use the Prime Number Theorem.

As we remarked earlier, Tchebycheff's motive for giving numerical information was to prove Bertrand's postulate, which can be interpreted as saying that for all $x > 1$ we have $\pi(2x) - \pi(x) > 0$. More refined results are known to-day. For example, it can be shown that for all $x \geqslant 21$ we have

$$\frac{3x}{5 \operatorname{Log} x} < \pi(2x) - \pi(x) < \frac{7x}{5 \operatorname{Log} x}$$

see Finsler [1], Rosser and Schoenfeld [3] and exercise 1.8.

More generally one can consider

$$\pi(x + y) - \pi(x)$$

and ask : " How large must y be, as a function of x, before the difference $\pi(x+y)-\pi(x)$ is positive. " This reduces to a study of the differences between consecutive prime numbers, a topic which has been the subject of extensive researches. Lack of space prevents us from discussing such problems and their solutions in this book. We shall merely mention some of the more interesting results and conjectures.

It has been *proved* that for all sufficiently large values of x there is at least one prime in the interval $[x, x + x^{7/12}]$, see Huxley [1], and it is *conjectured* that

$$\limsup_{n\to\infty} \frac{p_{n+1} - p_n}{(\operatorname{Log} n)^2} = 1 ,$$

which implies that for a fixed $\varepsilon > 0$ and all $x > x_0(\varepsilon)$ the interval $[x, x + (1+\varepsilon) \operatorname{Log}^2 x]$ coutains at least on prime, (see Cramér [2]).

In the opposite direction Bombieri and Davenport [2] have proved that

$$\liminf_{n\to\infty} \frac{P_{n+1} - P_n}{\operatorname{Log} P_n} \leqslant \frac{2+\sqrt{3}}{8} = 0.46650...$$

The sequence $\{ (P_{n+1} - P_n)/\operatorname{Log} P_n \}$ had been studied earlier by Ricci [1]. He showed that the set of limit points of the sequence has positive Lebesgue measure. However, the only known limit point is ∞. Of course, if the twin prime conjecture is true, then there are an infinity of integers n such that P_n and P_{n+2} are both prime and 0 would be a limit point of the sequence.

A generalisation of the twin prime conjecture, known as the k-tuple prime conjecture, has recently been shown to be relevant to the behaviour of $\pi(x)$. First, we state the conjecture.

" Let $k \geqslant 2$ be an integer and $a_1, ..., a_k$ integers which are coprime in pairs. If we define the function v by

$$v(m) = \operatorname{card} \{ a_i \,(\operatorname{mod} m) \text{ for } 1 \leqslant i \leqslant k \}$$

and suppose that for all $p \leqslant k$ we have $v(p) < p$, then one conjectures that there exist an infinity of integers n such that the k integers $a_1 + n, a_2 + n, ..., a_k + n$ are all prime. "

The twin prime conjecture is obviously a special case of the k-tuple prime conjecture. For more information on this conjecture and related topics see Halberstam [1]. Our interest in the conjecture stems from the fact Hensley and Richards [1] have shown that the k-tuple prime conjecture is in contradiction with another famous conjecture of prime number theory, namely that $\pi(x)$ is a sub-additive function, i.e.

$$\pi(x + y) \leqslant \pi(x) + \pi(y)$$

for all $x, y > 1$. As there is a considerable body of plausible reasoning in favour of the k-tuple prime conjecture one would conjecture that $\pi(x)$ is *not* a subadditive function.

Return now to the various forms of Tchebycheff's theorems. As we remarked in § 7 it is possible to obtain upper and lower bounds for $\pi(x)$, $\theta(x)$, $\psi(x)$ which are valid for all $x > 2$ by considering binomial coefficients. For example, exercise 1.9 shows that $\theta(x) < x \operatorname{Log} 4$ and with a little more work one can show that $\psi(x) < x \operatorname{Log} 3$ (see Hanson [1]). However, by using the deeper methods of chapters 2 and 4 together with considerable numerical computation one can prove much more precise results. It was in this way that Rosser and Schoenfeld [3] proved the following inequalities :

(*a*) For $x \geqslant 52$ we have

$$\frac{x}{\operatorname{Log} x}\left(1 + \frac{1}{2 \operatorname{Log} x}\right) < \pi(x) < \frac{x}{\operatorname{Log} x}\left(1 + \frac{3}{2 \operatorname{Log} x}\right),$$

(*b*) For $x \geqslant 41$ we have

$$x\left(1 - \frac{1}{\operatorname{Log} x}\right) < \theta(x) < x\left(1 + \frac{1}{2 \operatorname{Log} x}\right),$$

(*c*) For $x \geqslant 121$ we have

$$x\left(1 - \frac{1}{\operatorname{Log} x} + \frac{0{,}98}{x^{1/2}}\right) < \psi(x) < x\left(1 + \frac{1}{2 \operatorname{Log} x} + \frac{1{,}02}{x^{1/2}} + \frac{3}{x^{2/3}}\right),$$

(*d*) If P_n denotes the nth prime number, then for $n \geqslant 21$ we have :

$$n\left(\operatorname{Log} n + \operatorname{Log}\operatorname{Log} n - \frac{3}{2}\right) < p_n < n\left(\operatorname{Log} n + \operatorname{Log}\operatorname{Log} n - \frac{1}{2}\right).$$

It is easy to verify that the above inequality implies that for all $n \geqslant 1$ we have $p_n > \operatorname{Log} n$.

Lists of prime numbers have been prepared by several authors and we refer the reader to D. N. Lehmer [1] or to Baker and Gruenberger [1] for the enumeration of all primes up to about 10^7 and 10^8 respectively. An efficient computational procedure for computing the *exact* value of $\pi(x)$ for any given x is described in D. H. Lehmer [1].

GENERALISED PRIME NUMBERS

We now discuss an interesting abstraction of the multiplicative semi-group of positive integers which was introduced by Beurling [1] and has since proved to be a fascinating source of problems.

Let $\mathcal{T} = \{ P_i \}$ be a sequence of real numbers which satisfy

(i) $1 < P_1 \leqslant P_2 \leqslant \ldots,$

(ii) P_i tends to infinity with i.

Now consider the free multiplicative semigroup $Z(\mathcal{T})$ generated by the numbers P_i. The elements of $Z(\mathcal{T})$ consist of the set of real numbers of the form :

$$n = \prod_{r=1}^{\infty} p_r^{\nu_r} \tag{1.8}$$

where the ν_r are non-negative integers, only a finite number of which are non-zero.

The sequences $\mathcal{T}$ and $Z(\mathcal{T})$ mimic the sequences of primes and positive integers respectively. For this reason one calls $\mathcal{T}$ a set of generalised primes and $Z(\mathcal{T})$ a set of generalised integers.

We introduce the two functions

$$N(\mathcal{T}, x) = \sum_{\substack{n \leqslant x \\ n \in Z(\mathcal{T})}} 1 \quad \text{and} \quad \pi(\mathcal{T}, x) = \sum_{\substack{p \leqslant x \\ p \in \mathcal{T}}} 1$$

which reduce to $[x]$ and $\pi(x)$ in the special case of the natural numbers. It is reasonable to wonder about the relationship between $N(\mathcal{T}, x)$ and $\pi(\mathcal{T}, x)$. For example is it true that if $N(\mathcal{T}, x) \sim x$, then $\pi(\mathcal{T}, x) \sim x/\text{Log}\, x$ or at least $\pi(\mathcal{T}, x) = O(x/\text{Log}\, x)$?

Examples of generalised systems of integers arise naturally in number theory. An important example is as follows. Let K be an algebraic number field and $\mathcal{T}_1$ the sequence of positive integers, which are the norms of the prime ideals in the ring of integers of K. The norm of an integral ideal can be represented in the form (1.8) and consequently the norms of the integral ideals of K constitute a generalised system of integers. Weber proved that for this particular system

$$N(\mathcal{T}_1, x) = Ax + O(x^\theta) \tag{1.9}$$

where $A > 0$, $0 \leqslant \theta < 1$ and the implied " O " constant depend on K. Later-Landau [3] showed that

$$\pi(\mathcal{T}_1, x) = \int_2^x \frac{du}{\text{Log}\, u} + O\{ x \exp(- c \sqrt{\text{Log}\, x}) \} \tag{1.10}$$

where $c > 0$ and the implied " O " constant depend on K. An analysis of Landau's proof reveals that the only essential fact about K which is used is (1.9). It is with this observation in mind that we have arranged that the statements and proofs of all theorems about prime numbers in chapters 1, 2, 3 and 4 hold, with only trivial modification, for *any* generalised system of integers $Z(\mathcal{T})$ satisfying

$$N(\mathcal{T}, x) = Ax + O(x^\theta)\,, \tag{1.11}$$

where $A > 0$ and $0 \leqslant \theta < 1$.

There exist generalised systems satisfying (1.11) for which an estimate of type (1.10) is, apart from constants, best possible; see Hall [1]. However, for the prime ideal theorem it can be proved that

$$\pi(\mathfrak{I}_1, x) = \int_2^x \frac{du}{\operatorname{Log} u} + O\left(x \exp\left\{\frac{-c(\log x)^{3/5}}{(\log\log x)^{1/5}}\right\}\right)$$

by using the methods of chapter 11; see Mitsui [1].

If we weaken our hypothesis on the error term

$$R(\mathfrak{I}, x) = N(\mathfrak{I}, x) - Ax$$

from $R(\mathfrak{I}, x) = O(x^\theta)$ to $R(\mathfrak{I}, x) = O(x.\log^{-\gamma} x)$, then in certain circumstances one can still deduce that

$$\pi(\mathfrak{I}, x) = \frac{x}{\log x} + o\left(\frac{x}{\log x}\right). \tag{1.12}$$

For example, Beurling [1] proved that if $\gamma > 3/2$, then $\pi(\mathfrak{I}, x)$ satisfies (1.12) and Diamond [2] gave an example to show that if $\gamma \leqslant 3/2$, then (1.12) does not necessarily hold.

In the case when $\gamma \leqslant 3/2$ one can ask whether it is possible to deduce a weaker result than (1.12), namely

$$\pi(\mathfrak{I}, x) = O\left(\frac{x}{\operatorname{Log} x}\right). \tag{1.13}$$

It is conjectured that if $\gamma > 1$, then (1.13) is a consequence of

$$N(\mathfrak{I}, x) = x + O\left(\frac{x}{\log^\gamma x}\right)$$

For more information on generalised number systems and some references to the literature see Bateman and Diamond [3].

Exercises to chapter 1

1.1 Let f be a multiplicative function. Write

$$S(f) = \sum_{n=1}^{\infty} f(n) \quad \text{and} \quad P(f) = \prod_p \left\{ \sum_{k=0}^{\infty} f(p^k) \right\} .$$

Theorem 1.1 asserts that in certain circumstances $S(f) = P(f)$. By considering the two multiplicative functions f and g, which are defined on the prime powers by

$$f_1(p^k) = \frac{(-1)^{p-1}}{p^{3k/4}} \quad \text{and} \quad g(p) = 1 , \quad g(p^2) = -1 + \frac{1}{p^2} ,$$

$g(p^k) = 0$ for $k \geqslant 3$, show that condition (i) cannot be replaced by « $S(f)$ is convergent » and that condition (ii) cannot be replaced by « $P(f)$ is absolutely convergent ».

1.2 Is the condition $\sum_{n=1}^{\infty} | f(n) | < \infty$ necessary for the conclusion of theorem 1.1 to hold ? (see also exercise 7.9).

1.3 Let f be a completely multiplicative function. Prove that if $| f(p) | < 1$, then

(i) $\sum_p | f(p) | < \infty \Leftrightarrow \sum_{n=1}^{\infty} | f(n) | < \infty .$

In particular, if the series $\sum_p | f(p) |$ is convergent, then so is the series $\sum_p | f(p+1) |$.

(ii)* Is it possible to find a completely multiplicative function f such that the series $\sum_p | f(p+1) |$ is convergent, but the series $\sum_p | f(p) |$ does not converge ? (see Katai [1] for some analogous questions).

1.4 Let f be a completely multiplicative function satisfying $0 < f(n) < 1$ and $\sum_{n=1}^{\infty} f(n) < \infty$. We define a probability measure on the set of subsets of $\mathbb{N}$ as follows :

If $A \subset \mathbb{N}$, then

$$\Pr \{ m \in A \} = \frac{1}{\sum_{n=1}^{\infty} f(n)} \sum_{n \in A} f(n) .$$

Compute the following probabilities :

(i) Pr (n is divisible by k);
(ii) Pr (n is not divisible by k);
(iii) Pr (n is not divisible by p_1 and not divisible by p_2);
(iv) Pr (n is not divisible by any prime);
(v) Pr ($n = 1$).

Deduce that

$$\frac{1}{\sum_{n=1}^{\infty} f(n)} = \prod_p (1 - f(p)) .$$

1.5 For which values of α and β does the following series posses an Euler product :

$$\sum_{n=1}^{\infty} \frac{1}{(\alpha n + \beta)^s} \ ?$$

1.6 Prove that the series

$$\sum_{n=1}^{\infty} \left\{ \frac{\mu(n)}{\varphi(n)} \right\}^r$$

is absolutely convergent if $r > 1$. (This result will be used in chapter 9.)

1.7 Obtain the following asymptotic formula due to van Lint and Richert [1] : where γ is Euler's constant.

$$\sum_{n \leqslant x} \frac{|\mu(n)|}{\varphi(n)} = \operatorname{Log} x + \gamma + \sum_p \frac{\operatorname{Log} p}{p(p-1)} + o(1)$$

1.8 Use the following six exercises to prove that for all $n > 1$ we have

$$\frac{1}{3} \frac{n}{\operatorname{Log} n} < \pi(2\,n) - \pi(n) < \operatorname{Log} 4 \cdot \frac{n}{\operatorname{Log} n}$$

1.9 Prove that for all $n > 1$

$$\prod_{p<n} p \leqslant 4^n .$$

[Hint : Note that $C_{2k+1}^k < 4^k$ and $k + 1 \leqslant p \leqslant 2k + 1 \Rightarrow p \mid C_{2k+1}^k$, then by considering the cases k odd and even separately use induction to prove the result.]

1.10 If $n \geqslant 14$ show that

$$\pi(n) \leqslant \frac{1}{2} n - 1 .$$

1.11 Verify the following properties of the binomial coefficient $\binom{2n}{n}$:

(i) $\dfrac{4^n}{2\sqrt{n}} < \binom{2n}{n} < 4^n$, $\quad \dfrac{4^n}{2\sqrt{n}} < C_{2n}^n < 4^n$;

(ii) If p is a prime and $p \,\Big|\, \binom{2n}{n}$, then $p > \sqrt{2n} \Rightarrow p \,\Big\|\, \binom{2n}{n}$.

(iii) If p is a prime and $p^r \,\Big|\, \binom{2n}{n}$, then $p^r < 2n$.

(iv) If $n > 2$ and the prime p satisfies $\frac{2}{3} n < p \leqslant n$, then $p \nmid \binom{2n}{n}$.

1.12 Let $R_n = \prod_{n \leqslant p < 2n} p$ and write $\binom{2n}{n} = Q_n R_n$. Prove that

(i) $(Q_n, R_n) = 1$;

(ii) $Q_n < 4^{2/3n}(2n)^{1/2\sqrt{2n}}$.

1.13 Use the fact that $R_n \,\Big|\, \binom{2n}{n}$ to deduce the upper bound for $\pi(2n) - \pi(n)$ given in exercise 1.8.

1.14 Prove that if $n \geqslant 98$, then

$$\frac{4^{n/3}}{2\sqrt{n}(2\,n)^{1/2\sqrt{n}}} < R_n < (2\,n)^{\pi(2n)-\pi(n)}$$

and deduce the lower bound for $\pi(2\,n) - \pi(n)$ given in exercise 1.8.

1.15 Prove that 30 is the largest integer N with the property that every integer n satisfying

$$1 < n < N \quad \text{and} \quad (n, N) = 1$$

is a prime.

1.16 Prove the following result of Teufel [1].

(i) Let p_n denote the nth prime, where $p_1 = 1, p_2 = 2, p_3 = 3, \ldots$ Then for each integer $n \geqslant 1$ every integer $2\,k$ satisfying

$$0 \leqslant 2\,k \leqslant \sum_{\nu=0}^{2n-2} p_\nu$$

can be written in the form

$$2\,k = p_{2n-2} + \sum_{\nu=0}^{2n-2} \varepsilon_\nu\, p_\nu$$

where $\varepsilon_\nu \in \{-1, +1\}$ for $1 \leqslant \nu \leqslant 2\,n - 1$. (Hint : Use induction and Bertrand's postulate.)

(ii) For each $n > 1$ show that

$$p_{2n} = \sum_{\nu=0}^{2n-1} \varepsilon_\nu\, p_\nu \quad \text{où} \quad \varepsilon_\nu \in \{-1, +1\} \quad \text{with} \quad \varepsilon_\nu \in \{-1, +1\}$$

and

$$p_{2n+1} = 2\,\varepsilon_{2n}\, p_{2n} + \sum_{\nu=0}^{2n-1} \varepsilon_\nu\, p_\nu \quad \text{with} \quad \varepsilon'_\nu \in \{-1, +1\}\,.$$

1.17 Prove that each integer $n > 6$ can be written as a sum of distinct primes after making the following observations :

(i) If n satisfies $7 \leqslant n \leqslant 19$, then n can be written as a sum of distinct primes, each prime being less than 13.

(ii) If n satisfies $20 \leqslant n \leqslant 32$, then n can be written as a sum of distinct primes, each prime being less than 17.

(iii) Bertrand's postulate is true !

1.18 (i) Apply theorem 1, to the series

$$\sum_{p \leqslant x} \frac{\operatorname{Log} p}{p} \cdot \frac{1}{\operatorname{Log} p}$$

to deduce that

$$\sum_{p \leqslant x} \frac{1}{p} = \operatorname{Log}\operatorname{Log} x + a + \mathrm{O}\left(\frac{1}{\operatorname{Log} x}\right)$$

where $a = 0.2614972128\ldots$

(ii) Show that there exists an absolute constant $c > 0$ such that for all $x > 1$

$$\prod_{p \leqslant x} \left(1 - \frac{1}{p}\right) > \frac{c}{\log x} > 0 .$$

(iii) Show also that

$$\sum_{p \leqslant x} \frac{1}{p} = \int_1^x \frac{\pi(u)}{u^2}\, du + \mathrm{O}(1) .$$

(iv) Combine (i) and (iii) and prove that

$$\liminf_{x \to \infty} \pi(x) \frac{\log x}{x} \leqslant 1 \leqslant \limsup_{x \to \infty} \pi(x) \frac{\log x}{x} .$$

1.19 Suppose that the series $\sum_{n=1} a_n n^{-s}$ is convergent for $\mathfrak{R}(s) > \sigma_0$. Denote its sum by $f(s)$ and let $A(x) = \sum_{n \leqslant x} a_n$.

(i) Prove that

$$f(s) = s \int_1^\infty \frac{A(u)}{u^{1+s}}\, du .$$

(ii) Suppose that $A(x) > 0$ for all sufficiently large x and that when $s \to 1^+$ the product $(s - 1) f(s) \to l$, a finite limit. Prove that

$$\liminf_{x \to \infty} \frac{A(x)}{x} \leqslant l \leqslant \limsup_{x \to \infty} \frac{A(x)}{x} .$$

(iii) Suppose that $\lim \frac{A(x)}{x}$ exists and is equal to l. Prove that $\sigma_0 \leqslant 1$ and that $\lim_{s \to 1^+} (s-1) f(s)$ exists and is equal to l.

(iv) Does the existence of

$$\lim_{s \to 1^+} (s - 1) f(s)$$

imply the existence of

$$\lim_{x \to \infty} \frac{A(x)}{x} ?$$

1.20 Prove the following theorem of Mertens

$$\prod_{p \leqslant x} \left(1 - \frac{1}{p}\right) \sim \frac{e^{-\gamma}}{\operatorname{Log} x} \quad \text{and} \quad \prod_{p \leqslant x} \left(1 + \frac{1}{p}\right) \sim \frac{6\,e^{\gamma}}{\pi^2} \operatorname{Log} x$$

where γ to Euler's constant.

1.21 Show that

$$\liminf_{n \to \infty} \frac{\varphi(n) \operatorname{Log} \operatorname{Log} n}{n} = e^{-\gamma}.$$

1.22 Let f be a real valued function such that :

(i) $f(x) > 0$ for all $x > 1$,

(ii) $f(x)/\operatorname{Log} x$ decreases to zero as x tends to infinity. Prove that

$$\sum_p f(p) < \infty \Leftrightarrow \sum_{n=2}^{\infty} \frac{f(n)}{\operatorname{Log} n} < \infty .$$

If $\sum_p f(p)$ diverges, is it true that

$$\sum_{p \leqslant x} f(p) \sim \sum_{n \leqslant x} \frac{f(n)}{\operatorname{Log} n} \ ?$$

1.23 Let $\mathcal{A} = \{ a_n \}$ be a sequence of positive integers with the following property :

There exist real numbers θ and x_0 such that $0 \leqslant \theta < 1$ and for each $x > x_0$ the interval $(x, x + x^{\theta})$ contains at least one member of $\mathcal{A}$.

Prove that if $c \geqslant (1 - \theta)^{-1}$, then there exists a real number A, depending on $\mathcal{A}$ and c such that for every integer $n \geqslant 1$

$$[A^{c^n}] \in \mathcal{A} .$$

(The sequence of prime numbers have the above property and one can choose $\theta = \frac{7}{12}$.)

[Hint : Define by recursion the subsequence $\{ b_n \}$ of $\mathcal{A}$ as follows : b_1 is the least member of $\mathcal{A}$ satisfying $b_1 > x_0$ and for $n \geqslant 1$, b_{n+1} is the least member of $\mathcal{A}$ satisfying

$$b_n^c < b_{n+1} < b_n^c + b_n^{c\theta} .$$

Use the inequality

$$x^y + x^{y-1} < (1 + x)^y - 1$$

to deduce that

$$b_n^{c^{-n}} < b_{n+1}^{c^{-(n+1)}} < (1 + b_{n+1})^{c^{-(n+1)}} < (1 + b_n)^{c^{-n}}.$$

and hence

$$\lim_{n \to \infty} b_n^{c^{-n}}$$

exists and is equal to $A > 1$, with the consequence that

$$b_n \leqslant A^{c^n} < b_n + 1.]$$

1.24 Let $Z(\mathscr{T})$ be a generalised system of integers whose counting function $N(x)$ satisfies $N(x) = x + O(x)$. If the system $Z(\mathscr{T})$ is multiplicative, in the sense that $l_{mn} = l_m l_n$ for all positive integers m and n, then prove that $Z(\mathscr{T}) = \mathbb{Z}$.

1.25 Let $Z(\mathscr{T})$ be a generalised system of integers whose counting function satisfies

$$N(x) = x + O\left(\frac{x}{\operatorname{Log}^{\gamma} x}\right).$$

(i)* Prove that if $\gamma > 1$, then

$$\pi(\mathscr{T}, x) = O\left(\frac{x}{\operatorname{Log} x}\right).$$

(ii) Give examples to show that if $\gamma \leqslant 1$, then one does not necessarily have

$$\pi(\mathscr{T}, x) = O\left(\frac{x}{\operatorname{Log} x}\right).$$

1.26 Let $\{ a_n \}$, $\{ b_n \}$ be sequences of real numbers. Define the functions $A(x)$ and $B(x)$ by

$$A(x) = \sum_{n \leqslant x} a_n \quad \text{and} \quad B(x) = \sum_{n \leqslant x} b_n.$$

If $\{ c_n \}$ is the sequence defined by

$$c_r = \sum_{mn = r} a_m b_n.$$

then show that for any $y \geqslant 1$

$$C(x) = \sum_{n \leqslant x} c_n = \sum_{k \leqslant y} a_k B\left(\frac{x}{k}\right) + \sum_{j < x/y} b_j A\left(\frac{x}{j}\right) - A(y) B\left(\frac{x}{y}\right).$$

1.27 Denote by $d(n)$ the number of divisors of n. Prove that

$$\sum_{n \leqslant x} d(n) = x \operatorname{Log} x + (2\gamma - 1)x + O(x^{1/2})$$

1.28 If $\{x\}$ denotes the fractional part of x, prove that

$$\sum_{n \leqslant x} \left\{ \frac{x}{n} \right\} = (1 - \gamma)x + O(x^{1/2}).$$

1.29 An integer is square-free if it is not divisible by the square of any integer greater than 1. Denote by $N_2(x)$ the number of square-free integers less than x. Prove that

$$N_2(x) = \sum_{n \leqslant x} |\mu(n)| = \frac{6x}{\pi^2} + O(\sqrt{x}).$$

1.30 Use the Prime Number Theorem to deduce that

(i) $p_n \sim n \log n$,

(ii) $\displaystyle \lim_{x \to \infty} \frac{1}{x} \sum_{n \leqslant x} \frac{p_{n+1} - p_n}{\log n} = 1$,

(iii) $\displaystyle \liminf_{n \to \infty} \frac{p_{n+1} - p_n}{\log n} \leqslant 1$,

(iv) for any $\varepsilon > 0$ there exists an infinity of prime numbers p_n such that

$$p_{n+1} < (1 + \varepsilon) p_n.$$

1.31 Let $x > 0$ be a given real number. By using the Prime Number Theorem prove that there exists a sequence of primes $\{p(n)\}$ such that

$$\lim_{n \to \infty} \frac{p(n)}{n} = x.$$

Use the above result to show that the set of rationals

$$\{p/q,\ p,\ q \text{ both primes}\}$$

is dense in the positive real numbers.

1.32 If $Z(\mathcal{T})$ is a system of generalised integers whose counting function satisfies $N(x) = Ax + o(x)$, is it true that

$$\liminf \frac{\pi(x) \operatorname{Log} x}{x} \leqslant 1 \quad \text{and} \quad \limsup \frac{\pi(x) \operatorname{Log} x}{x} \geqslant 1\ ?$$

1.33 If $f(x)$ has a continuous second derivative in $Q \leqslant x \leqslant R$ and $\rho(x) = 1/2 - \{ x \}$, $\sigma(x) = \int_0^x \rho(t)\, dt$, then

$$\sum_{Q \leqslant n \leqslant R} f(n) = \int_Q^R f(x)\, dx + \rho(R) f(R) - \rho(Q) f(Q) -$$

$$- \sigma(R) f'(R) + \sigma(Q) f'(Q) + \int_Q^R \sigma(x) f''(x)\, dx .$$

1.34 If $\theta(z_0, z_1) = \sum_{z_0 < p \leqslant z_1} \log P$, then prove that

$$\theta(n, n/2) + \theta(n/3, n/4) + \theta(n/5, n/6) + \cdots = n \log 2 + O(\sqrt{n}) .$$

CHAPTER 2

The Prime Number Theorem

§1 INTRODUCTION

1.1 The primary objective of this chapter is to introduce the powerful methods of complex function theory into our study of prime numbers. As a specific application we shall prove the Prime Number Theorem, but the basic techniques of this chapter can be used to prove a great variety of asymptotic formulae for arithmetic functions.

As we saw in chapter 1 (theorem 1.5), the Prime Number Theorem is equivalent to any one of the statements

$$\pi(x) \sim \frac{x}{\log x}; \quad \Pi(x) \sim \frac{x}{\log x}; \quad \theta(x) \sim x; \quad \psi(x) \sim x.$$

It is not difficult to see that the problem of proving any of the above can be considered as a special case of the following general problem.

Given a Dirichlet series

$$f(s) = \sum_{n=1}^{\infty} \frac{a_n}{n^s},$$

then can one use knowledge about the behaviour of $f(s)$ to deduce information about the coefficient sum

$$S(x) = \sum_{n \leqslant x} a_n\ ?$$

For example, if $\Re(s) > 1$ then we know that

$$Z(s) = -\frac{\zeta'(s)}{\zeta(s)} = \sum_{n=1}^{\infty} \frac{\Lambda(n)}{n^s},$$

and we would like to deduce that as x tends to infinity

$$\psi(x) = \sum_{n \leqslant x} \Lambda(n) \sim x\ ?$$

There are several general techniques available for solving problems of this type. In this chapter we shall discuss three of them, namely

(i) the use of Cauchy's residue theorem,
(ii) the tauberian theorem of Ikehara,
(iii) the use of Plancherel's theorem.

1.2 Hadamard [1] and de la Vallée Poussin [1] independently in 1896 were the first to give a proof of the Prime Number Theorem; they used methods from complex function theory. Later their proofs were considerably simplified, largely by Landau. Soon other proofs of the Prime Number Theorem were found, but they all depended on the behaviour of $\zeta(s)$ in the complex plane and the proofs were indisputably analytic in nature.

It was natural to ask whether or not one could prove the Prime Number Theorem, which is, as we know, simply the statement that a certain real function tends to a limit, by a method not involving complex function theory. A great deal of effort was expended on this guest to no avail and by about 1930 many mathematicians had begun to feel that such a proof could not exist. This feeling was based upon the fact that it had been proved that the Prime Number Theorem was a consequence of the theorem that $\zeta(s)$ does not vanish on the line $\Re(s) = 1$ and conversely this theorem was a simple consequence of the Prime Number Theorem. From this relationship it was assumed that any proof of the Prime Number Theorem must implicitly use the fact that $\zeta(1 + it) \neq 0$ and that it was extremely unlikely that this latter assertion could be proved by purely real variable methods.

However, this heuristic conclusion was false, for Erdös [1] and Selberg [2] both published " elementary " proofs of the Prime Number Theorem in 1949. The " elementary " technique will be discussed in chapter 3. (Perhaps we ought to remark that the words " elementary proof " are often used by mathematicians in a technical sense to mean that the proof uses only very basic techniques and concepts. It does *not* mean that the proof is necessarily easy or short; often such proofs are long and complicated.)

1.3 The behaviour of $\zeta(s)$ in the complex plane will be of fundamental importance to a great deal of our later work. Before commencing our study of $\zeta(s)$ in § 2 we shall show how its behaviour is relevant to our first proof of the Prime Number Theorem. This will be done by discussing the structure of the proof in some detail and hopefully motivating for the reader the theorems about $\zeta(s)$ which are going to be proved in § 2.

We shall prove the Prime Number Theorem in the form :

As x tends to infinity, then $\psi(x) \sim x$. (2.1)

Technically it will be easier to deal with the function

$$\psi_1(x) = \int_1^x \psi(u)\, du,$$

rather than $\psi(x)$, because the irregularities of $\psi(x)$ are, to a certain extent, " smoothed out " in $\psi_1(x)$ and formal manipulations, such as the interchanging of limit operations, are easier to justify. Consequently we shall first prove :

As x tends to infinity, then $\psi_1(x) \sim \frac{1}{2} x^2$. (2.2)

The deduction of (2.1) from (2.2) will not be difficult; the non-trivial part of the argument is the proof of (2.2).

It will follow from the definition of Z(s) and by use of the Abel summation formula that for $\Re(s) > 1$

$$Z(s) = -\frac{\zeta'(s)}{\zeta(s)} = s \int_1^\infty \frac{\psi(u)}{u^{s+1}}\, du.$$

By an integration by parts we obtain

$$Z(s) = s(s+1) \int_1^\infty \frac{\psi_1(u)}{u^{s+2}}\, du,$$

We would like to « invert » this last relationship and write $\psi_1(x)$ as a function of Z(s). This is possible, either by the Mellin inversion theorem or by a direct calculation as we shall do in § 3; the result is

$$\frac{\psi_1(x)}{x^2} = \frac{1}{2\, i\pi} \int_{c-i\infty}^{c+i\infty} \frac{Z(s)\, x^{s-1}}{s(s+1)}\, ds,$$

where $c > 1$ is any fixed real number. Thus, in order to prove (2.2) we must show that as x tends to infinity the above integral tends to πi.

When one is faced with such an integral it is natural to try and evaluate it by choosing a suitable contour and then using Cauchy's residue theorem. What would be a good contour to choose ? Clearly there is no point in taking a contour strictly to the right of the line $\Re(s) = 1$, since the integrand has no singularities in this region. So we must go to the left of this line. Thus, our first step must be to show that $\zeta(s)$ and consequently Z(s) have analytic continuations to the left of the line $\Re(s) = 1$. This will be done as theorem 2.1, when we show that $\zeta(s)$ is holomorphic in the half-plane $\Re(s) > 0$ except for a simple pole at $s = 1$ with residue 1. Once this has been done we shall see that the only singularities of the integrand in the half plane $\Re(s) > 0$ are a simple pole at $s = 1$ with residue $\frac{1}{2}$ and simple poles at the zeros of $\zeta(s)$, if such exist.

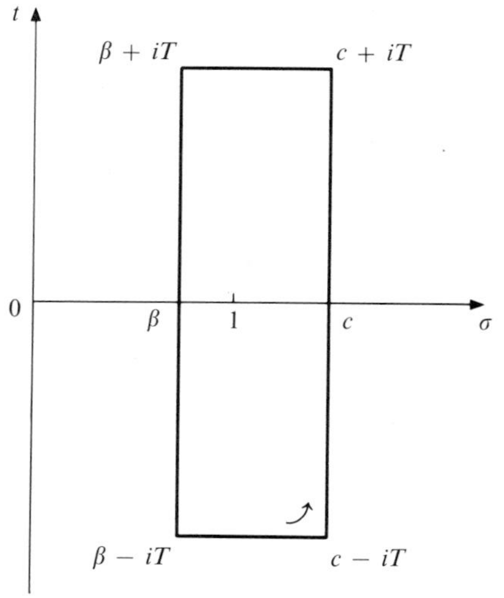

Fig. 1

If this were the best of all possible worlds we would be able to choose the simple contour illustrated in figure 1, where $\beta < 1$ is chosen such that for all values of T the rectangle contains no zeros of $\zeta(s)$. Let us suppose that such a β does exist. This is a very strong hypotheses, but it will help us to see the general principles behind the complex variable approach to the Prime Number Theorem.

If we write $A(s) = \mathrm{Z}(s)\, x^{s-1}/s(s+1)$, then by the residue theorem we would have

$$\frac{1}{2} = \frac{1}{2\,i\pi}\int_{c-iT}^{c+iT} A(s)\, ds + \frac{1}{2\,i\pi}\int_{c+iT}^{\beta+iT} A(s)\, ds + $$

$$+ \frac{1}{2\,i\pi}\int_{\beta+iT}^{\beta-iT} A(s)\, ds + \frac{1}{2\,i\pi}\int_{\beta-iT}^{c-iT} A(s)\, ds\,.$$

Now if we could prove that as T and x both tend to infinity the last three integrals all tend to zero we would have solved our problem. The second of the three integrals is easily dealt with, for it is

$$\frac{x^{\beta-1}}{2\,\pi i}\int_{\beta-iT}^{\beta+iT} \frac{x^{s-1}\,\mathrm{Z}(s)}{s(s+1)}\, ds = x^{\beta-1}\int_{\beta-iT}^{\beta+iT} \frac{\mathrm{Z}(s)}{s(s+1)}\, x^{it}\, ds = \mathrm{O}(x^{\beta-1})\,.$$

which tends to zero as x tends to infinity. The remaining two integrals are in absolute value less than

$$\frac{1}{2\,\pi}\int_{\beta}^{c} \frac{|\,\mathrm{Z}(\sigma + iT)\,|\, x^{\sigma}}{|\,(\sigma + iT)\,(\sigma + 1 + iT)\,|}\, d\sigma\,.$$

Hence if we can obtain a " good " upper bound for $|\,\mathrm{Z}(\sigma + iT)\,|$, valid uniformly for $\beta \leqslant \sigma \leqslant c$, then we ought to be able to show that for a fixed x the above integral tends to zero as T tends to infinity. Sufficiently good upper bounds for $|\,\mathrm{Z}(s)\,|$ will in fact be obtained in § 2 by finding upper bounds for $|\,\zeta'(s)\,|$ and $|\,\zeta(s)\,|^{-1}$.

In order to follow the above plan of attack we would have to find a $\beta < 1$ such that $\zeta(s)$ has no zeros in the strip $\beta \leqslant \Re(s) \leqslant c$. Unfortunately no such β is known. Riemann [1] conjectured that $\zeta(s)$ has no zeros in the half-plane $\Re(s) > \dfrac{1}{2}$.

This conjecture, known as the *Riemann Hypothesis*, is far from being resolved and is one of the most famous unsolved problems in mathematics. (The fact that $\zeta(s)$

does have an infinity of zeros on the line $\mathfrak{R}(s) = \frac{1}{2}$ will be proved in the *Notes* to chapter 5.)

We shall prove, as theorem 2.2 that $\zeta(s)$ has no zeros in the half-plane $\mathfrak{R}(s) \geqslant 1$, this is sufficient to prove the Prime Number Theorem. Improvements upon theorem 2.2 giving " zero free " regions for $\zeta(s)$ strictly to the left of the line $\mathfrak{R}(s) = 1$ are rather difficult to obtain. Such regions will be constructed in chapters 4 and 11.

Because of our relative lack of knowledge about the zeros of $\zeta(s)$ we must modify the contour in figure 1. The simplest type of contour which we shall arrange to contain the pole at $s = 1$ and no zero of $\zeta(s)$ is shown in figure 2. We now describe the contour in some detail. Let T, T_0 be real numbers satisfying $0 < T < T_0$. Since we will show that $\zeta(s)$ is holomorphic in the half plane $\mathfrak{R}(s) > 0$ except at $s = 1$ it follows that $\zeta(s)$ has only a finite number of zeros in the rectangle $\mathfrak{R}(s) > 0$, $|\mathfrak{Im}(s)| \leqslant T$. Thus, for each real number T we can choose a real number $\beta(T)$ so that $\zeta(s)$ has no zeros in the rectangle $\beta(T) \leqslant \mathfrak{R}(s) < 1$, $|\mathfrak{Im}(s)| \leqslant T$. With this choice of $\beta(T)$ there are no zeros of $\zeta(s)$ either on or within the contour shown in figure 2.

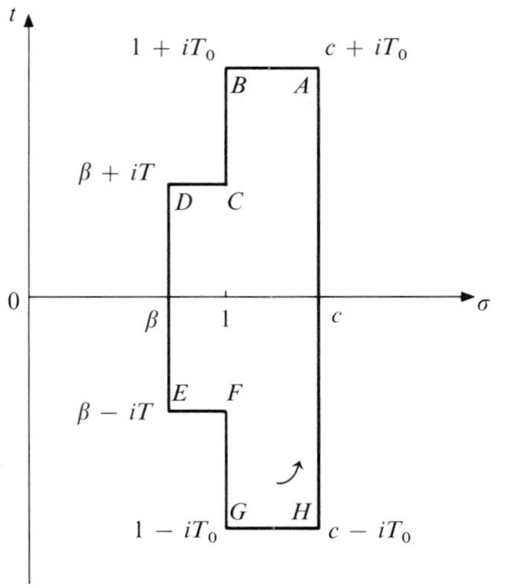

Fig. 2

In § 3 we carry out our earlier programme and show that for any given $\varepsilon > 0$ we have for all sufficiently large T_0, T and x

$$\left| \int_{\mathfrak{L}} \frac{Z(s)\, x^{s-1}}{s(s+1)}\, ds \right| < \varepsilon$$

where $\mathfrak{L}$ denotes the broken line $ABCDEFGH$. This means that

$$\frac{1}{2} = \lim_{x\to\infty} \lim_{T_0\to\infty} \frac{1}{2\,i\pi} \int_{c-iT_0}^{c+iT_0} \frac{Z(s)\, x^{s-1}}{s(s+1)}\, ds$$

$$= \lim_{x\to\infty} \frac{\psi_1(x)}{x^2}\,.$$

which is the asymptotic formula for $\psi_1(x)$, from which the Prime Number Theorem will be an easy consequence.

1.4 Contents of the chapter. In § 2 we commence our study of the Riemann zeta function by giving an analytic continuation into the half plane $\mathfrak{R}(s) > 0$ and establishing that $\zeta(s)$ has no zeros in the half plane $\Re(s) \geqslant 1$. Then we go on to obtain an upper bound for $|Z(s)|$ in various regions of the complex plane. As we have seen these results are of crucial importance to our proof of the Prime Number Theorem, which is given in § 3. This proof will be a detailed version of the argument sketched in § 1.2, section 4 is devoted to an analysis of the proof of the Prime Number Theorem and we discuss which of the properties of $\zeta(s)$ are really required in order to prove the Prime Number Theorem. The final two sections of the chapter deal rather briefly with the application of Tauberian theorems and Harmonic Analysis to prime number theory.

§ 2 $\zeta(s)$ IN THE HALF-PLANE $\Re(s) > 0$

2.1 We have seen in § 1 the importance of obtaining an analytic continuation of $\zeta(s)$ to the left of the line $\Re(s) = 1$. It can be shown that $\zeta(s)$ has an analytic continuation over the whole complex plane (see for example exercise 2.3 or theorem 5.2). In this chapter we content ourselves with an analytic continuation into the half-plane $\Re(s) > 0$.

Theorem 2.1 *The function $\zeta(s)$ can be analytically continued into the half plane $\Re(s) > 0$. In this half plane $\zeta(s)$ is holomorphic, except for a simple pole at $s = 1$ with residue* 1.

Proof. For $\Re(s) > 1$ we use the Abel summation formula (theorem 1.7), to deduce that

$$\zeta(s) = \sum_{n=1}^{\infty} \frac{1}{n^s} = s \int_1^{\infty} \frac{[u]}{u^{s+1}}\, du$$

$$= \frac{s}{s-1} - s \int_1^{\infty} \frac{\{u\}}{u^{s+1}}\, du \,. \tag{2.3}$$

Now observe that the final integral converges if $\Re(s) > 0$ and so defines a holomorphic in this half-plane. This gives the stated analytic continuation for $\zeta(s)$.

2.2 Zeros on the line $\Re(s) = 1$. From § 1.3 we know that it is of crucial importance to find regions of the complex plane which do not contain zeros of $\zeta(s)$. The simplest non-trivial result is theorem 2.2, which can be used to construct the contour shown in figure 2. This contour will then be used to prove the Prime Number Theorem. We shall return to the topic of " zero free regions " in chapters 4 and 11 where much deeper theorems will be proved.

Theorem 2.2 *The function $\zeta(s)$ has no zeros in the half-plane $\Re(s) \geqslant 1$.*

Proof. For $\Re(s) > 1$ we have the representation

$$\zeta(s) = \prod_p (1 - p^{-s})^{-1}$$

and it is clear that $\zeta(s)$ does not vanish in this open half-plane. Thus, it remains to show that $\zeta(s)$ has no zeros on the line $\Re(s) = 1$.

If $\sigma > 1$, then taking logarithms in the above representation we have, for all real t

$$\begin{aligned}\operatorname{Log}\{\zeta(\sigma + it)\} &= -\sum_p \operatorname{Log}(1 - p^{-\sigma - it}). \\ &= \sum_p \sum_{\nu=1}^{\infty} \nu^{-1} p^{-\nu(\sigma + it)}.\end{aligned}$$

Upon equating the real parts in the above equation we deduce that

$$\operatorname{Log} |\zeta(\sigma + it)| = \sum_p \sum_{\nu=1}^{\infty} \nu^{-1} p^{-\nu\sigma} \cos(\nu t \operatorname{Log} p). \tag{2.4}$$

Now we recall the following trivial inequality, valid for all real φ :

$$0 \leqslant 2(1 + \cos\varphi)^2 = 3 + 4\cos\varphi + \cos 2\varphi$$

Using the above inequality we see that for $\sigma > 1$ and all real t

$$\sum_p \sum_{\nu=1}^{\infty} \frac{1}{\nu} p^{-\nu\sigma} \{3 + 4\cos(\nu t \operatorname{Log} p) + \cos(2\nu t \operatorname{Log} p)\} \geqslant 0,$$

From this inequality and (2.4) we conclude that

$$3 \operatorname{Log} |\zeta(\sigma)| + 4 \operatorname{Log} |\zeta(\sigma + it)| + \operatorname{Log} |\zeta(\sigma + 2it)| \geqslant 0.$$

in other words, for $\sigma > 1$ and all real t :

$$|\zeta(\sigma)|^3 . |\zeta(\sigma + it)|^4 . |\zeta(\sigma + 2it)| \geqslant 1. \tag{2.5}$$

Suppose that $\zeta(s)$ has a zero at $s = 1 + it_0$. We must have $t_0 \neq 0$, since $s = 1$ is a pole of $\zeta(s)$. Choose $t = t_0$ in (2.5) and let σ tend to 1 from above. Since $\zeta(s)$ has a simple pole at $s = 1$ we have $\zeta(\sigma) = O(1/(\sigma - 1))$ and by hypothesis $\zeta(s)$ has a zero at $1 + it_0$, so $\zeta(\sigma + it_0) = O(\sigma - 1)$. Finally, since $\zeta(s)$ is holomorphic on the line $\Re(s) = 1$, except at $s = 1$, it follows that $\zeta(\sigma + i2t_0) = O(1)$. Consequently we have from (2.5)

$$\begin{aligned}1 \leqslant |\zeta(\sigma)|^3 |\zeta(\sigma + it_0)|^4 |\zeta(\sigma + 2it_0)| &= O\{(\sigma - 1)^{-3} (\sigma - 1)^4\} \\ &= O(\sigma - 1),\end{aligned}$$

which is a contradiction. Thus $\zeta(s)$ has no zeros on the line $\Re(s) = 1$. ■

2.3 Bounds for ζ, ζ' and Z. The following two theorems furnish upper bounds for $|\zeta(s)|^{\pm 1}$, $|\zeta'(s)|$ and $|Z(s)|$ in various regions of the complex plane. The results will be used often in later chapters. It is possible to refine our upper bounds at the expense of greatly complicating the argument. Such an improvement will be given in chapter 11.

Theorem 2.3 *Let θ satisfy $0 < \theta < 1$. For all $\sigma \geqslant \theta$ and all real t satisfying $|t| \geqslant 1$ we have*

(a) $$|\zeta(\sigma + it)| \leqslant \frac{7}{4}\frac{|t|^{1-\theta}}{\theta(1-\theta)},$$

(b) $$|\zeta'(\sigma + it)| \leqslant \frac{|t|^{1-\theta}}{\theta(1-\theta)}\left(\operatorname{Log}|t| + \frac{1}{\theta} + \frac{5}{4}\right).$$

Proof. (*a*) Since $\zeta(\sigma + it)$ and $\zeta(\sigma - it)$ are complex conjugates we can suppose that t is positive. Let x be a parameter to be chosen explicitly later satisfying $x \geqslant 1$.

We first establish an upper bound for $|\zeta(s)|$ in terms of x. If $s \neq 1$, then by the Abel summation formula (theorem 1.6) :

$$\sum_{n \leqslant x} \frac{1}{n^s} = [x]\, x^{-s} + s\int_1^x [u]\, u^{-s-1}\, du$$
$$= \frac{x^{1-s}}{1-s} - \frac{s}{1-s} - \{x\}\, x^{-s} - s\int_1^x \{u\}\, u^{-1-s}\, du. \tag{2.6}$$

If we now subtract (2.6) from (2.3) we see that for $\mathcal{R}(s) > 0$ and $s \neq 1$

$$\zeta(s) - \sum_{n \leqslant x}\frac{1}{n^s} = -\frac{x^{1-s}}{1-s} + \{x\}\, x^{-s} - s\int_x^\infty \{u\}\, u^{-1-s}\, du. \tag{2.7}$$

Upon writing $s = \sigma + it$, with $\sigma \geqslant 0$, $t \geqslant 1$ we deduce that

$$|\zeta(\sigma + it)| \leqslant \sum_{n \leqslant x}\left|\frac{1}{n^s}\right| + \left|\frac{x^{1-s}}{1-s}\right| + \left|\frac{\{x\}}{x^s}\right| + |s|\int_x^\infty \left|\frac{\{u\}}{u^{1+s}}\right| du,$$
$$\leqslant \sum_{n \leqslant x} n^{-\theta} + \frac{x^{1-\theta}}{t} + x^{-\theta} + |s|\int_x^\infty \frac{du}{u^{\sigma+1}}. \tag{2.8}$$

From (2.6), with $s = \theta$ we have

$$\sum_{n \leqslant x}\frac{1}{n^\theta} \leqslant \frac{x^{1-\theta}}{1-\theta}$$

and quite trivially

$$\frac{|s|}{\sigma} \leqslant \frac{\sigma + t}{\sigma} = 1 + \frac{t}{\sigma} \leqslant 1 + \frac{t}{\theta} = \frac{\theta + t}{\theta},$$

Thus, from (2.8) and the above remark we conclude

$$|\zeta(\sigma + it)| \leqslant \frac{x^{1-\theta}}{1 - \theta} + \frac{x^{1-\theta}}{t} + x^{-\theta} + \left(1 + \frac{t}{\theta}\right) x^{-\theta}.$$

If we now choose $x = t \geqslant 1$, then the above becomes

$$|\zeta(\sigma + it)| \leqslant \frac{t^{1-\theta}}{\theta(1 - \theta)} \left[1 + \frac{3\,\theta(1 - \theta)}{t}\right] \leqslant \frac{7}{4} \frac{t^{1-\theta}}{\theta(1 - \theta)}.$$

(*b*) We next obtain an upper bound for $|\zeta'(s)|$ in terms of the parameter x. In the region $\mathfrak{R}(s) > 0$, $s = 1$ equation (2.7) holds. Moreover, since the integral in (2.7) is absolutely uniformly convergent we may differentiate (2.7) with respect to S and obtain :

$$\zeta'(s) = -\sum_{n \leqslant x} n^{-s} \operatorname{Log} n + x^{1-s}\left(\frac{-1}{(s - 1)^2} + \frac{\operatorname{Log} x}{1 - s}\right)$$
$$- \{x\} x^{-s} \operatorname{Log} x - \int_x^{\infty} \frac{\{u\}}{u^{s+1}} (1 - s \operatorname{Log} u)\, du,$$

By making trivial estimates, as in (*a*), we find that for $\sigma \geqslant \theta$ and $t \geqslant 1$

$$|\zeta(\sigma + it)| \leqslant \sum_{n \leqslant x} n^{-\theta} \operatorname{Log} n + x^{1-\theta}\left(\frac{1}{t^2} + \frac{\operatorname{Log} x}{t}\right)$$
$$+ \frac{\operatorname{Log} x}{x^{\theta}} + \frac{1}{\theta x^{\theta}} + |s| \int_x^{\infty} \frac{\operatorname{Log} u}{u^{\sigma+1}}\, du.$$

The integral in the above expression is

$$\frac{\operatorname{Log} x}{\sigma x^{\sigma}} + \frac{1}{\sigma^2 x^{\sigma}}$$

and by using the inequalities

$$\frac{|s|}{\sigma} \leqslant 1 + t\theta^{-1} \quad \text{and} \quad \sum_{n \leqslant x} n^{-\theta} \operatorname{Log} n \leqslant \frac{x^{1-\theta}}{1 - \theta} \operatorname{Log} x,$$

we deduce that

$$|\zeta(\sigma + it)| \leqslant \frac{x^{1-\theta}}{1 - \theta} \operatorname{Log} x + x^{1-\theta}\left(\frac{1}{t^2} + \frac{\operatorname{Log} x}{t}\right)$$
$$+ \frac{\operatorname{Log} x}{x^{\theta}} + \frac{1}{\theta x^{\theta}} + \left(1 + \frac{t}{\theta}\right)\left(\frac{\operatorname{Log} x}{x^{\theta}} + \frac{1}{\theta x^{\theta}}\right).$$

If we again choose $x = t$ the above inequality simplifies and becomes

$$|\zeta'(\sigma + it)| \leqslant \frac{t^{1-\theta}}{\theta(1-\theta)}\left[\operatorname{Log} t + (1-\theta)\left(\frac{3\,\theta}{t}\operatorname{Log} t + \frac{2}{t} + \frac{\theta}{t^2} + \frac{1}{\theta}\right)\right].$$

Because $t \geqslant 1$ we certainly have $t^{-1}\operatorname{Log} t \leqslant 1$ and hence

$$|\zeta'(\sigma + it)| \leqslant \frac{t^{1-\theta}}{\theta(1-\theta)}\left[\operatorname{Log} t + \frac{1}{\theta} - 4\,\theta^2 + 2\,\theta + 1\right],$$

$$\leqslant \frac{t^{1-\theta}}{\theta(1-\theta)}\left[\operatorname{Log} t + \frac{1}{\theta} + \frac{5}{4}\right].$$ ∎

The final theorem gives estimates for $|\zeta(s)|^{\pm 1}$, $|\zeta'(s)|$ and $|Z(s)|$ in the half-plane $\mathcal{R}(s) \geqslant 1$. It is, of course, quite trivial that all these functions are bounded in the half-plane $\mathcal{R}(s) \geqslant 1 + \delta$ for any fixed $\delta > 0$. Our theorem is useful because it gives information about the behaviour of these functions on and in the neighbourhood of the line $\mathcal{R}(s) = 1$.

Theorem 2.4. *There exist absolute constants c_1, c_2, c_3, c_4, A and B such that for all $\sigma \geqslant 1$ and all real t satisfying $|t| \geqslant 8$ we have*

(*a*) $|\zeta(\sigma + it)| \leqslant c_1 \operatorname{Log}|t|$,

(b) $|\zeta'(\sigma + it)| \leqslant c_2(\operatorname{Log}|t|)^2$,

(c) $|\zeta(\sigma + it)|^{-1} \leqslant c_3(\operatorname{Log}|t|)^A$,

(d) $|Z(\sigma + it)| \leqslant c_4(\operatorname{Log}|t|)^B$.

Possible values for the constants are :

$$c_1 = 4\,e\,, \quad c_2 = 6\,e\,, \quad c_3 = 16(6\,e)^7\,, \quad c_4 = 16(6\,e)^8\,, \quad A = 7\,, \quad B = 9\,.$$

Proof. (*a*) If $|t| \geqslant e^2$ and we define $\theta(t)$ by $\theta(t) = 1 - (\operatorname{Log}|t|)^{-1}$, then $\frac{1}{2} \leqslant \theta(t) < 1$. Choosing $\theta = \theta(t)$ in theorem 2.3(*a*) we have for $\sigma \geqslant \theta(t)$; $|t| \geqslant e^2$

$$|\zeta'(\sigma + it)| \leqslant \frac{7}{4}\,\frac{e \operatorname{Log}|t|}{1 - (\operatorname{Log}|t|)^{-1}} < 4\,e \operatorname{Log}|t|$$

(*b*) Defining θ as above and using theorem 2.3(*b*) we have, for $\sigma \geqslant \theta(t)$, $|t| \geqslant e^2$:

$$|\zeta'(\sigma + it)| \leqslant \frac{e \operatorname{Log}|t|}{1 - (\operatorname{Log}|t|)^{-1}}\left[\operatorname{Log}|t| + \frac{5}{4} + (1 - (\operatorname{Log}|t|)^{-1})^{-1}\right]$$

$$\leqslant 2\,e(\operatorname{Log}|t|)^2 + \frac{5}{2}\,e \operatorname{Log}|t| + 4\,e \operatorname{Log}|t|$$

$$\leqslant \frac{21}{4}\,e(\operatorname{Log}|t|)^2 < 6\,e(\operatorname{Log}|t|)^2\,.$$

(*c*) We shall prove this part of the theorem in two steps : first the case $\Re(s) \geqslant 2$ and second the case $1 \leqslant \Re(s) \leqslant 2$.

From the Euler product representation of $\zeta(s)$ it is trivial that for $\Re(s) \geqslant 2$:

$$| \zeta(s) |^{-1} = \left| \sum_{n=1}^{\infty} \frac{\mu(n)}{n^s} \right| \leqslant \sum_{n=1}^{\infty} \frac{1}{n^2} = \frac{\pi^2}{6} < 6\,e\,.$$

We shall now suppose that $1 \leqslant \Re(s) \leqslant 2$. Let $\eta = \eta(t)$ be a real function, to be chosen explicitly later, satisfying $0 < \eta < 1$. From (2.5) we know that

$$| \zeta(\sigma + \eta) |^3 \; | \zeta(s + \eta) |^4 \; | \zeta(\sigma + \eta + 2\,it) | \geqslant 1\,.$$

This inequality will be used to deduce a lower bound for $| \zeta(s + \eta) |$ and from this lower bound we shall find a lower bound for $| \zeta(s) |$. The reason why we must do the proof this way is that (2.5) has not been proved when $\sigma = 1$ and as we cannot use it to deduce a lower bound for $| \zeta(1 + it) |$ directly.

From (2.3) we have for $1 < \sigma + \eta \leqslant 3$,

$$\begin{aligned} | \zeta(\sigma + \eta) | &= \left| \frac{\sigma + \eta}{\sigma + \eta - 1} - (\sigma + \eta) \int_1^{\infty} \frac{\{ u \}}{u^{\sigma+\eta+1}} \, du \right| \\ &\leqslant \frac{\sigma + \eta}{\sigma + \eta - 1} + (\sigma + \eta) \int_1^{\infty} \frac{du}{u^{\sigma+\eta+1}} = \frac{2(\sigma + \eta) - 1}{\sigma + \eta - 1} \\ &\leqslant \frac{5}{\sigma + \eta - 1} \leqslant \frac{5}{\eta} < \frac{6\,e}{\eta}\,. \end{aligned}$$

From part (*a*) we know that

$$| \zeta(\sigma + \eta + 2\,it) | \leqslant 4\,e \operatorname{Log} 2\,| t | \leqslant 6\,e \operatorname{Log} | t |$$

Using the above estimates for $| \zeta(\sigma + \eta) |$ and $| \zeta(\sigma + \eta + i\,2\,t) |$ we deduce that

$$1 \leqslant \left(\frac{6\,e}{\eta} \right)^3 | \zeta(s + \eta) |^4 \, 6\,e \operatorname{Log} | t | = \frac{(6\,e)^4}{\eta^3} (\operatorname{Log} | t |) \, | \zeta(s + \eta) |^4$$

and so

$$| \zeta(s + \eta) | \geqslant \frac{1}{6\,e} \eta^{3/4} (\operatorname{Log} | t |)^{-1/4}\,.$$

To deduce a lower bound for $| \zeta(s) |$ we observe that

$$\begin{aligned} | \zeta(s + \eta) - \zeta(s) | &= \left| \int_{\sigma}^{\sigma+\eta} \zeta'(u + it) \, du \right| \leqslant \int_{\sigma}^{\sigma+\eta} | \zeta'(u + it) | \, du \\ &\leqslant 6\,e\eta(\log | t |)^2 \end{aligned}$$

Thus we certainly have

$$\begin{aligned}|\zeta(s)| &\geqslant |\zeta(s+\eta)| - 6\,e\eta(\text{Log}\,|t|)^2 \\ &\geqslant \frac{1}{6\,e}\eta^{3/4}(\text{Log}\,|t|)^{-1/4} - 6\,e\eta(\text{Log}\,|t|)^2 .\end{aligned}$$

If we now choose

$$\eta = (2(6\,e)^2)^{-4}\,(\text{Log}\,|t|)^{-9}$$

then our lower bound for $|\zeta(s)|$ is

$$|\zeta(s)| \geqslant \frac{1}{16}(6\,e)^{-7}\,(\text{Log}\,|t|)^{-7}$$

and part (*c*) is proved.

(*d*) Since $Z(s) = -\zeta'(s)/\zeta(s)$ part (*d*) is an immediate consequence of parts (*b*) and (*c*). ■

For the moment we leave our study of $\zeta(s)$ and return to the proof of the Prime Number Theorem.

§ 3 THE PRIME NUMBER THEOREM

3.1 In this section we shall prove the Prime Number Theorem by carrying out in detail the programme outlined in § 1.2. We state the classical form of the result.

Theorem 2.5 *As x tends to infinity*

$$\pi(x) \sim \frac{x}{\text{Log}\,x} .$$

By theorem 1.5 we know that the Prime Number Theorem is equivalent to the statement :

As x tends to infinity, $\psi(x) \sim x$.

It is this latter result which we will eventually prove in § 3.4.

3.2 The first step in our programme was to establish the relationship

$$Z(s) = s(s+1)\int_0^\infty \frac{\psi_1(x)}{x^{s+1}}\,dx$$

and then « invert » this integral to express $\psi_1(x)$ as a function of Z and x. Before we do this it is convenient to introduce some notation and to prove a preliminary theorem. We define the function $E(x)$ by

$$E(x) = \begin{cases} 0 & \text{if} \quad x \leqslant 1 \\ x - 1 & \text{if} \quad x > 1 . \end{cases}$$

Theorem 2.6 (a) *If* $\Re(s) > 0$, *then we have*

$$\frac{1}{s(s+1)} = \int_0^\infty \frac{E(x)}{x^{s+2}}\, dx .$$

(b) *If* $c > 0$ *is any fixed real number, then for all* $x > 0$ *we have*

$$E(x) = \frac{1}{2\, i\pi} \int_{c-i\infty}^{c+i\infty} \frac{x^{s+1}}{s(s+1)}\, ds .$$

Proof. The proof of *(a)* is a trivial exercise. In order to prove *(b)* we shall consider the following two cases : (i) $x \geqslant 1$ and (ii) $0 < x < 1$.

Case (i) : The singularities of the function $x^{s+1}/s(s+1)$ are simple poles at $s = 0$ and $s = -1$ with residues x and -1 respectively.

Consider the contour shown in figure 3. It is formed by an arc Γ of the circle centre O, radius $R = (c^2 + T^2)^{1/2}$ and a segment of the line $\Re(s) = c$. By Cauchy's residue theorem we have, for $T > 1$,

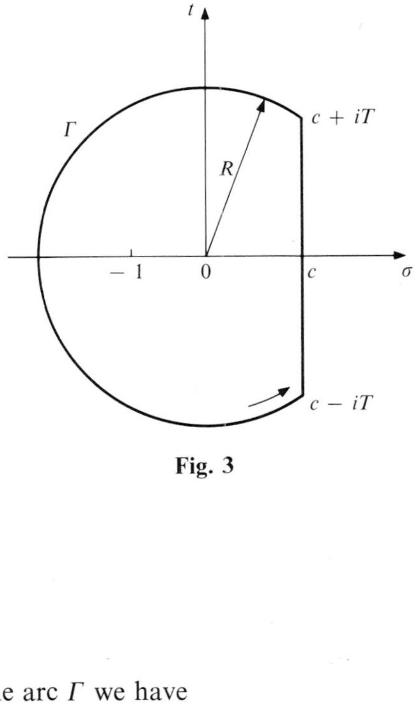

Fig. 3

$$x - 1 = \frac{1}{2\, i\pi} \int_{c-i\infty}^{c+i\infty} \frac{x^{1+s}}{s(s+1)}\, ds + \\ + \frac{1}{2\, i\pi} \int_{\Gamma'} \frac{x^{1+s}}{s(s+1)}\, ds .$$

As T tends to infinity the first integral tends to

$$\frac{1}{2\, i\pi} \int_{c-i\infty}^{c+i\infty} \frac{x^{1+s}}{s(s+1)}\, ds$$

Now let us consider the second integral. On the arc Γ we have

$$\left| \frac{x^{1+s}}{s(s+1)} \right| \leqslant \frac{x^{1+c}}{R(R-1)},$$

hence it follows that

$$\left| \int_{\Gamma} \frac{x^{1+s}}{s(s+1)} ds \right| \leqslant \frac{2\pi R x^{1+c}}{R.(R-1)} \to 0 .$$

which tends to zero as T tends to infinity. Thus we have shown that if $x > 1$, then

$$x - 1 = \frac{1}{2i\pi} \int_{c-i\infty}^{c+i\infty} \frac{x^{1+s}}{s(s+1)} ds.$$

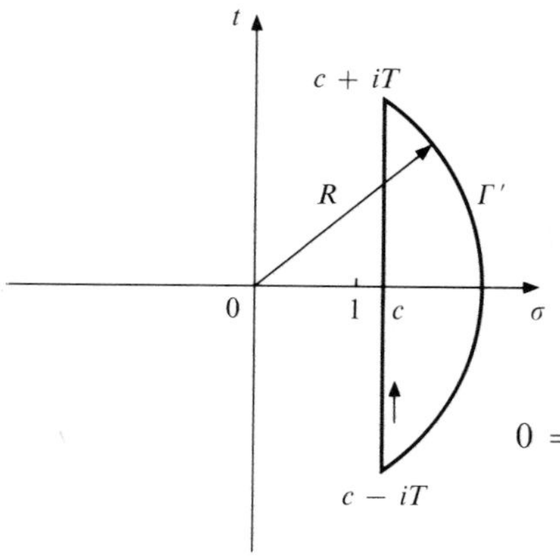

Fig. 4

Case (ii) : We now consider the integral of $x^{1+s}/s(s+1)$ around the contour shown in figure 4, which is formed by an arc of the circle centre O, radius $(c^2 + T^2)^{1/2}$ and a segment of the line $\mathcal{R}(s) = c$.

The function is holomorphic on and within this contour, hence by the residue theorem :

$$0 = \frac{1}{2i\pi} \int_{c-iT}^{c+iT} \frac{x^{1+s}}{s(s+1)} ds + \frac{1}{2i\pi} \int_{\Gamma'} \frac{x^{1+s}}{s(s+1)} ds .$$

As T tends to infinity the first integral tends to

$$\frac{1}{2i\pi} \int_{c-i\infty}^{c+i\infty} \frac{x^{1+s}}{s(s+1)} ds$$

and the second integral tends to zero since $0 < x < 1$. This completes the proof of the theorem. ■

We are now in a position to prove the " inversion " theorem in the following form.

Theorem 2.7 (a) *If* $\mathcal{R}(s) > 1$, *then*

$$Z(s) = s(s+1) \int_1^{\infty} \psi_1(x) \frac{dx}{x^{s+2}} .$$

(b) *For all* $x \geqslant 1$ *and any* $c > 1$ *we have*

$$\psi_1(x) = \frac{1}{2i\pi} \int_{c-i\infty}^{c+i\infty} \frac{Z(s)}{s(s+1)} x^{s+1} ds .$$

Proof. (*a*) By the definition of Z(*s*), if $\mathfrak{R}(s) > 1$, then

$$Z(s) = \sum_{n=1}^{\infty} \frac{\Lambda(n)}{n^s}.$$

By the Abel summation formula (theorem 1.7), and the fact that $\psi(x) = O(x)$, we have

$$Z(s) = s \int_1^{\infty} \psi(x) \frac{dx}{x^{s+1}}. \tag{2.9}$$

Integrate the above integral by parts and obtain

$$Z(s) = s\left[\frac{\psi_1(x)}{x^{s+1}}\right]_1^{\infty} + s(s+1)\int_1^{\infty} \psi_1(x) \frac{dx}{x^{s+2}}.$$

Because $\psi(x) = O(x)$, it follows that $\psi_1(x) = O(x^2)$ and since $\mathfrak{R}(s) > 1$, the last expression is zero. This proves (*a*).

(*b*) If $c > 1$, then from the definition of Z(*s*) we have

$$\int_{c-i\infty}^{c+i\infty} Z(s) \frac{x^{1+s}}{s(s+1)} ds = \sum_{n=1}^{\infty} n\Lambda(n) \int_{c-i\infty}^{c+i\infty} \frac{(x/n)^{s+1}}{s(s+1)} ds.$$

since the series

$$\int_{c-i\infty}^{c+i\infty} \sum_{n=1}^{\infty} \left| \frac{\Lambda(n)}{n^s} \cdot \frac{x^{s+1}}{s(s+1)} ds \right|$$

is convergent we can appeal to the theorem of dominated convergence to justify interchanging the summation and integration in the above equation to obtain :

$$\int_{c-i\infty}^{c+i\infty} Z(s) \frac{x^{1+s}}{s(s+1)} ds = \int_{c-i\infty}^{c+i\infty} \sum_{n=1}^{\infty} \frac{\Lambda(n)}{n^s} \cdot \frac{x^{s+1}}{s(s+1)} ds.$$

Using theorem 2.6(*b*) it now follows that

$$\int_{c-i\infty}^{c+i\infty} Z(s) \frac{x^{1+s}}{s(s+1)} ds = 2\, i\pi \sum_{n=1}^{\infty} n\Lambda(n)\, E\left(\frac{x}{n}\right),$$

$$2\, i\pi \sum_{n \leqslant x} \Lambda(n)\,(x - n).$$

Applying the Abel summation formula (theorem 1.6), to the last sum we see that it is equal to $\psi_1(z)$. Thus, we have shown that if $c > 1$, then

$$\psi_1(x) = \frac{1}{2\pi i} \int_{c-i\infty}^{c+i\infty} \frac{Z(s)\, x^{1+s}}{s(s+1)} ds. \qquad \blacksquare$$

3.3 We are now in a position to obtain an asymptotic formula for $\psi_1(x)$ along the lines indicated in § 1.

Theorem 2.8 *As x tends to infinity*

$$\psi_1(x) \sim \frac{1}{2}x^2 .$$

Proof. For any $c > 1$ and $x \geqslant 1$ it follows from theorem 2.7(*b*) that

$$\frac{\psi_1(x)}{x^2} = \frac{1}{2\,i\pi}\int_{c-i\infty}^{c+i\infty} \frac{Z(s)}{s(s+1)}\, x^{s-1}\, ds .$$

We shall show that as x tends to infinity the above integral tends to $i\pi$. This will be done by integrating the function

$$A(s) = \frac{Z(s)\, x^{s-1}}{s(s+1)}$$

around the contour $\mathcal{C}(T, T_0)$ shown in figure 2 of § 1.3.

The only singularity of $A(s)$ on or within the contour $\mathcal{C}(T, T_0)$ is a simple pole at $s = 1$ with residue $\frac{1}{2}$. If we denote the broken line $BCDEFG$ by $\mathcal{L}(T, T_0)$, then the residue theorem gives

$$\pi i = \int_{HA} A(s)\, ds + \int_{AB} A(s)\, ds + \int_{\mathcal{L}(T,T_0)} A(s)\, ds + \int_{GH} A(s)\, ds .$$

For a fixed value of x it follows from theorem 2.4(*d*) that both the integrals

$$\int_A^B A(s)\, ds \quad \text{and} \quad \int_G^H A(s)\, ds$$

are $O(T_0^{-2}(\text{Log } T_0)^B)$. Thus, as T_0 tends to infinity both these integrals tend to zero and we have

$$\begin{aligned}\frac{1}{2} &= \frac{1}{2\,\pi i}\int_{c-i\infty}^{c+i\infty} A(s)\, ds + \frac{1}{2\,\pi i}\int_{\mathcal{L}(T,\infty)} A(s)\, ds \\ &= \frac{\psi_1(x)}{x^2} + \frac{1}{2\,\pi i}\int_{\mathcal{L}(T,\infty)} A(s)\, ds .\end{aligned}$$

All that remains now is to show that given any $\varepsilon > 0$, then for all sufficiently large values of T and x

$$\left| \int_{\mathcal{L}(T,\infty)} A(s)\, ds \right| < \varepsilon .$$

It is obvious that

$$\left| \int_{\mathcal{L}(T,\infty)} A(s)\, ds \right| \leqslant \left| \int_{-\infty}^{F} A(s)\, ds \right| + \left| \int_{\mathcal{M}(T)} A(s)\, ds \right| + \left| \int_{c}^{\infty} A(s)\, ds \right|,$$

where $\mathcal{M}(T)$ denotes the broken line $FEDC$. From theorem 2.4(d) we know that on the two segments $[-\infty, F]$ and $[c, \infty]$ we certainly have

$$A(1 + it) = \mathrm{O}(t^{-2} \operatorname{Log}^B t)\,.$$

Hence both the integrals

$$\int_{-\infty}^{F} A(s)\, ds \quad \text{and} \quad \int_{c}^{\infty} A(s)\, ds$$

are $\mathrm{O}(T^{-1})$. Thus, given any $\varepsilon > 0$, there exists $T(\varepsilon)$ such that for all $T > T(\varepsilon)$ we have

$$\left| \int_{-\infty}^{F} A(s)\, ds \right| + \left| \int_{c}^{\infty} A(s)\, ds \right| < \frac{1}{2}\varepsilon\,.$$

We now study the remaining integral. Let $T > T(\varepsilon)$ be chosen and denote by $K(T)$ the maximum value of $|\mathrm{Z}(s)/s(s+1)|$ on the broken line $\mathcal{M}(T)$. It now follows that

$$\begin{aligned} \left| \int_{\mathcal{M}(T)} A(s)\, ds \right| &\leqslant \int_{\mathcal{M}(T)} \left| \frac{\mathrm{Z}(s)}{s(s+1)} x^{s-1}\, ds \right| \\ &\leqslant K(T) \left\{ 2 \int_{\beta}^{1} x^{\sigma-1}\, d\sigma + \int_{-T}^{+T} x^{\beta-1}\, dt \right\} \\ &= 2\, K(T) \left\{ \frac{1}{\operatorname{Log} x} - \frac{x^{\beta-1}}{\operatorname{Log} x} + T x^{\beta-1} \right\}. \end{aligned}$$

For all sufficiently large x, say $x > x(T, \varepsilon)$ the above expression is less than $\frac{1}{2}\varepsilon$. Thus we have shown that for all sufficiently large values of T and x

$$\left| \int_{\mathcal{L}(T,\infty)} A(s)\, ds \right| < \varepsilon\,,$$

which implies that $\psi_1(x) \sim \frac{1}{2}x^2$. ■

3.4 Proof of the Prime Number Theorem. The final step in the argument is to deduce that $\psi_1(x) \sim \frac{1}{2} x^2$ implies $\psi(x) \sim x$.

Let $h = h(z)$ be a function of x, to be chosen explicitly later, satisfying for all $x > 0$ the condition $0 < h(x) < x$.

Because ψ is a positive, monotone increasing function of x we have

$$\frac{1}{h} \int_{x-h}^{x} \psi(u)\, du \leqslant \psi(x) \leqslant \frac{1}{h} \int_{x}^{x+h} \psi(u)\, du ,$$

or equivalently

$$\frac{\psi_1(x) - \psi_1(x-h)}{h} \leqslant \psi(x) \leqslant \frac{\psi_1(x+h) - \psi_1(x)}{h} .$$

Using the fact that $\psi_1(x) = \frac{1}{2} x^2 + o(x^2)$ we see the above expression is

$$1 - \varepsilon/2 + o(1) \leqslant \frac{\psi(x)}{x} \leqslant 1 + \varepsilon/2 + o(1)$$

and as ε is arbitrary small we conclude that $\psi(x) \sim x$.

§ 4 ANALYSIS OF THE PROOF OF THE PRIME NUMBER THEOREM

4.1 It is of some methodological interest to look at our proof of the Prime Number Theorem and ask whether it is really necessary to methods which appear to be far removed from the simple limit operation involved in the theorem. The principal tools which we used were :

(*a*) Cauchy's residue theorem,

(*b*) an upper bound for $| Z(1 + it) |$,

(*c*) the fact that $\zeta(s)$ is holomorphic in the region $\mathfrak{R}(s) > 0$, $s \neq 1$ and has no zeros on the line $\mathfrak{R}(s) = 1$.

We shall now examine (*a*), (*b*), (*c*) in turn to see if they are really necessary ingredients in a proof of the Prime Number Theorem.

4.2 The use of Cauchy's theorem can be easily eliminated from the proof. It is only used in the proofs of theorems 2.6 and 2.8.

Theorem 2.6 is a simple consequence of the following " inversion " theorem for Fourier Transforms :

$$\textit{If} \quad g(\xi) = \int_{-\infty}^{+\infty} f(t)\, e^{-i\xi t}\, dt, \quad \textit{then} \quad f(t) = \frac{1}{2\pi} \int_{-\infty}^{+\infty} g(\xi)\, e^{i\xi t}\, d\xi . \tag{2.10}$$

provided that both f and g are in $\mathfrak{L}^1$.

If we write $x = e^\xi$ and $s = c + it$, then the formulae of theorem 2.6 become

$$\frac{1}{(c + it)(c + 1 + it)} = \int_{-\infty}^{+\infty} \frac{E(e^\xi)}{e^{\xi(c+1)}} e^{-i\xi t} \, d\xi$$

and

$$\frac{E(e^\xi)}{e^{\xi(c+1)}} = \frac{1}{2\pi} \int_{-\infty}^{+\infty} \frac{1}{(c + it)(c + 1 + it)} e^{i\xi t} \, dt \, ;$$

which is a special case of the general theorem.

Now we show how use of Cauchy's theorem can be avoided in the proof of theorem 2.8. If we replace s by $s + 1$ in theorem 2.7, then for any $c > 0$ and all $x \geqslant 1$ we have

$$\frac{\psi_1(x)}{x^2} - \frac{1}{2}\left(1 - \frac{1}{x}\right) = \frac{1}{2\, i\pi} \int_{c-i\infty}^{c+i\infty} \frac{1}{s+1} \left\{ \frac{Z(s+1)}{s+2} - \frac{1}{2\, s} \right\} x^s \, ds \, .$$

Thus, in order to prove theorem 2.8 we must show that the above integral tends to zero when x tends to infinity. This is an easy consequence of the following general result.

Theorem 2.9 *Suppose that $c > 0$ and the function f defined for all $x \geqslant 1$ by*

$$\chi(x) = \int_{c-i\infty}^{c+i\infty} g(s)\, x^s \, ds$$

is independent of c and $g(s)$ satisfies :

(i) *$g(s)$ is continuous on the half-plane $\mathfrak{R}(s) \geqslant 0$,*
(ii) *$| g(s) | \leqslant K(1 + | t |)^{-3/2}$ uniformly in σ on the half-plane $\mathfrak{R}(s) \geqslant 0$. Then*

$$\lim_{x \to \infty} f \, \chi(x) = 0 \, .$$

Proof. Write $x = e^\omega$ and $s = c + it$, then

$$\chi(e^\omega)\, e^{-\omega c} = i \int_{-\infty}^{+\infty} g(c + it)\, e^{i\omega t} \, dt \, .$$

Now take successively $c = \sigma$ and $c = 2\sigma$ in the above equation. A subtraction then yields

$$\chi(e^\omega)\,(e^{-\omega\sigma} - e^{-2\omega\sigma}) = i \int_{-\infty}^{+\infty} [g(\sigma + it) - g(2\sigma + it)]\, e^{i\omega t} \, dt \, . \qquad (2.11)$$

Thus, by using (ii), we certainly have

$$|\chi(e^{\omega})(e^{-\omega\sigma} - e^{-2\omega\sigma})| \leqslant \int_{-\infty}^{+\infty} |g(\sigma + it) - g(2\sigma + it)|\, dt = G(\sigma) .$$

Hypotheses (i) and (ii) imply that the above integral is uniformly convergent on every compact subset of the half-plane $\mathcal{R}(s) \geqslant 0$, hence

$$\lim_{\sigma \to 0^+} \int_{-\infty}^{+\infty} |g(\sigma + it) - g(2\sigma + it)|\, dt = \int_{-\infty}^{+\infty} |g(\sigma + it) - g(2\sigma + it)|\, dt = 0 .$$

If we choose $\sigma = \omega^{-1}$ in (2.11), then

$$\lim_{\omega \to \infty} \{ \chi(e^{\omega})(e^{-1} - e^{-2}) \} = 0 ,$$

and the theorem is proved. ■

If we take

$$g(s) = \frac{1}{s+1} \left\{ \frac{Z(s+1)}{s+2} - \frac{1}{2s} \right\}$$

then theorems 2.1 and 2.2 imply that $g(s)$ satisfies condition (i) of the theorem and theorem 2.4(d) implies that condition (ii) is satisfied. Hence we can conclude, without use of Cauchy's theorem, that $\psi_1(x) \sim \frac{1}{2} x^2$.

4.3 It is more difficult to eliminate both theorem 2.4(d) and the Cauchy residue theorem from a proof of the Prime Number Theorem. One possible way of doing this is to appeal to the following general theorem of Ikehara, which will be proved in § 5.

Theorem 2.10 *Suppose that the function F has the following properties :*

(i) *In the half-plane $\mathcal{R}(s) > 1$ the function F has the representation*

$$F(s) = \int_0^{\infty} A(u)\, e^{-su}\, du ,$$

where A is a positive, monotone increasing function.

(ii) *In the region $\mathcal{R}(s) \geqslant 1$, $s \neq 1$ the function F has the representation*

$$G(s) = F(s) - \frac{1}{s-1}$$

where $G(s)$ is continuous on the half-plane $\mathcal{R}(s) \geqslant 1$.

Then we have

$$\lim_{u \to \infty} A(u)\, e^{-u} = 1\,.$$

The Prime Number Theorem is now a simple consequence of the above theorem, for if we write $A(u) = \psi(e^u)$, then for $\mathcal{R}(s) > 1$ we have by (2.9) :

$$F(s) = \frac{Z(s)}{s} = \int_1^\infty \psi(e^u)\, e^{-su}\, du\,.$$

and from theorems 2.1 and 2.2 it follows that

$$\frac{Z(s)}{s} - \frac{1}{s-1} = G(s)$$

is continuous on the half-plane $\mathcal{R}(s) \geqslant 1$. Thus the conditions of theorem 2.10 are satisfied and we can conclude that

$$\lim_{u \to \infty} \psi(e^u)\, e^{-u} = 1 \quad \text{or} \quad \psi(x) \sim x\,.$$

The only properties of $\zeta(s)$ used in this proof of the Prime Number Theorem are given by theorems 2.1 and 2.2.

4.4 The elimination of theorem 2.2 from a proof of the Prime Number Theorem is possible and chapter 3 will be principally devoted to the exposition of the so-called " elementary " technique. Historically this was the most difficult proof to find. Indeed it was long thought to be impossible for as we remarked in § 1.1, if one assumes theorem 2.1 and the Prime Number Theorem, then one can deduce theorem 2.2. The proof is as follows. For $\mathcal{R}(s) > 1$ we have

$$Z(s) - \frac{s}{s-1} = s \int_1^\infty \frac{\psi(x) - x}{x^{s+1}}\, dx\,.$$

and using the Prime Number Theorem in the form $\psi(x) = x + R(x)$, where $R(x) = o(x)$ it follows that

$$\left| Z(s) - \frac{s}{s-1} \right| \leqslant |\, s\,| \int_1^\infty \frac{|\, R(x)\,|}{x^{\sigma+1}}\, dx = o\left(\frac{1}{\sigma - 1}\right).$$

In particular, if s tends to $1 + it_0$, where $t_0 \neq 0$, then

$$Z(s) = o\left(\frac{1}{\sigma - 1}\right).$$

Thus $Z(s)$ does not have a pole at $1 + it_0$ and consequently $\zeta(s)$ does not have a zero at $1 + it_0$, which is theorem 2.2.

§ 5 IKEHARA'S THEOREM

5.1 In this section we shall prove a slightly stronger version of theorem 2.10. Ikehara's theorem and theorem 2.11 belong to an important class of analytic results known, for historical reasons, as *tauberian theorems.* This class forms a rich and fascinating branch of analysis with numerous applications to number theory. For an introduction to the subject see Weiner [1], Pitt [1] and Delange [1].

Theorem 2.11 *Suppose that the function F has the following properties :*

(i) *In the half-plane* $\mathfrak{R}(s) > 1$ *the function F has the representation*

$$F(s) = \int_0^\infty A(u)\, e^{-us}\, du$$

where A is a positive, monotone increasing function.

(ii) *There exists* $T > 0$ *such that in the region* $\mathfrak{R}(s) \geqslant 1$, $s \neq 1$, $|\mathfrak{Im}(s)| \leqslant T$ *the function F has the representation*

$$F(s) - \frac{1}{s-1} = G(s)$$

where $G(s)$ *is continuous on the strip* $\mathfrak{R}(s) \geqslant 1$, $|\mathfrak{Im}(s)| \leqslant T$.

Then there exists two functions P_1 *and* P_2 *such that*

(a) $\lim\limits_{x\to\infty} P_1(x) = \lim\limits_{x\to\infty} P_2(x) = 1$,

(b) $0 < P_1(T) \leqslant \liminf\limits_{u\to\infty} A(u)\, e^{-u} \leqslant \limsup\limits_{u\to\infty} A(u)\, e^{-u} \leqslant P_2(T) < \infty$.

The functions P_1 *and* P_2 *can be determined explicitly.*

5.2 It is clear that Ikehara's theorem is a simple consequence of the above theorem, since we can take T as large as we please.

The applications of theorems 2.10 and 2.11 to number theory arise as follows. If in $\mathfrak{R}(s) > 1$ the function $f(s)$ has the representation

$$f(s) = \sum_{n=1}^{\infty} \frac{a_n}{n^s}$$

where $a_n \geqslant 0$, then provided that $S(x) = \sum\limits_{n\leqslant x} a_n = o(x^\delta)$ for every $\delta > 1$, we can use the Abel summation formula to deduce

$$f(s) = s \int_1^\infty \frac{S(x)}{x^{s+1}}\, dx\ .$$

A simple change of variable then yields

$$F(s) = \frac{f(s)}{s} = \int_0^\infty A(u)\, e^{-su}\, du$$

where $S(e^u) = A(u)$. We can now try and prove that $F(s)$ satisfies condition (ii) of the theorem and so deduce something about the asymptotic behaviour of $S(x)$.

With the hypotheses of theorem 2.11 one cannot, in general, deduce a stronger conclusion. For example, if we take

$$F(s) = \frac{1}{s}\left\{\zeta(s) + \frac{1}{2}\zeta(s-i) + \frac{1}{2}\zeta(s+i)\right\} = \frac{1}{s}\sum_{n=1}^{\infty} \frac{1+\cos(\operatorname{Log} n)}{n^s}$$

then it is easy to see that the hypotheses of theorem 2.11 are only satisfied for $T < 1$. So we cannot conclude that $\lim_{x\to\infty} S(x)/x = 1$. This is not supprising, for a simple application of the Abel summation formula shows that

$$S(x) = x + \frac{1}{\sqrt{2}}\, x \cos\left(\operatorname{Log} x - \frac{\pi}{4}\right) + o(x)\,,$$

thus

$$1 - \frac{1}{\sqrt{2}} = \liminf_{x\to\infty} \frac{S(x)}{x} \neq \limsup_{x\to\infty} \frac{S(x)}{x} = 1 + \frac{1}{\sqrt{2}}\,.$$

Many variations of theorem 2.11 are known, some of which are given in § 5.7. We shall not however make any use of them in later chapters.

5.3 Proof of theorem 2.11. On the half strip $\mathcal{R}(s) > 0$, $|\mathcal{I}m(s)| \leqslant T$ we have

$$F(s+1) - \frac{1}{s} = G(s+1)$$

and so $G(s+1)$ has the integral representation

$$G(\sigma + 1 + it) = \int_0^\infty (A(u)\, e^{-u} - 1)\, e^{-u\sigma}\, e^{-iut}\, du\,.$$

on the half strip. The basic idea behind the proof is to observe that for $\sigma > 0$, $F(s+1) - s^{-1}$ is the Fourier transform of $(A(u)\, e^{-u} - 1)\, e^{-u\sigma}$. It is now natural to try to mimic the inversion theorem for Fourier transforms and to try to express $(A(u)\, e^{-u} - 1)\, e^{-u\sigma}$ in terms of $G(s+1)$. One can then hope to use the hypotheses on G to deduce information about $A(u)\, e^{-u} - 1$.

The first difficulty in carrying out the above programme is that $G(\sigma + 1 + it)$, considered as a function of t, is only defined for $|t| \leqslant T$ and so does not have a Fourier transform ! To circumvent this difficulty we shall introduce a real, continuous function k with support $[-1, +1]$, then work with the function

$$G(\sigma + 1 + it)\, k\left(\frac{t}{T}\right),$$

which is now defined on $(-\infty, +\infty)$.

The precise choice of k will not be very important for most of the proof. It only becomes relevant in §§ 5.5 and 5.6 when we explicitly exhibit the functions P_1 and P_2. However, for technical convenience we shall assume that k is such that its Fourier transform, k^*, is absolutely integrable on $(-\infty, \infty)$. That is

$$k^*(y) = \int_{-\infty}^{+\infty} k(x)\, e^{-ixy}\, dx \quad \text{and} \quad \int_{-\infty}^{+\infty} |k^*(y)|\, dy < \infty .$$

We remark that from (2.10) we have

$$\int_{-\infty}^{+\infty} k^*(y)\, dy = 2\pi k(0) . \tag{2.12}$$

For the specific choices of k to be made in §§ 5.5 and 5.6 it will be apparent that the hypotheses on k^* are satisfied.

5.4 Following the idea of mimicking the Fourier inversion theorem we consider, for any fixed $\sigma > 0$

$$\int_{-\infty}^{+\infty} G(\sigma + 1 + it)\, k\left(\frac{t}{T}\right) e^{i\omega t}\, dt ,$$

From the integral representation of $G(\sigma + 1 + it)$ we have

$$\int_{-\infty}^{+\infty} \int_{0}^{\infty} (A(u)\, e^{-u} - 1)\, e^{-u\sigma}\, k\left(\frac{t}{T}\right) e^{i(\omega t - ut)}\, du\, dt .$$

Because the integral on the right hand side is absolutely uniformly convergent on all compact subsets of the domain $-\infty < t < \infty$, $u > 0$ we can interchange the order of integration and obtain

$$\int_{-\infty}^{+\infty} G(\sigma + 1 + it)\, k\left(\frac{t}{T}\right) e^{i\omega t}\, dt = \int_{0}^{\infty} (A(u)\, e^{-u} - 1)\, e^{-u\sigma}\, Tk^*(uT - \omega T)\, du .$$

Now we utilise the idea used in the proof of theorem 2.9 and write the above equation with 2 σ in place of σ. A subtraction then yields

$$H_\sigma^*(\omega) - H_{2\sigma}^*(\omega) = \int_{-\infty}^{+\infty} \{ G(\sigma + 1 + it) - G(2\sigma + 1 + it) \} k\left(\frac{t}{T}\right) e^{i\omega t} \, dt$$

$$= \int_0^\infty (A(u)\, e^{-u} - 1)(e^{-u\sigma} - e^{-2u\sigma})\, Tk^*(uT - \omega T)\, du\,. \tag{2.13}$$

We look at the left hand side of the above equation. Since k is continuous on its support it follows that k is bounded, say by K. Thus we have

$$\left| \int_{-\infty}^{+\infty} \{ G(\sigma + 1 + it) - G(2\sigma + 1 + it) \} k\left(\frac{t}{T}\right) e^{i\omega t} \, dt \right| \leqslant$$

$$\leqslant K \int_{-T}^{+T} | G(\sigma + 1 + it) - G(2\sigma + 1 + it) | \, dt\,.$$

The continuity of $G(s + 1)$ in the half strip $\mathfrak{R}(s) \geqslant 0$, $|\mathfrak{Im}(t)| \leqslant T$ implies that the right hand side of the inequality tends to zero when σ tends to zero. Hence, if we write $\sigma = \omega^{-1}$ and denote the left hand side of (2.13) by $L(\omega)$, then

$$L(\omega) = \int_0^\infty (A(u)\, e^{-u} - 1)(e^{-u/\omega} - e^{-2u/\omega})\, Tk^*(uT - \omega T)\, du$$

and

$$L(\omega) \to 0 \quad \text{when} \quad \omega \to \infty\,.$$

For convenience later we now introduce some new notation. First we write $u = \omega + y/T$. The expression for $L(\omega)$ becomes

$$L(\omega) = \int_{-\omega T}^{+\infty} \left\{ A\left(\omega + \frac{y}{T}\right) e^{-\omega - y/T} - 1 \right\} \{ e^{-1 - y/\omega T} - e^{-2 - 2y/\omega T} \}\, k^*(y)\, dy\,.$$

Second we define the functions $\alpha(u)$ and $Q(u)$ by

$$\alpha(u) = A(u)\, e^{-u} \quad \text{and} \quad Q(u) = \begin{cases} e^{-u} - e^{-2u} & \text{if} \quad u \geqslant 0 \\ 0 & \text{if} \quad u < 0\,. \end{cases}$$

For use later we observe that the maximum of $Q(u)$ occurs at $u = \operatorname{Log} 2$ and that for $u > \operatorname{Log} 2$ the function $Q(u)$ is monotone decreasing.

With the above notation the integral becomes

$$L(\omega) = \int_{-\omega T}^{+\infty} \alpha\left(\omega + \frac{y}{T}\right) Q\left(1 + \frac{y}{\omega T}\right) k^*(y)\, dy - \int_{-\infty}^{+\infty} Q\left(1 + \frac{y}{\omega T}\right) k^*(y)\, dy\,. \tag{2.14}$$

Because the second of the above integrals is absolutely uniformly convergent on all compact subsets of $(0, \infty)$ it follows that

$$\lim_{\omega\to\infty} \int_{-\infty}^{+\infty} Q\left(1 + \frac{y}{\omega T}\right) k^*(y)\, dy = \int_{-\infty}^{+\infty} \lim_{\omega\to\infty} Q\left(1 + \frac{y}{\omega T}\right) k^*(y)\, dy$$

$$= Q(1) \int_{-\infty}^{+\infty} k^*(y)\, dy\,.$$

Thus we have

$$\int_{-\infty}^{+\infty} Q\left(1 + \frac{y}{\omega T}\right) k^*(y)\, dy = Q(1) \int_{-\infty}^{+\infty} k^*(y)\, dy + K(\omega)\,,$$

where $K(\omega) \to 0$ when $\omega \to \infty$ and (2.14) becomes

$$\int_{-\omega T}^{+\infty} \alpha\left(\omega + \frac{y}{T}\right) Q\left(1 + \frac{y}{\omega T}\right) k^*(y)\, dy = Q(1) \int_{-\infty}^{+\infty} k^*(y)\, dy + L(\omega) + K(\omega). \tag{2.15}$$

This completes the formal manipulations. What we are going to do now is make specific choices for k in such a way that we can deduce upper and lower bounds for lim sup $\alpha(u)$ and lim inf $\alpha(u)$ respectively.

5.5 An upper bound for lim sup $\alpha(u)$. Our choice of k here is

$$k(x) = \begin{cases} 1 - |x| & \text{if } |x| \leqslant 1 \\ 0 & \text{if } |x| > 1\,. \end{cases}$$

A simple calculation shows that

$$k^*(y) = \int_{-\infty}^{+\infty} k(x)\, e^{-ixy}\, dx = \left(\frac{\sin \frac{1}{2} y}{\frac{1}{2} y}\right)^2 .$$

Let μ be a fixed real number to be chosen explicitly later. We also suppose that ω is so large that

$$0 < \mu \leqslant \omega T \quad \text{and} \quad 1 - \frac{\mu}{\omega T} > \operatorname{Log} 2\,.$$

Since α, Q and k^* are all non-negative functions we have

$$\int_{-\omega T}^{+\infty} \alpha\left(\omega + \frac{y}{T}\right) Q\left(1 + \frac{y}{\omega T}\right) k^*(y)\, dy \geqslant \int_{-\mu}^{+\mu} \alpha\left(\omega + \frac{y}{T}\right) Q\left(1 + \frac{y}{\omega T}\right) k^*(y)\, dy\,.$$

Recall that $A(u) = \alpha(u)\, e^u$ is monotone increasing for all $u > 0$, thus if $-\mu \leqslant y \leqslant \mu$, then

$$\alpha\left(\omega + \frac{y}{\omega T}\right) \geqslant \alpha\left(\omega - \frac{y}{\omega T}\right) e^{-2\mu/T}\,.$$

Also we recall that $Q(u)$ is monotonic decreasing for $u > \text{Log}\, 2$, hence if $-\mu \leqslant y \leqslant \mu$, then

$$Q\left(1 + \frac{y}{\omega T}\right) \geqslant Q\left(1 + \frac{\mu}{\omega T}\right).$$

It now follows that

$$\int_{-\mu}^{+\mu} \alpha\left(\omega + \frac{y}{T}\right) Q\left(1 + \frac{y}{\omega T}\right) k^*(y)\, dy \geqslant e^{-2\mu/T} \alpha\left(\omega - \frac{\mu}{T}\right) Q\left(1 + \frac{\mu}{\omega T}\right) \int_{-\mu}^{+\mu} k^*(y)\, dy\,.$$

We write

$$J(\eta) = \int_{-\eta}^{+\eta} k^*(y)\, dy = \int_{-\eta}^{+\eta} \left(\frac{\sin \frac{1}{2} y}{\frac{1}{2} y}\right)^2 dy\,.$$

then by (2.12) $= 2\,\pi$ and so

$$\alpha\left(\omega - \frac{\mu}{T}\right) \leqslant \frac{Q(1)}{Q\left(1 + \frac{\mu}{\omega T}\right)} \frac{2\,\pi}{J(\mu)} e^{2\mu/T} + \frac{L(\omega) + K(\omega)}{Q\left(1 + \frac{\mu}{\omega T}\right) J(\mu)} e^{2\mu/T}$$

Consequently we have

$$\limsup_{u \to \infty} \alpha(u) = \limsup_{\omega \to \infty} \alpha\left(\omega - \frac{\mu}{T}\right) \leqslant \frac{2\,\pi}{J(\mu)} e^{2\mu/T}\,.$$

If we now choose $\mu = \sqrt{T}$ and define the function P_2 by

$$P_2(x) = \frac{2\pi}{J(\sqrt{x})} e^{2/\sqrt{x}} .$$

then $P_2(x) \to 1$ as $x \to \infty$ and $\limsup_{u \to \infty} \alpha(u) \leqslant P_2(T)$.

5.6 A lower bound for lim inf $\alpha(u)$. Let v be an integral multiple of 2π to be chosen explicitly later. Our choice for k in (2.15) is now

$$k(x) = \begin{cases} 1 - |x| + \dfrac{\sin vx}{v} & \text{for } |x| \leqslant 1 \\ 0 & \text{for } |x| > 1 . \end{cases}$$

A simple integration gives

$$k^*(y) = \int_{-\infty}^{+\infty} k(x)\, e^{-ixy}\, dx = \left(\frac{\sin \frac{1}{2} y}{\frac{1}{2} y} \right)^2 \frac{v^2}{v^2 - y^2} ,$$

We note that $k^*(y)$ is positive on the interval $[-v, +v]$ and negative on the intervals $(-\infty, -v]$ and $[v, \infty)$. If we now suppose that ω is so large that $\omega T \geqslant v$ and $1 - v/\omega T > \operatorname{Log} 2$, then

$$\int_{-v}^{+v} \alpha\left(\omega + \frac{y}{T}\right) Q\left(1 + \frac{y}{\omega T}\right) k^*(y)\, dy \geqslant$$
$$\int_{-\omega T}^{+\infty} \alpha\left(\omega + \frac{y}{T}\right) Q\left(1 + \frac{y}{\omega T}\right) k^*(y)\, dy .$$

From the fact that $\alpha(u)\, e^u$ is monotone increasing and $Q(u)$ monotone decreasing for $u > \operatorname{Log} 2$ we conclude

$$e^{2v/T} \alpha\left(\omega + \frac{v}{T}\right) Q\left(1 - \frac{v}{\omega T}\right) \int_{-v}^{+v} k^*(y)\, dy \geqslant$$
$$\int_{-v}^{+v} \alpha\left(\omega + \frac{y}{T}\right) Q\left(1 + \frac{y}{\omega T}\right) k^*(y)\, dy .$$

Let us write

$$I_v(\eta) = \int_{-\eta}^{+\eta} \left(\frac{\sin \frac{1}{2} y}{\frac{1}{2} y} \right)^2 \frac{v^2}{v^2 - y^2}\, dy = \int_{-\eta}^{+\eta} k^*(y)\, dy .$$

By (2.12) we know that $I_\nu(\infty) = 2\pi$. Hence from (2.15)

$$\alpha\left(\omega - \frac{\nu}{T}\right) \geqslant \frac{Q(1)}{Q\left(1 - \frac{\nu}{\omega T}\right)} \frac{2\pi}{I_\nu(\nu)} e^{-2\nu/T} + \frac{L(\omega) + K(\omega)}{Q\left(1 - \frac{\nu}{\omega T}\right) I_\nu(\nu)} e^{-2\nu/T},$$

Consequently we can conclude that

$$\liminf_{u\to\infty} \alpha(u) = \liminf_{\omega\to\infty} \alpha\left(\omega + \frac{\nu}{T}\right) \geqslant \frac{2\pi}{I_\nu(\nu)} e^{-2\nu/T}.$$

If we now choose $\nu = \nu(x) = 2\pi \max\{1, \sqrt{[x]}\}$ and define the function P_1 by

$$P_1(x) = \frac{2\pi}{I_{\nu(x)}(\nu(x))} e^{-2\nu(x)/x}.$$

then we have shown

$$\liminf_{u\to\infty} \alpha(u) \geqslant P_1(T).$$

All that remains is to prove that $P_1(x) \to 1$ when $x \to \infty$. This is equivalent to proving that if ν is a multiple of 2π, then

$$\lim_{\nu\to\infty} I_\nu(\nu) = 2\pi = \int_{-\infty}^{+\infty} \left(\frac{\sin\frac{1}{2}y}{\frac{1}{2}y}\right)^2 dy.$$

We shall prove this fact by using Lebesgue's dominated convergence theorem to show that as ν tends to infinity the following difference tends to zero :

$$D_\nu = \int_{-\nu}^{+\nu} \left(\frac{\sin\frac{1}{2}y}{\frac{1}{2}y}\right)^2 \frac{\nu^2}{\nu^2 - y^2}\,dy - \int_{-\nu}^{+\nu} \left(\frac{\sin\frac{1}{2}y}{\frac{1}{2}y}\right)^2 dy = \int_{-\nu}^{+\nu} \frac{4\sin^2\left(\frac{1}{2}y\right)}{\nu^2 - y^2}\,dy$$

If we define the sequence of functions $\{\varphi_\nu\}$ by

$$\varphi_\nu(y) = \begin{cases} 4\dfrac{\left(\sin\frac{1}{2}y\right)^2}{\nu^2 - y^2} & \text{if} \quad |y| \leqslant \nu \\ 0 & \text{if} \quad |y| > \nu. \end{cases}$$

then, because v is a multiple of 2π, we have $0 \leqslant \varphi_v(y) \leqslant 1$ and the dominated convergence theorem implies that

$$\lim_{v \to \infty} \int_{-v}^{+v} \frac{4 \sin^2 \left(\frac{1}{2} y\right)}{v^2 - y^2} dy = \lim_{v \to \infty} \int_{-\infty}^{+\infty} \varphi_v(y)\, dy = \int_{-\infty}^{+\infty} \lim_{v \to \infty} \varphi_v(y)\, dy = 0 .$$

Thus we have

$$\lim_{v \to \infty} I_v(v) = \int_{-\infty}^{+\infty} \left(\frac{\sin \frac{1}{2} y}{\frac{1}{2} y} \right)^2 dy = 2\pi .$$

and the proof of the theorem is complete. ■

5.7 Variations of theorem 2.11 Many variations of Ikehara's and theorem 2.11 are known, see for example Delange [1]. The ennunciations and proofs of such extensions tend to be rather elaborate. We shall state one extension which is useful in the study of generalised prime numbers.

Theorem 2.12 *Suppose that the function* $F(s)$ *has the following properties :*

(i) *There exists* $\beta > 0$ *such that for* $\mathfrak{R}(s) > \beta$ $F(s)$ *has the representation*

$$F(s) = \int_0^\infty A(u)\, e^{-us}\, du$$

where $A(u)$ *is a positive monotone increasing function.*

(ii) *There exist three real constants* $\alpha > -1$, $c > 0$ *and* $q > 1$ *such that the function*

$$G(s) = F(s) - c \frac{\Gamma(\alpha + 1)}{(s - \beta)^{\alpha + 1}}$$

satisfies the condition

$$\int_{-T}^{+T} | G(\sigma + it) - G(q(\sigma - \beta) + \beta + it) |\, dt = \mathrm{o}\left\{ \frac{1}{(\sigma - \beta)^\alpha} \right\}$$

when $\sigma \to \beta^+$.

Then there exists two functions $\mathrm{P}_1(x)$ *and* $\mathrm{P}_2(x)$, *such that*

(a) $\lim_{x \to \infty} \mathrm{P}_1(x) = \lim_{x \to \infty} \mathrm{P}_2(x) = 1$,

$$\text{(b)}\ 0 < c\mathrm{P}_1\left(\frac{T}{\beta}\right) \leqslant \liminf \frac{A(u)}{u^\alpha\, e^{\beta u}} \leqslant \limsup \frac{A(u)}{u^\alpha\, e^{\beta u}} \leqslant c\mathrm{P}_2\left(\frac{T}{\beta}\right) < \infty\,.$$

The proof of the above result is more or less the same as the proof of theorem 2.11. The main difference is that instead of changing σ to $2\,\sigma$ and subtracting to obtain (2.13) one changes σ to $q\sigma$ and then subtracts. The details of the proof are left as an exercise for the reader.

Condition (ii) is rather complicated, so we now give more special, but more easily applied alternative hypotheses :

$$\text{(ii)}'\quad \int_{-T}^{+T} |\, G'(\sigma + it)\,|\, dt = \mathrm{o}\left\{\frac{1}{(\sigma - \beta)^{\alpha+1}}\right\}$$

when $\sigma \to \beta^+$.

$$\text{(ii)}''\ F(s) = \frac{H(s)}{(s - \beta)^{\alpha+1}} \quad \text{et} \quad H(\beta) = c\Gamma(\alpha + 1)\,.$$

where H is holomorphic on the half-strip $\Re(s) \geqslant \beta, |\,\Im m(s)\,| \leqslant T$ *and* $H(\beta) = c\Gamma(\alpha+1)$.

With either of the above hypotheses in place of (ii) one still obtains the same conclusion.

§ 6 HELSON'S METHOD

6.1 We have already studied two important techniques for attacking the general problem mentioned in § 1.1. Now we shall briefly discuss a third very simple and elegant method due to Helson [1]. His principal tool is the use of Plancherel's theorem, namely :

If f and g are square integrable functions such that

$$g(y) = \int_{-\infty}^{+\infty} f(x)\, e^{-ixy}\, dx\,,$$

then

$$\int_{-\infty}^{+\infty} |\, f(x)\,|^2\, dx = \frac{1}{2\,\pi} \int_{-\infty}^{+\infty} |\, g(y)\,|^2\, dy\,.$$

A proof of this theorem can be found in Titchmarsh [1]. The general aim of Helson's method is as follows. Given a function $\varphi(s)$ defined for $\Re(s) > 1$ by

$$\varphi(s) = \sum_{n=1}^{\infty} \frac{a_n}{n^s}$$

then, provided certain regularity conditions are satisfied, one would like to show that for every $k > 0$

$$S(x) = \sum_{n \leqslant x} a_n = o\{ x/(\operatorname{Log} x)^k \} .$$

6.2 We shall first give the general structure of the method, then, as an illustrative example, a detailed proof that

$$M(x) = \sum_{n \leqslant x} \mu(x) = o(x/\operatorname{Log} x) .$$

The fact that this implies the Prime Number Theorem will be seen in chapter 3.

First assume that $S(x) = \sum_{n \leqslant x} a_n = o(x^\delta)$ for every $\delta > 1$. An application of the Abel summation formula then yields

$$\frac{\varphi(s)}{s} = \int_1^\infty \frac{S(x)}{x^{s+1}} dx = \int_0^\infty S(e^u) e^{-su} du .$$

Suppose now that we can differentiate the above equation k times so

$$(-1)^k \frac{d^k}{ds^k} \left\{ \frac{\varphi(s)}{s} \right\} = \int_{-\infty}^{+\infty} u^k S(e^u) e^{-su} du .$$

If we define F by

$$F(u) = \begin{cases} S(e^u) & \text{for} \quad u \geqslant 0 \\ 0 & \text{for} \quad u < 0 , \end{cases}$$

and

$$G(\sigma + it) = (-1)^k \frac{d^k}{ds^k} \left\{ \frac{\varphi(s)}{s} \right\}$$

then

$$G(\sigma + it) = \int_{-\infty}^{+\infty} \{ u^k S(e^u) e^{-\sigma u} \} e^{-itu} du .$$

Assuming we can prove that for $\sigma > 1$ both $G(\sigma + it)$ and $u^k F(u) e^{-\sigma u}$ are square integrable, then we can appeal to Plancherel's theorem to conclude

$$\int_{-\infty}^{+\infty} | u^k S(e^u) e^{-\sigma u} |^2 du = \frac{1}{2\pi} \int_{-\infty}^{+\infty} | G(\sigma + it) |^2 dt .$$

Furthermore if, as $\sigma \to 1^+$ the right hand integral is bounded it follows that there exists K such that for all $\sigma > 1$

$$\int_{-\infty}^{+\infty} | u^k F(u) e^{-u\sigma} |^2 \, du < K .$$

As $\sigma \to 1^+$ the above integrand is a positive, increasing function of σ which tends to $| u^k F(u) e^{-u} |^2$. Thus, by Beppo Levi's theorem

$$\limsup_{\sigma \to 1^+} \int_{-\infty}^{+\infty} | u^k \mathrm{S}(e^u) e^{-\sigma u} |^2 \, du \leqslant K \int_{-\infty}^{+\infty} | u^k \mathrm{S}(e^u) e^{-u} |^2 \, du \leqslant K < \infty .$$

It is now not very surprising that from the existence of the last integral one can usually prove

$$\lim_{u \to \infty} | u^k \mathrm{S}(e^u) e^{-u} | = 0 ,$$

which is, of course, equivalent to

$$\mathrm{S}(x) = \mathrm{o}(x/\mathrm{Log}^k x) .$$

6.3 We now apply Helson's method to study $M(x)$.

Theorem 2.13 *As x tends to infinity*

$$M(x) = \sum_{n \leqslant x} \mu(n) = \mathrm{o}\left(\frac{x}{\mathrm{Log}^{2/3} x}\right).$$

Proof. We define the function F for all real values of x by

$$F(x) = \begin{cases} M(e^x) & \text{if } x \geqslant 0 \\ 0 & \text{if } x < 0 . \end{cases}$$

Upon noting that $| F(x) | \leqslant e^x$ for all $x \geqslant 0$ we deduce from the Abel summation formula (theorem 1.7), that if $\mathfrak{R}(s) > 1$, then

$$\frac{1}{\zeta(s)} = \sum_{n=1}^{\infty} \frac{\mu(n)}{n^s} = s \int_{-\infty}^{+\infty} F(u) e^{-su} \, du .$$

which we write as

$$\frac{1}{s\zeta(s)} = \int_{-\infty}^{+\infty} uF(u) e^{-\sigma u} e^{-itu} \, du .$$

Since the integral is absolutely and uniformly convergent we may differentiate the above equation with respect to s to obtain

$$\frac{s\zeta'(s) + \zeta(s)}{s^2 \zeta^2(s)} = \int_{-\infty}^{+\infty} uF(u)\, e^{-\sigma u}\, e^{-itu}\, du\,.$$

If $\sigma \geqslant 1$ is a fixed real number we define the function G_σ for all real t by

$$G_\sigma(t) = \frac{(\sigma + it)\,\zeta'(\sigma + it) + \zeta(\sigma + it)}{(\sigma + it)^2\, \zeta^2(\sigma + it)}\,.$$

Theorem 2.4 implies that as $|t| \to \infty$ we have uniformly in σ

$$G_\sigma(t) = \mathrm{O}\left\{\frac{\mathrm{Log}^{16}|t|}{|t|}\right\}.$$

Furthermore, when $|t| \to 0$

$$G_\sigma(t) = \mathrm{O}(1)\,.$$

Thus for $\sigma \geqslant 1$, $G_\sigma(t)$ is square integrable and, for $\sigma > 1$, $uF(u)\,e^{-\sigma u}$ is also square integrable.

For $\sigma > 1$, we apply Plancherel's theorem to conclude

$$\int_{-\infty}^{+\infty} |uF(u)\, e^{-\sigma u}|^2\, du = \frac{1}{2\pi}\int_{-\infty}^{+\infty} |G_\sigma(t)|^2\, dt\,.$$

Upon letting $\sigma \to 1^+$ the right hand side of the above equation tends to a finite limit and there exists an absolute constant K such that for $\sigma > 1$

$$H(\sigma) = \int_{-\infty}^{+\infty} |uF(u)\, e^{-\sigma u}|^2\, du < K$$

As $\sigma \to 1^+$ the above integrand is a positive, increasing function of σ which tends to $|uF(u)\,e^{-u}|^2$. Thus by Beppo Levi's theorem

$$\limsup_{\sigma \to 1^+} \int_{-\infty}^{+\infty} |uF(u)\, e^{-\sigma u}|^2\, du = \int_{-\infty}^{+\infty} |uF(u)\, e^{-u}|^2\, du < \infty\,.$$

It remains to show that

$$\lim_{u \to \infty} uF(u)\, e^{-u} = 0\,.$$

This is easily done, for suppose the contrary. Then there exists an infinite sequence of integers $\{ n_j \}$ and a real number $\eta > 0$ such that

$$| F(\operatorname{Log} n_j) | \frac{\operatorname{Log} n_j}{n_j} > \eta \quad \text{for} \quad j = 1, 2, \ldots$$

Consequently,

$$\int_{\operatorname{Log} n_j}^{\operatorname{Log}(1+n_j)} | ue^{-u} F(u) |^2 \, du = F(\operatorname{Log} n_j)^2 \int_{\operatorname{Log} n_j}^{\operatorname{Log}(1+n_j)} u^2 e^{-2u} \, du \geqslant$$

$$\geqslant F(\operatorname{Log} n_j)^2 \frac{(\operatorname{Log} n_j)^2}{e^2 n_j^2} > \eta^2 e^{-2} ,$$

which implies

$$\int_{-\infty}^{+\infty} | uF(u) e^{-u} |^2 \, du = \infty ,$$

a contradiction, and the theorem is proved. ■

Exercises to chapter 2

2.1 Suppose that $xf(x)$ is continuous and increasing for $x > a$ and that as x tends to infinity

$$\int_a^x f(t)\,dt \sim Ax^m ,$$

where $m > 0$. Prove that $f(x) \sim mAx^{m-1}$ as x tends to infinity.

2.2 Define the function f for $\mathfrak{R}(s) > 0$ by

$$f(s) = \sum_{n=1}^{\infty} \frac{(-1)^{n-1}}{n^s} .$$

(i) For s satisfying $\mathfrak{R}(s) > 1$ relate $f(s)$ to $\zeta(s)$, then deduce an analytic continuation of $\zeta(s)$ into the half plane $\mathfrak{R}(s) > 0$.

(ii) Show that for real s satisfying $0 < s < 1$ we have $\zeta(s) < 0$.

2.3 (i) When is

$$(x-1)^{1-s} - x^{1-s} = \sum_{n=0}^{\infty} \binom{n+s-1}{n+1} x^{-n-s} \,?$$

Use the above equation to deduce that

$$1 = \sum_{n=0}^{\infty} \binom{n+s-1}{n+1} (\zeta(n+s) - 1) .$$

(ii) Prove that

$$(1 - 2^{1-s})\,\zeta(s) = \sum_{n=1}^{\infty} \frac{\Gamma(s+n)}{\Gamma(s)\,n\,!} \frac{\zeta(n+s)}{2^{n+s}} .$$

(iii) Deduce that $\zeta(s)$ has an holomorphic continuation over the whole complex plane, except for a simple pole at $s = 1$.

2.4 Suppose that $f(s) = \sum_{n=1}^{\infty} h(n)\, n^{-s}$ has abscissa of convergence a and simple pole at $s = a$. Prove that the Laurent expansion of f about $s = a$ is given by

$$f(s) = \frac{c}{s-a} + \sum_{r=0}^{\infty} (-1)^r \frac{c_r}{r\,!} (s-a)^r$$

where the coefficients c_r are given by

$$c_r = \lim_{x \to \infty} \left\{ \sum_{n \leqslant x} n^{-a} h(n) (\operatorname{Log} n)^r - \frac{c}{r+1} (\operatorname{Log} x)^{r+1} \right\}.$$

2.5 Prove that $\zeta(s)$ has no zeros in the region $|s-1| \leqslant 1$ as follows. Write

$$\zeta(s)(s-1) = 1 + \sum_{r=0}^{\infty} \gamma_r (s-1)^{r+1},$$

deduce that

$$|(s-1)\zeta(s)| > 1 - \sum_{r=0}^{\infty} |\gamma_r|$$

in the region $|s-1| \leqslant 1$ and then prove that $\sum_{r=0}^{\infty} |\gamma_r| < 1$.

2.6 The Hurwitz zeta function $\zeta(s, \alpha)$ is defined for all $\alpha \in [0, 1)$ and $\Re(s) > 1$ by

$$\zeta(s, \alpha) = \sum_{n=0}^{\infty} \frac{1}{(n+\alpha)^s} \qquad \Re(s) > 1.$$

show that $\zeta(s, \alpha)$ has an analytic continuation to the left of the line $\Re(s) = 1$ and has a simple pole at $s = 1$. Furthermore, use the technique of the previous exercise to deduce that $\zeta(s, \alpha)$ has no zeros in the disc $|s-1| < 1$.

2.7 Prove that for $\Re(s) > 1$ we have

$$\sum_p \frac{1}{p^s} = \sum_{n=1}^{\infty} \frac{\mu(n)}{n} \operatorname{Log} \zeta(ns).$$

2.8 If $\Re(s) > 1$ show that

$$\operatorname{Log} \zeta(s) = s \int_2^{\infty} \frac{\pi(x)}{x(x^s - 1)}\, dx.$$

Can you use the above equation to give a direct proof that $\pi(x) \sim x/\operatorname{Log} x$ when x tends to infinity ?

2.9 Denote by $\pi_n(x)$ the number of integers less than or equal to x which are the product of precisely n prime factors.

(i) Prove that

$$\pi_n(x) = \frac{x}{\operatorname{Log} x} \frac{(\operatorname{Log} \operatorname{Log} x)^{n-1}}{(n-1)!} + O\left\{ \frac{x}{\operatorname{Log} x} (\operatorname{Log} \operatorname{Log} x)^{n-2} \right\}.$$

2.10 (*a*) Show that :

(i) $$\zeta^2(s) = \sum_{n=1}^{\infty} \frac{d(n)}{n^s}, \qquad \mathfrak{R}(s) > 1,$$

where $d(n)$ denotes the number of divisors of n, if $\mathfrak{R}(s) > 1$;

(ii) $$\frac{\zeta(s)}{\zeta(2s)} = \sum_{n=1}^{\infty} \frac{|\mu(n)|}{n^s}, \qquad \mathfrak{R}(s) > 1,$$

where $\mu(n)$ is the Möbius function, if $\mathfrak{R}(s) > 1$;

(iii) $$\frac{\zeta^2(s)}{\zeta(2s)} = \sum_{n=1}^{\infty} \frac{2^{\nu(n)}}{n^s}, \qquad \mathfrak{R}(s) > 1,$$

where $\nu(n)$ denotes the number of different prime factors of n, if $\mathfrak{R}(s) > 1$;

(iv) $$\frac{\zeta^4(s)}{\zeta(2s)} = \sum_{n=1}^{\infty} \frac{\{d(n)\}^2}{n^s}, \qquad \mathfrak{R}(s) > 1;$$

if $\mathfrak{R}(s) > 1$;

(v) $$\frac{\zeta(2s)}{\zeta(s)} = \sum_{n=1}^{\infty} \frac{\lambda(n)}{n^s}, \qquad \mathfrak{R}(s) > 1,$$

where $\lambda(n)$ is Liouville's function (defined as $\lambda(n) = (-1)^r$ if n has r prime factors), if $\mathfrak{R}(s) > 1$;

(vi) $$\frac{\zeta(s-1)}{\zeta(s)} = \sum_{n=1}^{\infty} \frac{\varphi(n)}{n^s}, \qquad \mathfrak{R}(s) > 2,$$

where $\varphi(n)$ is Euler's function, for $\mathfrak{R}(s) > 2$;

(vii) $$\zeta(s)\,\zeta(s-a) = \sum_{n=1}^{\infty} \frac{\sigma_a(n)}{n^s},$$

where $\sigma_a(n)$ denotes the sum of the ath powers of the divisors of n, if $\mathfrak{R}(s) > \max\{1, \mathfrak{R}(a)+1\}$.

(*b*) Obtain asymptotic formulae for the above coefficient sums.

2.11 Let $f(s) = \sum_{n=1}^{\infty} a_n n^{-s}$ be holomorphic for $\Re(s) > 1$. Suppose that :
(i) For $n = 1, 2, \ldots, a_n = O(g(n))$, where g is a non-decreasing function.
(ii) As $\sigma \to 1^+$, then

$$\sum_{n=1}^{\infty} \frac{|a_n|}{n^\sigma} = O\left\{\frac{1}{(\sigma - 1)^\alpha}\right\}.$$

Prove that if $c > 0$, $\sigma + c > 1$, $s = \sigma + it$, $T > 0$ and x is *not* an integer, then

$$\sum_{n<x} \frac{a_n}{n^s} = \frac{1}{2\,i\pi}\int_{c-iT}^{c+iT} f(s+\omega)\frac{x^\omega}{\omega}\,d\omega + O\left\{\frac{x^c}{T(\sigma + c - 1)^\alpha}\right\}$$
$$+ O\left\{g(2\,x)\frac{x^{1-\sigma}}{T}\operatorname{Log} x\right\} + O\left\{\frac{g(x)\,x^{-\sigma}}{T}\langle x \rangle\right\}$$

where $N = [x]$. If x is an integer the corresponding result is

$$\sum_{n<x} \frac{a_n}{n^s} + \frac{1}{2}\frac{a_x}{n^s} = \frac{1}{2\,\pi i}\int_{c-iT}^{c+iT} f(s+\omega)\frac{x^\omega}{\omega}\,d\omega + O\left\{\frac{x^c}{T(\sigma + c - 1)^\alpha}\right\} +$$
$$+ O\left\{\frac{g(2\,x)\,x^{1-\sigma}}{T}\operatorname{Log} x\right\} + O\left\{\frac{g(x)\,x^{-\sigma}}{T}\right\}.$$

2.12 (i) Prove that $\sum_{p \leqslant x} p = \frac{x^2}{2 \operatorname{Log} x} + o\left(\frac{x^2}{\operatorname{Log} x}\right)$.
(ii) Let f be a real valued function defined on $\mathbb{R}^+$ with the following properties : (*a*) $f(x) > 0$; (*b*) f is monotone increasing; (*c*) for each $u \in \mathbb{R}^+$ the limit

$$\lim_{x\to\infty} \frac{f(ux)}{f(x)}$$

exists and is equal to $\varphi(u)$. Prove that

$$\sum_{p \leqslant x} f(p) \sim \frac{xf(x)}{(1+\beta)\operatorname{Log} x}$$

where $\beta = \operatorname{Log}(\varphi(e))$.

2.13 (i) Suppose that for $\Re(s) > 1$ the function F is defined by

$$F(s) = \sum_{n=1}^{\infty} \frac{a_n}{n^s}$$

where $a_n \geqslant 0$ for all n. If $F(s)$ has no singularities on the line $\mathcal{R}(s) = 1$ except for a simple pole at $s = 1$ with residue 1, then show that

$$\sum_{n \leqslant x} a_n \sim x.$$

(ii) If $G(s)$ is defined by

$$G(s) = \sum_{n=1}^{\infty} \frac{b_n}{n^s}$$

where the sequence $\{ b_n \}$ satisfies $b_n = \mathrm{O}(a_n)$, and the only singularity of $G(s)$ on the line $\mathcal{R}(s) = 1$ is a simple pole at $s = 1$ with residue α, then prove that

$$\sum_{n \leqslant x} b_n \sim \alpha x .$$

(Hint : Consider the function $\dfrac{cF(s) + G(s)}{c + \alpha}$, where c is " large " and the cases when $\{ b_n \}$ is a real or a complex sequence separately.)

2.14 Prove that $M(x) = \sum_{n \leqslant x} \mu(n) = \mathrm{o}(x)$ by considering the function $\zeta(s) + \zeta^{-1}(s)$.

2.15 Use Helson's method to prove that as x tends to infinity, then for any fixed positive integer k we have

$$M(x) = \sum_{n \leqslant x} \mu(n) = \mathrm{o}\left(\frac{x}{\operatorname{Log}^k x}\right),$$

$$g(x) = \sum_{n \leqslant x} \frac{\mu(n)}{n} = \mathrm{o}\left(\frac{1}{\operatorname{Log}^k x}\right),$$

$$\psi(x) = \sum_{n \leqslant x} \Lambda(n) = x + \mathrm{o}\left(\frac{x}{\operatorname{Log}^k x}\right).$$

2.16 A Ramanujan sum $c_q(n)$ is defined as

$$c_q(n) = \sum_{\substack{k=1 \\ (k,q)=1}}^{q} \exp\left(2\, i\pi \frac{kn}{q}\right).$$

(These sums often arise naturally in Number Theory, see for example chapter 9.) Prove the following assertions :

(i) If $(a, b) = 1$, then

$$c_a(n)\, c_b(n) = c_{ab}(n) .$$

(ii) $c_q(n) = \sum_{d \mid (q,n)} \mu\left(\frac{q}{d}\right) d$.

(iii) $c_q(n) = \mu(q_n) \frac{\varphi(q)}{\varphi(q_n)}$. where $q_n = q/(q, n)$.

(iv) If $\Re(s) > 0$ and $\sigma_s(n) = \sum_{d/n} d^s$, then

$$\sum_{q=1}^{\infty} \frac{c_q(n)}{q^{s+1}} = \frac{1}{\zeta(s+1)} \frac{\sigma_s(n)}{n^s} .$$

(v) $\sum_{q=1}^{\infty} \frac{c_q(n)}{q} = 0$.

(vi) $\sum_{n=1}^{\infty} \frac{c_q(n)}{n} = - \Lambda(q)$.

CHAPTER 3

An elementary proof of the Prime Number Theorem

§1 INTRODUCTION

1.1 In this chapter we shall give, as a consequence of a refinement of theorem 1.10 an elementary proof of the Prime Number Theorem. The words " elementary proof " have, of course, the technical meaning which was mentioned in the introduction to chapter 2.

This chapter is a direct continuation of § 6 of chapter 1 and we shall assume nothing from chapter 2.

There is a point of terminology which the reader might find used in some earlier books on prime number theory without explanation. Prior to the discovery of an elementary proof of the Prime Number Theorem by Erdös [1] and Selberg [2] it had become customary to divide theorems about prime numbers into two classes :

(*a*) the *elementary theorems*, i.e. theorems which could be proved by elementary algebraic manipulation and rudimentary real variable analysis, and

(*b*) the *transcendental theorems*, i.e. those theorems which could only be proved by complex variable methods.

Some curiosities did arise, for example in the formal equation

$$\prod_p \left(1 - \frac{1}{p}\right) = \sum_{n=1}^{\infty} \frac{\mu(n)}{n}$$

it was an elementary theorem that the left hand side diverged to zero and a transcendental theorem that the right hand side converged to zero ! The classification led quite naturally to the notion of " equivalence " or " depth " between transcendental theorems. Two theorems were " equivalent " or of the same " depth " if either could be deduced from the other by elementary methods. For example, Landau showed that the Prime Number Theorem was " equivalent " to

$$M(x) = \sum_{n \leqslant x} \mu(n) = o(x) \quad \textit{as } x \textit{ tends to infinity}$$

and to the statement

$$\sum_{n=1}^{\infty} \frac{\mu(n)}{n} = 0 \, .$$

Since " transcendental " theorems had a tendency to become " elementary " theorems with the passage of time the classification was not fixed and when, in 1949, the Prime Number Theorem itself became " elementary " the whole point of the classification vanished.

1.2 We shall see in § 1.3 that the Prime Number Theorem follows directly from theorem 3.1 which is a refinement of theorem 1.10. The proof of theorem 3.1 depends upon the theorem that $M(x) = o(x)$. It is to the proof of this latter fact that the bulk of the chapter is devoted.

Theorem 3.1 *Let (f, F) be a Möbius pair of functions with F a real valued function of bounded variation in every finite interval $[1, x]$. Suppose that as x tends to infinity*

$$F(x) = Ax(\operatorname{Log} x)^2 + Bx(\operatorname{Log} x) + Cx + O(x^{\beta}) \, ,$$

where A, B, C, β are real constants and $0 \leqslant \beta < 1$. Then as x tends to infinity we have

$$f(x) = 2\,Ax \operatorname{Log} x + (B - 2\,A\gamma)\,x + o(x) \, ,$$

where γ is Euler's constant.

1.3 The Prime Number Theorem is an immediate consequence of theorem 3.1, for if we take $f(x) = \psi(x)$ and $F(x) = \operatorname{Log}([x]\,!)$, then from chapter 1, § 5.2 we know that $F(x) = x \operatorname{Log} x - x + O(\operatorname{Log} x)$ and as F is monotonic it is of bounded variation in every finite interval. Thus theorem 3.1 is applicable and we conclude that $\psi(x) = x + o(x)$, which as we know (theorem 1.5), implies that

$$\pi(x) = \frac{x}{\operatorname{Log} x} + o\left(\frac{x}{\operatorname{Log} x}\right) .$$

1.4 Proof of theorem 3.1. We shall now prove theorem 3.1 under the assumption that as x tends to infinity

$$M(x) = \sum_{n \leqslant x} \mu(n) = o(x)$$

As in the proof of theorem 1.10 we shall first prove the theorem in the particular case $A = B = C = 0$ and reduce the general case to this special case.

Let η be a fixed real number in the interval (0, 1). Then for $x > \eta^{-1}$ we have, since (f, F) is a Möbius pair,

$$\begin{aligned} f(x) &= \sum_{n \leqslant x} \mu(n)\, F\left(\frac{x}{n}\right) = \sum_{n \leqslant \eta x} \mu(n)\, F\left(\frac{x}{n}\right) + \sum_{\eta x < n \leqslant x} \mu(n)\, F\left(\frac{x}{n}\right) \\ &= \Sigma_0 + \Sigma_1 . \end{aligned}$$

We consider the sums Σ_1 and Σ_2 separately.

(i) *The sum* Σ_1. Under the assumption $A = B = C = 0$, and taking K to be the " O "-constant for F, the following inequalities are quite trivial :

$$\begin{aligned} |\Sigma_0| &\leqslant \sum_{n \leqslant \eta x} \left| F\left(\frac{x}{n}\right) \right| \leqslant \sum_{n \leqslant \eta x} K.\left(\frac{x}{n}\right)^{\beta} \leqslant Kx^{\beta} \int_1^{\eta x} u^{-\beta}\, du \\ &\leqslant Kx^{\beta} \frac{(\eta x)^{1-\beta}}{1-\beta} = K_1\, x\eta^{1-\beta} . \end{aligned}$$

Thus we certainly have

$$\frac{|\Sigma_1|}{x} \leqslant K_1\, \eta^{1-\beta} .$$

(ii) *The sum* Σ_2. Because F is a function of bounded variation in every finite interval we can write

$$F(x) = \mathrm{P}(x) - Q(x) ,$$

where P and Q are both positive, monotonic increasing functions. The sum Σ_2 can now be written as

$$\begin{aligned} \Sigma_1 &= \sum_{\eta x < n \leqslant x} \mu(n)\, \mathrm{P}\left(\frac{x}{n}\right) - \sum_{\eta x < n \leqslant x} \mu(n)\, Q\left(\frac{x}{n}\right) \\ &= \Sigma_{\mathrm{P}} - \Sigma_Q . \end{aligned}$$

By theorem 1.8

$$\left| \sum_{\eta x < n \leqslant x} \mu(n)\, \mathrm{P}\left(\frac{x}{n}\right) \right| \leqslant 2\, \mathrm{P}\left(\frac{1}{\eta}\right) \max_{\eta x < n \leqslant x} | M(n) |$$
$$\leqslant 2\, \mathrm{P}\left(\frac{1}{\eta}\right) x\varepsilon(\eta x),$$

where the function ε is defined by

$$\varepsilon(\mathrm{X}) = \underset{x \geqslant \mathrm{X}}{\text{upper bound}} \left\{ \frac{| M(x) |}{x} \right\}.$$

Similarly we have

$$\left| \sum_{\eta x < n \leqslant x} \mu(n)\, Q\left(\frac{x}{n}\right) \right| \leqslant 2\, Q\left(\frac{1}{\eta}\right) x\varepsilon(\eta x).$$

We know that

$$\frac{| f(x) |}{x} \leqslant \frac{| \sum_1 |}{x} + \frac{| \sum_2 |}{x}$$

and from our estimates for $\sum_1$ and $\sum_2$ we obtain

$$\frac{| f(x) |}{x} \leqslant K_1\, \eta^{1-\beta} + 2\left\{ \mathrm{P}\left(\frac{1}{\eta}\right) + Q\left(\frac{1}{\eta}\right) \right\} \varepsilon(x\eta).$$

Now we appeal to theorem 3.8, which asserts that $M(x) = \mathrm{o}(x)$, to conclude that for fixed $\eta > 0$, $\varepsilon(x\eta) \to 0$ as $x \to \infty$. Whence it follows that

$$\limsup_{x \to \infty} \left| \frac{f(x)}{x} \right| \leqslant K_1\, \eta^{1-\beta}.$$

Since η can be taken as small as we like we must have $f(x) = \mathrm{o}(x)$. The general case follows exactly as in the proof of theorem 1.10. ■

It remains to show that $M(x) = \mathrm{o}(x)$. The proof is not easy.

1.5 Structure of the proof that $M(x) = \mathrm{o}(x)$. The starting point of our proof will be the existence of the inequality

$$| M(x) |\, (\mathrm{Log}\, x)^2 \leqslant \sum_{n \leqslant x} c(k) \left| M\left(\frac{x}{k}\right) \right| + \mathrm{O}(x\, \mathrm{Log}\, x) \tag{3.1}$$

where $\{ c(k) \}$ is a sequence of positive integers about which we will know very little, other than the asymptotic formula for

$$\sum_{k \leqslant x} \frac{c(k)}{k}. \tag{3.2}$$

The asymptotic formula will show that the sequence does not increase very rapidly. We will then use information about $M(x)$ to obtain upper bounds for the terms $\left| M\left(\frac{x}{k}\right) \right|$ and these, combined with our information about the sequence $\{ c(k) \}$, will enable us to deduce from (3.1) that for any $\varepsilon > 0$ and all $x > x_0(\varepsilon)$ we have $| M(x) | < \varepsilon x$ i.e. that $M(x) = o(x)$.

The information about upper bounds for $| M(x) |$ will be given as theorem 3.7. The result is completely self contained and very straightforward to prove. From theorem 3.7 and the inequality (3.1) the deduction that $M(x) = o(x)$ is reasonably easy and fairly " natural ". However, the proof of (3.1) presents more of a problem. It is not a difficult result to prove *per se*; the difficulty lies in making the result seem " natural " and not something which is pulled from a magicians bag of tricks which just happens to work.

We have chosen to follow the rules of the " elementary proof " game and give a strictly elementary proof of the inequality. Our proof will be a consequence of some properties of Möbius pairs of functions to be proved in § 2. However we shall not give any motivation for the theorems to be proved in § 2 other than the general one that they lead to the proof of (3.1), which will be used in § 3 to prove that $M(x) = o(x)$. Instead, we reserve for section 4 some explanation as to why elementary proofs of the Prime Number Theorem work. We hope that our " explanation " will make the proof more " natural " for the reader.

§ 2 MÖBIUS PAIRS OF SEQUENCES

2.1 In this section we shall specialise the notion of Möbius pairs of functions to Möbius pairs of sequences. Then we shall prove two general theorems about Möbius pairs of sequences, the object being to prove theorem 3.4 which gives $b(k)$ and $c(k)$ in the following two Möbius pairs :

$$(\mu(k) \log^2 k\, ,\, b(k)) \quad \text{and} \quad (c(k)\, ,\, \log^2 k)\,.$$

We end the section by proving the inequality (3.1) and obtaining an asymptotic formula for the sum (3.2).

2.2 Let f, F be a pair of functions defined for all $x \geqslant 1$. We recall that (f, F) is a Möbius pair of functions if f and F are related by

$$F(x) = \sum_{n \leqslant x} f\left(\frac{x}{n}\right).$$

In the special case when f and F are arithmetic functions, that is their support is $\mathbb{N}$, we shall say that (f, F) is a *Möbius pair of sequences.* The above equation then reduces to

$$F(n) = \sum_{m \mid n} f\left(\frac{n}{m}\right).$$

Our first two results are useful trivialities.

Theorem 3.2 *Let φ be any function defined for all $x \geqslant 1$. If (f, F) is a Möbius pair of sequences, then*

$$\left(\sum_{n \leqslant x} f(n)\, \varphi\left(\frac{x}{n}\right),\ \sum_{n \leqslant x} F(n)\, \varphi\left(\frac{x}{n}\right)\right)$$

is a Möbius pair of functions.

Proof. Define the functions g and G by

$$g(x) = \sum_{n \leqslant x} f(n)\, \varphi\left(\frac{x}{n}\right) \quad \text{and} \quad G(x) = \sum_{n \leqslant x} F(n)\, \varphi\left(\frac{x}{n}\right).$$

Then we have

$$\begin{aligned} G(x) &= \sum_{n \leqslant x} F(n)\, \varphi\left(\frac{x}{n}\right) = \sum_{n \leqslant x} \sum_{km = n} f(m)\, \varphi\left(\frac{x}{km}\right) \\ &= \sum_{km \leqslant x} f(m)\, \varphi\left(\frac{x}{km}\right) = \sum_{m \leqslant x} \sum_{k \leqslant x/m} f(m)\, \varphi\left(\frac{x}{mk}\right) = \sum_{m \leqslant x} g\left(\frac{x}{m}\right), \end{aligned}$$

which shows that (g, G) is a Möbius pair of functions. ■

Theorem 3.3 *Suppose that (f, F) is a Möbius pair of sequences. Define the sequences $\{ g(k) \}$, $\{ H(k) \}$ by $g(k) = f(k) \operatorname{Log} k$ and $H(k) = F(k) \operatorname{Log} k$. Suppose that G and h are such that (g, G) and (h, H) are Möbius pairs of sequences, then*

(a) $G(k) = F(k) \operatorname{Log} k - \sum_{mn = k} F(m)\, \Lambda(n)$

(b) $h(k) = f(k) \operatorname{Log} k + \sum_{mn = k} f(m)\, \Lambda(n)$.

Proof. (*a*) Since (f, F) and (g, G) are Möbius pairs,

$$G(k) = \sum_{mn=k} g(m) = \sum_{mn=k} f(m) \operatorname{Log} m$$

$$\begin{aligned} F(k) \operatorname{Log} k - G(k) &= \sum_{mn=k} f(m) \operatorname{Log}(mn) - \sum_{mn=k} f(m) \operatorname{Log} m \\ &= \sum_{mn=k} f(m) \operatorname{Log} n = \sum_{mn=k} f(m) \sum_{ts=r} \Lambda(t) \\ &= \sum_{mts=k} f(m) \Lambda(t) = \sum_{rt=k} F(r) \Lambda(t) . \end{aligned}$$

The proof of (*b*) is similar. ∎

2.3 In our proof that $M(x) = o(x)$ we shall make use of the special sequences $\{ b(k) \}$ and $\{ c(k) \}$ which are defined below. These sequences, which are rather strange at first sight, arise as half of Möbius pairs of more natural arithmetic functions. The sequences are defined by

$$b(k) = - \Lambda(k) \operatorname{Log} k + \sum_{mn=k} \Lambda(m) \Lambda(n)$$

and

$$c(k) = \Lambda(k) \operatorname{Log} k + \sum_{mn=k} \Lambda(m) \Lambda(n) .$$

We now have the following result.

Theorem 3.4 *With the above notation the following are Möbius pairs of sequences :*

(a) $(\mu(k) \operatorname{Log} k, - \Lambda(k))$,

(b) $(\mu(k) \operatorname{Log}^2 k, b(k))$,

(c) $(c(k), \operatorname{Log}^2 k)$.

Proof. Parts (*a*) and (*b*) follow by applying theorem 3.3(*a*) twice, starting from the Möbius pair of sequences $(\mu(n), \delta(n))$ and part (*c*) follows from theorem 3.3(*b*) on starting with the Möbius pair of sequences $(\Lambda(n), \operatorname{Log} n)$. ∎

We are now in a position to prove the inequality (3.1).

Theorem 3.5 *As x tends to infinity we have*

$$| M(x) | (\operatorname{Log} x)^2 \leqslant \sum_{k \leqslant x} c(k) \left| M\left(\frac{x}{k}\right) \right| + 2 x \operatorname{Log} x .$$

Proof. From theorem 3.4(*b*) and theorem 3.2 with $\varphi(u) \equiv 1$ we see that

$$\left(\sum_{k \leqslant x} \mu(k)\,(\operatorname{Log} k)^2,\ \sum_{k \leqslant x} b(k)\right)$$

is a Möbius pair of functions. Hence by the Möbius inversion formula

$$\sum_{k \leqslant x} \mu(k)\,(\operatorname{Log} k)^2 = \sum_{kn \leqslant x} \mu(n)\, b(k)\ .$$

The right hand side is equal to

$$\sum_{k \leqslant x} b(k)\, M\left(\frac{x}{k}\right)\ ;$$

and upon using the Abel summation formula we see that it is equal to

$$M(x)\,(\operatorname{Log} x)^2 - 2\int_1^x \frac{M(u)}{u} \operatorname{Log} u\, du\ .$$

Thus we have

$$M(x)\,(\operatorname{Log} x)^2 = \sum_{k \leqslant x} b(k)\, M\left(\frac{x}{k}\right) + 2\int_1^x \frac{M(u)}{u} \operatorname{Log} u\, du\ .$$

The trivial estimate $|\, M(u)\, | \leqslant u$ and the inequality $|\, b(k)\, | \leqslant c(k)$ now leads to the theorem. ■

During the course of our proof that $M(x) = o(x)$ we shall need some information about the rate of growth of the sequence $\{\, c(k)\, \}$ and it is convenient to prove the result now.

Theorem 3.6 *As x tends to infinity we have :*

(a) $$\sum_{k \leqslant x} \frac{c(k)}{k} = (\operatorname{Log} x)^2 - 2\,\gamma \operatorname{Log} x + O(1)$$

where γ is Euler's constant.

(b) *If H is the interval $(u, v]$, then*

(b) $$\sum_{u < k \leqslant v} \frac{c(k)}{k} = \{\, \operatorname{Log}(uv) - 2\,\gamma\, \} \operatorname{Log}\left(\frac{v}{u}\right) + O(1)\ .$$

Proof. Define the function f by

$$f(x) = \sum_{k \leqslant x} \frac{x}{k}\, c(k) - x(\operatorname{Log} x)^2\ ,$$

Let F be the Möbius transform of f, namely

$$F(x) = \sum_{n \leqslant x} \sum_{kn \leqslant x} \frac{x}{kn} c(k) - \sum_{n \leqslant x} \frac{x}{n} \left(\operatorname{Log} \frac{x}{n}\right)^2 .$$

Using theorem 3.4(c), which asserts that $(c(k), \operatorname{Log}^2 k)$ is a Möbius pair of sequences, we write

$$\begin{aligned} F(x) &= \sum_{m \leqslant x} \frac{x}{m} \left\{ (\operatorname{Log} m)^2 - \left(\operatorname{Log} \frac{x}{m}\right)^2 \right\} \\ &= \sum_{m \leqslant x} \frac{x}{m} \left\{ (\operatorname{Log} x)^2 - 2 \operatorname{Log} \left(\frac{x}{m}\right) \operatorname{Log} x \right\} . \\ &= \sum_{n \leqslant x} \frac{x}{n} \left\{ \operatorname{Log}^2 x - 2 \operatorname{Log} \left(\frac{x}{n}\right) \cdot \operatorname{Log} x \right\} . \end{aligned}$$

Using the notation and results of § 5.2 of chapter 1 we can write the above sum as

$$\begin{aligned} F(x) &= x(\operatorname{Log} x)^2 \, \mathrm{U}(x) - 2 \, x \operatorname{Log} x \, . \, \mathrm{V}(x) \\ &= x(\operatorname{Log} x)^2 \left(\operatorname{Log} x + \gamma + \mathrm{O}\left(\frac{1}{x}\right) \right) \\ &\qquad - 2 \, x \operatorname{Log} x \left(\frac{1}{2} (\operatorname{Log} x)^2 + \gamma \operatorname{Log} x + \delta + \mathrm{O}\left(\frac{1}{x}\right) \right) \\ &= - \gamma x (\operatorname{Log} x)^2 - 2 \, \delta x \operatorname{Log} x + \mathrm{O}(\operatorname{Log}^2 x) . \end{aligned}$$

Now we apply theorem 1.10 and conclude that

$$f(x) = - 2 \, \gamma x \operatorname{Log} x + \mathrm{O}(x) .$$

This proves the first part of the theorem and the second is a simple consequence. ■

§ 3 THE SUM $M(x)$

3.1 Our aim in this section is to give the final steps in the proof that $M(x) = \mathrm{o}(x)$. The proof of this result is, in principal quite straightforward. One starts with the inequality

$$| M(x) | (\operatorname{Log} x)^2 \leqslant \sum_{k \leqslant x} c(k) \left| M\left(\frac{x}{k}\right) \right| + \mathrm{O}(x \operatorname{Log} x) \tag{3.3}$$

and then uses it in conjunction with information about the sequence $\{ c(k) \}$ to deduce that $M(x) = o(x)$. However, one has to be rather careful. If one uses only the trivial upper bound for $\left|M\left(\frac{x}{k}\right)\right|$, namely $\left|M\left(\frac{x}{k}\right)\right| \leqslant \frac{x}{k}$, one obtains

$$| M(x) | (\operatorname{Log} x)^2 \leqslant \sum_{k \leqslant x} c(k) \frac{x}{k} + O(x \operatorname{Log} x)$$

and after theorem 3.6 we can conclude that

$$| M(x) | (\operatorname{Log} x)^2 \leqslant x \{ (\operatorname{Log} x)^2 - 2 \gamma \operatorname{Log} x + O(1) \} + O(x \operatorname{Log} x),$$

from which we can only deduce that

$$\limsup_{x \to \infty} \frac{| M(x) |}{x} \leqslant 1 .$$

Thus, in order to deduce a non-trivial result from (3.3) we must use more information about $M(x)$. The key fact will be given by theorem 3.7, which asserts that if $\Delta > 1$ is any fixed real number, then every " large " interval I of the form $[\rho, \rho e^{\Delta})$ contains a subinterval i upon which $| M(x) |$ is less than $K\Delta^{-1}$, where K is an absolute constant.

Consequently when we use (3.3) we shall divide the interval $[1, x]$ into disjoint subintervals $I_0, \ldots, I_N$ of the required type, then we will find the subintervals $i_n \subseteq I_n$ upon which $| M(x) |$ is small.

We will use the " good " upper bound for $| M(x/k) |$ whenever $x/k \in i_n$ for $n = 0, \ldots, N$ and the trivial upper bound whenever $x/k \notin i_n$, $n = 0, \ldots, N$. In conjunction with theorem 3.5(*b*) we will be able to deduce that for all $x > x_0(\Delta)$ an upper bound for the right hand side of (3.3) is

$$\frac{c_1}{\Delta} x (\operatorname{Log} x)^2 ,$$

where c_1 is an absolute constant. Since Δ can be as large as we please it will follow that

$$\limsup_{x \to \infty} \left| \frac{M(x)}{x} \right| = 0 .$$

3.2 The additional fact about $| M(x) |$ which will be needed for the proof of theorem 3.8 will now be proved.

Theorem 3.7 *There exists an absolute constant K with the following property. Let* $\Delta > 1$ *be any fixed real number and suppose that* δ *satisfies* $\Delta^{-1} \leqslant \delta \leqslant 1$. *Then every interval I of the form* $[\rho, \rho e^{\Delta})$, *with* $\rho \geqslant e^{\Delta}$, *contains a subinterval* $i = [\sigma, \sigma e^{\delta})$ *throughout which we have*

$$\left| \frac{M(u)}{u} \right| \leqslant K.\delta .$$

Proof. Let $K_1 \geqslant 1$ be an absolute constant, to be chosen explicitly later. We show first that there exists $u_0 \in I$ such that

$$| M(u_0) | < \frac{K_1}{\Delta} u_0 . \tag{3.4}$$

For suppose not, then the step function $M(u)/u$ is of fixed sign in the interval I, because its "jumps" are of magnitude $1/u$ and $1/u \leqslant 1/e^{\Delta} \leqslant K_1/\Delta$. Consequently we have

$$\left| \int_I \frac{M(u)}{u^2} du \right| = \int_I \frac{| M(u) |}{u^2} du \geqslant \frac{K_1}{\Delta} \int_I \frac{du}{u} = K_1 . \tag{3.5}$$

We shall now show that there exists an absolute constant K_2 such that for all $x \geqslant 1$ we have

$$\left| \int_1^x \frac{M(u)}{u^2} du \right| \leqslant K_2 \tag{3.6}$$

and consequently for any interval I it follows that

$$\left| \int_I \frac{M(u)}{u^2} du \right| \leqslant 2 K_2 . \tag{3.7}$$

Thus, if we choose $K_1 > \max \{ 1, 2 K_2 \}$, then (3.7) implies that (3.5) is false, hence (3.4) must hold for some $u_0 \in I$.

The proof of (3.6) is easy. Let $F(x) = x - 1$ and suppose that f is such that (f, F) is a Möbius pair of functions. We have, by definition and the Abel summation formula that

$$f(x) = \sum_{n \leqslant x} \mu(n) \left(\frac{x}{n} - 1 \right) = x \int_1^x \frac{M(u)}{u^2} du .$$

From theorem 1.10 we also know that $|f(x)| \leqslant K_2 x$, where K_2 is an absolute constant and we have the inequality (3.6).

To complete the proof of the theorem we choose any interval $i \subseteq I$ of the form $i = [\sigma, \sigma e^{\delta})$ which contains u_0. For any $u \in i$ we trivially have

$$| M(u) - M(u_0) | \leqslant | u - u_0 |$$

whence it follows that

$$\begin{aligned} | M(u) | &\leqslant | M(u_0) | + | u - u_0 | \\ &\leqslant \frac{K_1}{\Delta} u_0 + | u - u_0 | \leqslant \frac{K_1}{\Delta} \sigma e^{\delta} + \sigma e^{\delta} - \sigma \\ &\leqslant \{ K_1 e^{\delta} + (e^{\delta} - 1) \delta^{-1} \} \delta\sigma . \end{aligned}$$

since $\Delta^{-1} \leqslant \delta$. Because $\delta \leqslant 1$ we also have

$$K_1 e^{\delta} + (e^{\delta} - 1)\,\delta^{-1} \leqslant K_1\, e + e - 1 = K\,.$$

Hence for any $u \in i = [\sigma, \sigma e^{\delta})$ we have

$$\left|\frac{M(u)}{u}\right| \leqslant \left|\frac{M(u)}{\sigma}\right| \leqslant K\delta$$

and the proof of the theorem is complete. ■

3.3 We are now is a position to prove the principal result of this section and thus complete our proof of theorem 3.1.

Theorem 3.8 *As x tends to infinity we have*

$$M(x) = \sum_{n \leqslant x} \mu(n) = \mathrm{o}(x)\,.$$

Proof. Let $\Delta > 1$ be a fixed real number. Define for each positive integer n, intervals I_n and $J_n(x)$ as follows :

$$I_n = [e^{n\Delta}, e^{(n+1)\Delta}[\text{ and } J_n = J_n(x) =]xe^{-(n+1)\Delta}, xe^{-n\Delta}]\,.$$

Let N be the integer satisfying $x \in I_N$. Since $x/k \in I_n$ if and only if $k \in J_n(x)$ we see that the inequality of theorem 3.5 can be written as

$$|\,M(x)\,|\,(\mathrm{Log}\,x)^2 \leqslant \sum_{n=0}^{N} \sum_{k \in J_n} c(k) \left| M\left(\frac{x}{k}\right) \right| + \mathrm{O}(x.N\Delta)\,. \tag{3.8}$$

Let $v(u)$ be any integer valued function of u with the following properties :

$$0 < v(u) \leqslant u\,; \quad v(u) \to \infty \quad \text{and} \quad v(u)/u \to 0 \quad \text{as} \quad u \to \infty\,.$$

(For example we could take $v(u)=[\sqrt{u}]$.) In order to obtain a non-trivial upper estimate for the right hand side of (3.8) we shall split the summation over n into three parts :

$$0 \leqslant n < v(N)\,; \quad v(N) \leqslant n < N\,; \quad n = N\,.$$

For n in the range $[0, v(N))$ and $n = N$ we shall use the trivial estimate for $M(x)$. Thus for these n we have :

$$\sum_{k \in J_n} c(k) \left| M\left(\frac{x}{k}\right) \right| \leqslant x \sum_{k \in J_n} \frac{c(k)}{k}\,.$$

For each n in the range $[\nu(N), N)$ we appeal to theorem 3.7 to deduce the existence of a subinterval $i_n = [\sigma_n, \sigma_n e^{\delta}) \subseteq I_n$ in which we have

$$| M(u) | \leqslant K\delta u .$$

Whenever $x/k \in i_n$ we shall use this estimate in (3.8) in place of the trivial estimate for $| M(x/k) |$.

If we define the intervals $j_n(x) \subseteq J_n(x)$ by

$$j_n(x) =]x\sigma_n^{-1} e^{-\delta}, x\sigma_n^{-1}]$$

and note that $x/k \in i_n$ if and only if $k \in j_n(x)$, then for n satisfying $\nu(N) \leqslant n < N$ we have, on denoting by m_n the upper bound of $| M(u) |/u$ for $u \in I_n$,

$$\sum_{k \in J_n} c(k) \left| M\left(\frac{x}{k}\right) \right| \leqslant x m_n \sum_{k \in J_n - j_n} \frac{c(k)}{k} + K\delta x \sum_{k \in j_n} \frac{c(k)}{k} . \tag{3.9}$$

We shall use theorem 3.6 to obtain estimates for the sums :

$$\sum_{k \in J_n} \frac{c(k)}{k}, \qquad \sum_{k \in J_n - j_n} \frac{c(k)}{k} \text{ and } \sum_{k \in j_n} \frac{c(k)}{k} .$$

On taking $H = J_n = (xe^{-(n+1)\Delta}, xe^{-n\Delta}]$ in theorem 3.6 and on noting that $e^{N\Delta} \leqslant x < e^{(N+1)\Delta}$ it follows that

$$\sum_{k \in J_n} \frac{c(k)}{k} = 2\left(N - n + \frac{1}{2}\right) \Delta^2 - 2\,\gamma\Delta + O(1) .$$

Next taking $H = j_n(x) = (x\sigma_n^{-1} e^{-\delta}, x\sigma_n^{-1}] \subseteq J_n(x)$ we obtain

$$2\,\Delta(N - n - 1)\,\delta + O(1) \leqslant \sum_{k \in j_n} \frac{c(k)}{k} \leqslant 2\,\Delta(N - n + 1)\,\delta + O(1) . \tag{3.10}$$

From the above estimates it follows immediately that

$$\sum_{k \in J_n - j_n} \frac{c(k)}{k} \leqslant 2\left(N - n + \frac{1}{2}\right) \Delta^2 - 2\,\gamma\Delta - 2\,\Delta(N - n - 1)\,\delta + O(1) . \tag{3.11}$$

The completion of the proof is now just a matter of mopping up. First using the trivial estimate for $M(x)$ when $n < \nu$ or $n = N$ we have

$$\begin{aligned} \sum_{n < \nu(N)} \sum_{k \in J_n} c(k) \left| M\left(\frac{x}{k}\right) \right| &\leqslant \sum_{n < \nu(N)} x \left\{ 2\left(N - n + \frac{1}{2}\right) \Delta^2 + O(\Delta) \right\} \\ &= O(x\Delta^2\, N\nu(N)) , \end{aligned} \tag{3.12}$$

and also

$$\sum_{k \in J_N} c(k) \left| M\left(\frac{x}{k}\right) \right| = O(x\Delta^2) . \tag{3.13}$$

Finally, from (3.9), (3.10) and (3.11) we have the estimate

$$\sum_{n=\nu(N)}^{N-1} \sum_{k \in J_n} c(k) \left| M\left(\frac{x}{k}\right) \right|$$
$$\leqslant \sum_{n=\nu(N)}^{N-1} 2\, x m_n \left\{ \left(N - n + \frac{1}{2}\right) \Delta^2 - (N - n - 1)\, \Delta\delta - \gamma\Delta + O(1) \right\}$$
$$+ \sum_{n=\nu(N)}^{N-1} 2\, K\delta x \left\{ (N - n + 1)\, \Delta\delta + O(1) \right\}$$
$$\leqslant \sum_{n=\nu(N)}^{N-1} 2(N - n + 1)\, \Delta x \left\{ m_n(\Delta - \delta) + K\delta^2 \right\} + O(N\Delta^2 x)\,.$$

Denoting by m_ν^* the upper bound of m_n for n satisfying $\nu \leqslant n < N$ and summing the arithmetic progression :

$$\sum_{\nu \leqslant n < N} \sum_{k \in J_n} c(k) \left| M\left(\frac{x}{k}\right) \right| \leqslant N^2\, \Delta x m_\nu^*(\Delta - \delta) + K\delta^2\, \Delta x N^2 + O(N\Delta^2 x)\,. \tag{3.14}$$

Combining the inequalities (3.8), (3.12), (3.13), (3.14) and noting that Log $x \geqslant N\Delta$ we deduce that

$$|\, M(x)\, |\, (\Delta N)^2 \leqslant N^2\, \Delta x m_\nu^*(\Delta - \delta) + K\delta^2\, N^2\, \Delta x + O(x\nu N\Delta^2)\,.$$

Hence, on dividing by $(N\Delta^2)\, x$ and taking the upper bound for $M(x)/x$ with $x \in I_N$ we obtain

$$m_N \leqslant m_\nu^* \left(1 - \frac{\delta}{\Delta}\right) + K\frac{\delta^2}{\Delta} + O\left\{ \frac{\nu(N)}{N} \right\}.$$

If we let $\lambda = \limsup m_n = \lim m_\nu^*$ as n tends to infinity it follows that

$$\lambda \leqslant \lambda\left(1 - \frac{\delta}{\Delta}\right) + K\frac{\delta^2}{\Delta}$$

which implies that $\lambda \leqslant K\delta$, or in other words

$$\limsup_{x \to \infty} \left| \frac{M(x)}{x} \right| \leqslant K\delta\,.$$

On taking δ as small as possible, namely $\delta = \Delta^{-1}$ it follows that, since Δ can be as large as we like, we must have

$$\limsup_{x \to \infty} \left| \frac{M(x)}{x} \right| = 0$$

and the theorem is proved. ■

§ 4 WHY CAN ELEMENTARY PROOFS EXIST ?

4.1 Perhaps this question belongs more to a book on philosophy than to one on mathematics, but since many eminent mathematicians held the conviction that an elementary proof of the Prime Number Theorem could not exist, the question does deserve some consideration. The presentation of an elementary proof as a *fait accompli*, whilst logically irreproachable, does not really satisfy ones aesthetic feelings or give any real insight into the relationship between the elementary proof and the properties of ζ in the complex plane.

Let us put ourselves in the position of somebody seeking an elementary proof of the Prime Number Theorem. First of all what are our restrictions on methods going to be ? To some extent this a matter of personal choice, but the generally accepted notion seems to be that one should confine ones attention to the algebraic manipulation of arithmetic functions, such as Λ, ψ, μ, M, π, etc. to-gether with the concept and very basic properties of limit operations. One could give a precise axiomatic description of what " elementary number theory " should consist of, but in our opinion this is not worth the effort involved and we prefer to be vague and informal about " elementary number theory ".

Assuming that we have agreed upon what is to constitute " elementary number theory ", then we can certainly state the Prime Number Theorem within the framework of " elementary number theory " as

$$\pi(x) = \sum_{p \leqslant x} 1 \sim \frac{x}{\operatorname{Log} x}$$

As we saw in chapter 1 it is easy to prove, within the confines of " elementary number theory " that :

$$\left(\pi(x) \sim \frac{x}{\operatorname{Log} x} \right) \Leftrightarrow (\psi(x) \sim x) \Leftrightarrow (\theta(x) \sim x)$$

The big problem being can we prove any of the above assertions using only the self-imposed restrictions of " elementary number theory ".

Now let us consider the known " non-elementary " proofs of the Prime Number Theorem. The proof which uses the least number of facts about ζ is the one given in § 4.3 of chapter 2 as a consequence of Ikehara's tauberian theorem. The properties which we needed were :

(i) ζ has a simple pole at $s = 1$ with residue 1 and is otherwise holomorphic in some open set containing the half plane $\mathcal{R}(s) \geqslant 1$.

(ii) ζ does not vanish upon the half plane $\mathcal{R}(s) \geqslant 1$. More precisely we used (i) and the following equivalent formulation of (ii).

(iii) The function $Z(s) = -\dfrac{\zeta'(s)}{\zeta(s)}$ has no poles on the half plane $\Re(s) \geqslant 1$, except at $s = 1$ where there is a simple pole with residue 1.

Thus, three questions naturally present themselves :

(*a*) Can we formulate the above analytic facts about ζ within the framework of " elementary number theory " ? That is, can we write down a set of statements in terms of arithmetical functions and simple limit operations which imply the truth of (i) and (ii) ?

(*b*) If the answer to (*a*) is yes, then can we prove the statements within the confines of " elementary number theory " ?

(*c*) Assuming that (*a*) and (*b*) receive affirmative answers, then can we mimic the tauberian argument within " elementary number theory " and so deduce an elementary proof of the Prime Number Theorem ?

We shall only concern ourselves with questions (*a*) and (*b*) since historically they were the more difficult to answer.

Consider the following situation. Let f be an analytic function, which for $\Re(s) > 1$ has a representation of the form :

$$f(s) = \sum_{n=1}^{\infty} \frac{a_n}{n^s}.$$

By considering the arithmetic function defined by the sequence $\{ a_n \}$ and perhaps other related arithmetic functions we wish to make statements within the confines of " elementary number theory " which imply that f has certain analytic properties. This is a reasonable hope because f is completely determined for $\Re(s) > 1$ by the sequence $\{ a_n \}$ and one would expect that its analytic properties in the neighbourhood of the line $\Re(s) = 1$ could be expressed, in a fairly simple way, in terms of the sequence $\{ a_n \}$.

The simplest analytic statement one can make about f is :

f is holomorphic in an open set containing the half plane $\Re(s) \geqslant 1$, except for a simple pole at $s = 1$ with residue r (3.15)

What kind of elementary statements imply the above analytic fact ? By an application of the Abel summation formula we see that the statement

$$\sum_{n \leqslant x} a_n = rx + O(1) \tag{3.16}$$

implies the truth of (3.15). Thus if the above elementary statement can be proved by techniques from elementary number theory, then we will have captured the analytic property of f.

For the Riemann zeta function we have $a_n = 1$ for all n so (3.16) is a triviality to prove by elementary number theory. For the function $Z(s)$, on the other hand, things are not as simple. We shall see in chapter 6 that, for Z(s), the statement (3.16) is false. Thus we must search further.

As we just saw, the statement

$$\sum_{n \leqslant x} 1 = x + O(1)$$

implies that $\zeta(s)$ satisfies (3.15). Consequently $Z(s) = -\zeta'(s)/\zeta(s)$ is holomorphic in an open set containing $\mathfrak{R}(s) \geqslant 1$ except for poles on the line $\mathfrak{R}(s) = 1$. So what we wish to find now is an elementary statement which implies that a function which is holomorphic except for poles on the line $\mathfrak{R}(s) = 1$ can have, in fact, only one simple pole at $s = 1$.

Certainly the statement

$$\sum_{n \leqslant x} a_n = x + o(x)$$

is elementary and, as we saw in chapter 2, § 4, implies that the related Dirichlet series has no poles on $\mathfrak{R}(s) = 1$ except for a simple pole at $s = 1$. Unfortunately, verifying this statement for Z(s) is the Prime Number Theorem and we are back where we started.

What is really needed are statements which hold for Z(s) and which we can prove by using elementary techniques. Essentially only one result is known; this is given as theorem 3.9 below. Every elementary proof of the Prime Number Theorem is based upon this result.

As we shall see, the hypotheses of the theorem can all be formulated as elementary statements when $f(s) = Z(s)$.

Theorem 3.9 (i) *Suppose that f is holomorphic in an open set containing the half plane $\mathfrak{R}(s) \geqslant 1$ except for poles on the line $\mathfrak{R}(s) = 1$.*

(ii) *In the region $\mathfrak{R}(s) > 1$ the function f has the representation*

$$f(s) = \sum_{n=1}^{\infty} \frac{a_n}{n^s}.$$

(iii) *Define $g(s)$ for $\mathfrak{R}(s) \geqslant 1$ by $g(s) = -f'(s) + f^2(s)$. In the region $\mathfrak{R}(s) > 1$ write*

$$g(s) = \sum_{n=1}^{\infty} \frac{b_n}{n^s},$$

where $b_n = a_n \operatorname{Log} n + \sum_{rs=n} a_r a_s$ and suppose that

$$B(x) = \sum_{n \leqslant x} b_n \sim 2x \operatorname{Log} x .$$

(iv) *The coefficients a_n are all non-negative.*

Then we have :

(a) *Hypotheses* (i), (ii), (iii) *imply that* f *has only simple poles on the line* $\mathfrak{R}(s)=1$ *and, except at* $s = 1$, *the residue at these poles is* -1.

(b) *Hypotheses* (i), (ii), (iii) *and* (iv) *imply that the only pole of* $f(s)$ *on the line* $\mathfrak{R}(s) = 1$ *is a simple pole at* $s = 1$ *with residue* 1.

Proof. (*a*) First we prove that $g(s)$ does not have any multiple poles on the line $\mathfrak{R}(s) = 1$, except at $s = 1$. By the Abel summation formula (theorem 1.7), we have for $\mathfrak{R}(s) > 1$

$$\begin{aligned} g(s) &= \sum_{n=1}^{\infty} \frac{b_n}{n^s} = s \int_1^{\infty} \frac{B(u)}{u^{s+1}}\, du \\ &= s \int_1^{\infty} \frac{2 \operatorname{Log} u}{u^s}\, du + \mathrm{o}\left(|s| \int_1^{\infty} \frac{\operatorname{Log} u}{u^{\sigma}}\, du \right) \\ &= \frac{2s}{(s-1)^2} + \mathrm{o}\left(\frac{|s|}{(\sigma - 1)^2} \right). \end{aligned}$$

If $s_0 = 1 + it_0$ is a pole of $g(s)$, then upon taking $s = \sigma + it_0$ in the above equation and letting $\sigma \to 1^+$ we see that $g(s)$ cannot have a double pole, except at $s = 1$.

Since $g(s) = -f'(s) + f^2(s)$ it now follows that $f(s)$ can only have simple poles on the line $\mathfrak{R}(s) = 1$.

Suppose f has a pole at $s_0 = 1 + it_0$ with residue $r(s_0)$. In the neighbourhood of s_0 we have the representation

$$f(s) = \frac{r(s_0)}{s - s_0} + F(s)$$

with $F(s)$ holomorphic. Consequently, $g(s)$ has the representation

$$g(s) = \frac{r(s_0) + r^2(s_0)}{(s - s_0)^2} + \frac{2\, r(s_0)\, F(s)}{s - s_0} + F^2(s) - F'(s)\,.$$

If $s_0 \neq 1$, then $g(s)$ does not have a double pole at s_0 and we must have $r(s_0) + r^2(s_0) = 0$, so $r(s_0) = -1$. At $s = 1$ we know $g(s)$ has a double pole with coefficient 2, thus $r(1) + r^2(1) = 2$, which implies that either $r(1)=1$ or $r(1)=-2$.

(*b*) Our first use of hypothesis (iv) is to prove $r(1) = 1$. Because $a_n \geqslant 0$ and since $f(s)$ has a pole at $s = 1$,

$$\lim_{s \to 1^+} \sum_{n=1}^{\infty} \frac{a_n}{n^s} = +\infty\,.$$

Hence, the residue of the pole at $s = 1$ must be positive. We have shown that the only possible values for $r(1)$ are $+1$ and -2. Thus we know $r(1) = 1$.

We next show that $f(s)$ has no other poles on the line $\mathfrak{R}(s) = 1$. Our proof of this fact is a variation of Hadamard's original proof that $\zeta(s)$ has no zeros on the line $\mathcal{R}(s) = 1$.

For $\mathcal{R}(s) > 1$ we have

$$f(s) = f(\sigma + it) = \sum_{n=1}^{\infty} a_n \frac{\cos (t \operatorname{Log} n)}{n^\sigma} + i \sum_{n=1}^{\infty} a_n \frac{\sin (t \operatorname{Log} n)}{n^\sigma}$$
$$= R(\sigma, t) + iI(\sigma, t) .$$

The use of Cauchy's inequality then gives us :

$$\left\{ \sum_{n=1}^{\infty} a_n \frac{\cos (t \operatorname{Log} n)}{n^\sigma} \right\}^2 = \left\{ \sum_{n=1}^{\infty} a_n^{1/2} \frac{\cos (t \operatorname{Log} n)}{n^{\sigma/2}} \cdot \frac{a_n^{1/2}}{n^{\sigma/2}} \right\}^2$$
$$\leqslant \left\{ \sum_{n=1}^{\infty} a_n \frac{\cos^2 (t \operatorname{Log} n)}{n^\sigma} \right\} \left\{ \sum_{n=1}^{\infty} \frac{a_n}{n^\sigma} \right\}$$
$$= \frac{1}{2} \left\{ \sum_{n=1}^{\infty} \frac{a_n}{n^\sigma} (1 + \cos (2\, t \operatorname{Log} n)) \right\} \left\{ \sum_{n=1}^{\infty} \frac{a_n}{n^\sigma} \right\} .$$

We have shown, for $\sigma > 1$ and any real t,

$$R^2(\sigma, t) \leqslant \frac{1}{2} \{ R(\sigma, 0) + R(\sigma, 2\, t) \} R(\sigma, 0) . \tag{3.17}$$

We now relate the behaviour of $f(s)$ in the neighbourhood of a pole to the behaviour of $R(\sigma, t)$. If $f(s)$ has a pole at $s_0 = 1 + it_0$, then in the neighbourhood of s_0

$$f(s) = \frac{r(s_0)}{s - s_0} + F(s) = R(\sigma, t) + iI(\sigma, t) .$$

Write $s = \sigma + it_0$ and then let $\sigma \to 1^+$. Upon equating the real and imaginary parts, we see

$$\frac{r(s_0)}{\sigma - 1} \sim R(\sigma, t_0) .$$

We know that $r(1) = 1$ and $r(s_0) = -1$ if $s_0 \neq 1$. Using the above asymptotic relation in conjunction with (3.17) we deduce, for any $\varepsilon > 0$ and all σ sufficiently close to 1,

$$\frac{(1 - \varepsilon)^2}{(\sigma - 1)^2} \leqslant \frac{1}{2} \left\{ \frac{1 + \varepsilon}{\sigma - 1} + R(\sigma, 2\, t_0) \right\} \frac{1 + \varepsilon}{\sigma - 1} .$$

and consequently

$$\frac{1 - 6\, \varepsilon + \varepsilon^2}{1 + \varepsilon} \frac{1}{\sigma - 1} \leqslant R(\sigma, 2\, t_0) .$$

If ε is sufficiently small it follows that $R(\sigma, 2\, t_0)$ tends to ∞ when $\sigma \to 1^+$. Thus $f(s)$ has a pole at $1 + i\, 2\, t_0$ with a *positive* residue. This contradicts (*a*). ■

Translating the hypotheses of theorem 3.9 into elementary statements about $Z(s)$ is easy. As we shall show, the statements which we write down are true but proving them within the confines of elementary number theory is non-trivial.

The elementary statements which imply that $s = 1$ is the only pole of $Z(s)$ on the line $\Re(s) = 1$ are :

$$\sum_{n \leqslant x} 1 = x + O(1)\,; \quad \Lambda(n) \geqslant 0\,;$$

$$\sum_{n \leqslant x} \Lambda(n) \operatorname{Log} n + \sum_{rs \leqslant x} \Lambda(r)\,\Lambda(s) \sim 2\,x \operatorname{Log} x\,.$$

Obviously only the third statement presents a problem. Before attempting to prove it by elementary, methods, we verify that it is true.

By the Abel summation formula and Tchebycheff's theorem,

$$\sum_{n \leqslant x} \Lambda(n) \operatorname{Log} n + \sum_{rs \leqslant x} \Lambda(r)\,\Lambda(s) = \psi(x) \operatorname{Log} x + O(x) + \sum_{n \leqslant x} \Lambda(n)\,\psi\left(\frac{x}{n}\right).$$

We next apply the Prime Number Theorem to see

$$\begin{aligned} &\psi(x) \operatorname{Log} x + \sum_{n \leqslant x} \Lambda(n)\,\psi\left(\frac{x}{n}\right) + O(x) \\ &\quad = x \operatorname{Log} x + o(x \operatorname{Log} x) + \sum_{n \leqslant x} \Lambda(n) \left\{ \frac{x}{n} + o\left(\frac{x}{n}\right) \right\} \\ &\quad = x \operatorname{Log} x + o(x \operatorname{Log} x) + x \sum_{n \leqslant x} \frac{\Lambda(n)}{n} + o\left\{ x \sum_{n \leqslant x} \frac{\Lambda(n)}{n} \right\}. \end{aligned}$$

Recall that in chapter 1 (1.7) we showed $\sum \frac{\Lambda(n)}{n} = \operatorname{Log} x + O(1)$. The result follows.

It remains to prove the asymptotic formula within elementary number theory. This follows from the more general theorem which we state next.

Theorem 3.10 *The statement*

$$\psi(x) \operatorname{Log} x + \sum_{kr \leqslant x} \Lambda(k)\,\Lambda(r) = 2\,x \operatorname{Log} x + O(x)$$

can be proved within elementary number theory.

Proof. The following facts from chapter 1 will be freely used without special comment :

$$\operatorname{Log} n = \sum_{d \mid n} \Lambda(d)\,, \qquad \Lambda(n) = -\sum_{d \mid n} \mu(d) \operatorname{Log} d\,,$$

$$\sum_{k \leqslant x} \frac{1}{k} = \operatorname{Log} x + \gamma + O\left(\frac{1}{x}\right), \qquad \sum_{d \mid n} \mu(d) = \begin{cases} 1 \text{ si } n = 1 \\ 0 \text{ si } n > 1 \end{cases}$$

$$\psi(x) = O(x) \text{ and } \sum_{n \leqslant x} \frac{\Lambda(n)}{n} = \operatorname{Log} x + O(1).$$

Consider first the sum

$$\sum_{kr \leqslant x} \Lambda(k)\,\Lambda(r) = \sum_{n \leqslant x} \sum_{kr = n} \Lambda(k)\,\Lambda(r)$$

We will transform the inner sum into a more convenient form. We have

$$\begin{aligned} \sum_{kr = n} \Lambda(k)\,\Lambda(r) &= -\sum_{r \mid n} \Lambda(r) \sum_{d \mid n/r} \mu(d) \operatorname{Log} d \\ &= -\sum_{d \mid n} \mu(d) \operatorname{Log} d \sum_{r \mid n/d} \Lambda(r) \\ &= -\sum_{d \mid n} \mu(d) \operatorname{Log} d \operatorname{Log}\left(\frac{n}{d}\right) \\ &= \Lambda(n) \operatorname{Log} n + \sum_{d \mid n} \mu(d) \operatorname{Log}^2 d. \end{aligned} \tag{3.18}$$

Now suppose $n > 1$ and consider the expression

$$\begin{aligned} \sum_{d \mid n} \mu(d) \operatorname{Log}^2\left(\frac{x}{d}\right) &= \sum_{d \mid n} \mu(d) (\operatorname{Log}^2 x - 2 \operatorname{Log} x \operatorname{Log} d + \operatorname{Log}^2 d) \\ &= -2 \operatorname{Log} x \sum_{d \mid n} \mu(d) \operatorname{Log} d + \sum_{d \mid n} \mu(d) \operatorname{Log}^2 d \\ &= 2\,\Lambda(n) \operatorname{Log} x + \sum_{d \mid n} \mu(d) \operatorname{Log}^2 d \end{aligned} \tag{3.19}$$

From (3.18) and (3.19) we see that

$$\sum_{kr = n} \Lambda(r)\,\Lambda(k) = -2\,\Lambda(n) \operatorname{Log} x + \Lambda(n) \operatorname{Log} n + \sum_{d \mid n} \mu(d) \operatorname{Log}^2\left(\frac{x}{d}\right).$$

Summing the above equation over all $n \leqslant x$ we obtain

$$\sum_{kr \leqslant x} \Lambda(k)\,\Lambda(r) = -2\,\psi(x) \operatorname{Log} x + \sum_{n \leqslant x} \Lambda(n) \operatorname{Log} n + \sum_{n \leqslant x} \sum_{d \mid n} \mu(d) \operatorname{Log}^2\left(\frac{x}{d}\right),$$

hence

$$\begin{aligned} &\psi(x) \operatorname{Log} x + \sum_{kr \leqslant x} \Lambda(r)\,\Lambda(k) \\ &\qquad = -\psi(x) \operatorname{Log} x + \sum_{n \leqslant x} \Lambda(n) \operatorname{Log} n + \sum_{n \leqslant x} \sum_{d \mid n} \mu(d) \operatorname{Log}^2\left(\frac{x}{d}\right). \end{aligned}$$

An application of the Abel summation formula and use of the fact $\psi(x) = O(x)$ shows that

$$\sum_{n \leqslant x} \Lambda(n) \operatorname{Log} n = \psi(x) \operatorname{Log} x + O(x)$$

Thus, to complete the proof of the theorem we must show that

$$S(x) = \sum_{n \leqslant x} \sum_{d \mid n} \mu(d) \operatorname{Log}^2 \left(\frac{x}{d}\right) = 2\, x \operatorname{Log} x + O(x)\,.$$

If γ is Euler's constant, then

$$S(x) - \gamma^2 = \sum_{n \leqslant x} \sum_{d \mid n} \mu(d) \left\{ \operatorname{Log}^2 \left(\frac{x}{d}\right) - \gamma^2 \right\}$$

$$= \sum_{d \leqslant x} \mu(d) \left[\frac{x}{d}\right] \left\{ \operatorname{Log}^2 \left(\frac{x}{d}\right) - \gamma^2 \right\}.$$

The error introduced by replacing $[x/d]$ by x/d is

$$O\left(\sum_{d \leqslant x} \left\{ \operatorname{Log}^2 \left(\frac{x}{d}\right) - \gamma^2 \right\} \right) = O(x)\,.$$

So

$$S(x) = x \sum_{d \leqslant x} \frac{\mu(d)}{d} \left\{ \log^2 \left(\frac{x}{d}\right) - \gamma^2 \right\} + O(x)$$

$$= x \sum_{d \leqslant x} \frac{\mu(d)}{d} \left\{ \log \left(\frac{x}{d}\right) - \gamma \right\} \left\{ \log \left(\frac{x}{d}\right) + \gamma \right\} + O(x)$$

$$= x \sum_{d \leqslant x} \frac{\mu(d)}{d} \left\{ \log \left(\frac{x}{d}\right) - \gamma \right\} \left\{ \sum_{k \leqslant x/d} \frac{1}{k} + O\left(\frac{d}{x}\right) \right\} + O(x)$$

$$= x \sum_{d \leqslant x} \sum_{k \leqslant x/d} \frac{\mu(d)}{kd} \left\{ \log \left(\frac{x}{d}\right) - \gamma \right\} + O\left\{ x \sum_{d \leqslant x} \frac{1}{d} \left(\log \left(\frac{x}{d}\right) - \gamma \right) \frac{d}{x} \right\}$$

$$+ O(x)$$

$$= x \sum_{dk \leqslant x} \frac{\mu(d)}{kd} \left\{ \operatorname{Log} \left(\frac{x}{d}\right) - \gamma \right\} + O(x)$$

$$= x \sum_{n \leqslant x} \frac{1}{n} \sum_{d \mid n} \mu(d) \left\{ \operatorname{Log} \left(\frac{x}{d}\right) - \gamma \right\} + O(x)$$

$$= x(\operatorname{Log} x - \gamma) + x \sum_{n \leqslant x} \frac{\Lambda(n)}{n} + O(x)$$

$$= 2\, x \operatorname{Log} x + O(x)\,.$$

This completes the proof of theorem 3.10 ■

We have found satisfactory answers to questions (*a*) and (*b*). There are many ways of answering the third question; indeed all the various elementary proofs of the Prime Number Theorem without error term differ only in their tauberian argument. They all use theorem 3.10 which is known as the Selberg inequality.

It is not immediately obvious where the expression $-\,Z^1(s) + Z^2(s)$ occurred in our proof of the Prime Number Theorem. In accordance with the rules of elementary number theory, the connection between formal manipulation of finite sums and analytic properties of Dirichlet series was deliberately obscurred. We reconsider our proof now and make these relationships visible.

Recall that in § 2.3 we defined

$$c(k) = \Lambda(k) \operatorname{Log} k + \sum_{mn=k} \Lambda(m)\, \Lambda(n) .$$

From a trivial comparison of coefficients of the Dirichlet series,

$$\sum_{k=1}^{\infty} \frac{c(k)}{k^s} = -\, Z'(s) + Z^2(s) .$$

Proving the above fact within elementary number theory was theorems 3.2, 3.3, 3.4.

To simplify the tauberian argument we chose to prove the inequality

$$\sum_{k \leqslant x} \frac{c(k)}{k} = (\operatorname{Log} x)^2 - 2\,\gamma \operatorname{Log} x + O(1) ,$$

where γ is Euler's constant. Writing

$$\sum_{k \leqslant x} c(k) = \sum_{k \leqslant x} k\, \frac{c(k)}{k} .$$

and applying the Abel summation formula, we see that our inequality implies the more usual Selberg inequality.

For other elementary proofs of the Prime Number Theorem see Breusch [2], Kalecki [1], [2], Levinson [1], Nevanlinna [1].

Notes to chapter 3

THE INGHAM TAUBERIAN THEOREM

The following slightly stronger version of theorem 3.1 was proved by Ingham [3].

Theorem 3.11 *Let f be a positive, non-decreasing function which is defined for all $x \geqslant 1$ and is such that as x tends to infinity*

$$\sum_{n \leqslant x} f\left(\frac{x}{n}\right) = ax \operatorname{Log} x + bx + o(x).$$

Then we can conclude that

(a) $f(x) = ax + o(x)$

and

(b) $\displaystyle\int_1^{\infty} \frac{f(x) - ax}{x^2}\, dx = b - a\gamma$

where γ is Euler's constant.

The deduction of (*a*) was made by using Wiener's tauberian method, together with the fact that ζ does not vanish on the line $\mathcal{R}(s) = 1$, but the deduction of (*b*) from the hypotheses and (*a*) was completely elementary (see exercise 3.8).

Later Erdös and Ingham [2] considered an interesting variant of theorem 3.9. Let $\{ a_n \}$ be a sequence of real numbers such that $1 < a_1 < a_2 < \ldots$ and the series $\sum_n a_n^{-1}$ is convergent. If f is defined for all $x \geqslant 1$, bounded in every finite interval and satisfies the asymptotic relation

$$f(x) + \sum_{a_i \leqslant x} f\left(\frac{x}{a_i}\right) = \left(1 + \sum_i a_i^{-1}\right) x + o(x). \tag{3.20}$$

then we can conclude, by rather simple arguments that :

(*a*) If $\sum_i a_i^{-1} < 1$, then $f(x) \sim x$.

(*b*) If $\sum_i a_i^{-1} = 1$, then for some $c \in [0, \infty)$ we have

$$1 - c \leqslant \liminf_{x \to \infty} \frac{f(x)}{x} \leqslant \limsup_{x \to \infty} \frac{f(x)}{x} \leqslant 1 + c.$$

However if we add the extra hypothesis that f is non-decreasing then we can conclude that if $\sum_i a_i^{-1} = 1$ and the sequence $\{ a_n \}$ is *not* of the form $a_n = a^{r_n}$ for some $a > 1$ and the r_n odd integers, then $f(x) = x + o(x)$. In this exceptional case it is known by means of examples that we do not always have $c = 0$. The proofs of these results are completely elementary, but to understand the result one must adopt a slightly more sophisticated attitude. This Erdös and Ingham do by considering the "zeta function" Z, defined for $\mathcal{R}(s) \geqslant 1$ by

$$Z(s) = 1 + \sum_n a_n^{-s}$$

and proving that a necessary and sufficient condition that a function which satisfies (3.20) also satisfies $f(x) \sim x$ is that $Z(s)$ has no zeros on the line $\mathcal{R}(s) = 1$.

A more direct generalisation was found by Segal [2], who proved the following result.

Theorem 3.12 *Suppose that f is a positive, non-decreasing function such that as x tends to infinity*

$$\sum_{n \leqslant x} f\left(\frac{x}{n}\right) = xg(x) + o\{ x^2 g'(x) \}$$

where the function g is positive, twice continuously differentiable in $[1, \infty)$ *and is such that :*

(i) $g'(x) > 0$ *for* $x \in [1, \infty)$,

(ii) $xg'(x)$ *is eventually non-increasing,*

(iii) *for some integer k we have* $x(\text{Log } x)^k g'(x) = h(x)$ *is non-decreasing from some point on and* $\liminf h(x) = \infty$.

Then we can conclude that

$$f(x) \sim x^2 g'(x) .$$

It would be interesting to know to what extent the theorem remains true under less restrictive hypotheses on the function g. Clearly some conditions on g are required by considering the simple example $f(x) = 6x^2/\pi^2$ and $g(x) = x$. Segal conjectures that condition (iii) is not necessary for the truth of the theorem, but a proof seems rather difficult to find.

Using the elementary technique of Selberg, Karamata [1], [2] proved : *If* $\{ a_n \}$ *is a sequence of real numbers with the property that if f is defined by*

$$f(x) = \sum_{n \leqslant x} a_n$$

and satisfies

$$ax \text{ Log } x + bx + o(x) = \sum_{n \leqslant x} f\left(\frac{x}{n}\right) = \sum_{n \leqslant x} a_n \left[\frac{x}{n}\right],$$

then we always have

$$\int_1^x \frac{f(t) - at}{t}\, dt = o(x) .$$

If, in addition, we suppose that $a_n \geqslant -M$ *for all n, where* $M \geqslant 0$ *is a constant, then we can deduce that*

$$f(x) = ax + o(x) .$$

Earlier Shapiro [1] had proved the following, related but simpler theorem. *Let* $\{ a_n \}$ *be a sequence of real numbers such that :*

(i) $a_n \geqslant 0$ *for all n and* (ii)

$$\sum_{n \leqslant x} a_n \left[\frac{x}{n} \right] = x \operatorname{Log} x + O(x) .$$

Then there exist strictly positive real numbers α, β *such that*

$$\alpha x < \sum_{n \leqslant x} a_n < \beta x .$$

for all sufficiently large values of x. This theorem contains as a special case theorem 1.13, due to Tchebycheff. For a proof of Shapiro's theorem see exercise 3.10.

The theorem of Karamata mentioned above is very closely related to another theorem of Ingham which was proved in [3], namely the following result :

Let $\{ a_n \}$ *be a sequence of real numbers and suppose that there exists a constant* $K \geqslant 0$ *such that for all integers n we have* $na_n \geqslant -K$. *If the series*

$$\sum_{n \leqslant x} a_n \frac{n}{x} \left[\frac{x}{n} \right]$$

tends to a limit as x tends to infinity, then the series

$$\sum_{n \leqslant x} a_n$$

tends to the same limit.

THE FUNCTION $M(x)$

The behaviour of the function $M(x) = \sum_{n \leqslant x} \mu(n)$ is, as we shall see in later chapters, intimately related to the distribution of prime numbers and to the function $\psi(x)$. Although we have shown, by elementary methods in this chapter and by analytic methods in chapter 2, that $M(x) = o(x)$ it is not easy to replace the " $o(x)$ " by some simple, explicit function. By using analytic methods and a considerable amount of numerical computation (for example the fact that the first $3{,}5 \times 10^6$ zeros of ζ are on the line $\Re(s) = \frac{1}{2}$) Schoenfeld [1] was able to show that *for* $x \geqslant 6$

$$| M(x) | \leqslant \frac{ax}{(\operatorname{Log} x)^\alpha}$$

where one can take the pair $(a\,;\,\alpha)$ *to be any one of the couples* : $\left(0.47\,;\,\frac{2}{3}\right)$, $(2.9\,;\,1)$, $\left(5.5\,;\,\frac{10}{9}\right)$.

In chapter 4 we shall deduce the " better " upper bound

$$M(x) = \mathrm{O}(x \exp(-\,\alpha \sqrt{\operatorname{Log} x}))\,,$$

where α can be any number in the range $0 < \alpha < \frac{1}{15}$ and the implied " O " constant is a function of α. However, we shall not compute any numerical value for this constant; there is no doubt that it would be extremely large.

From improvements in estimates for $| M(x) |$ one can, in certain circumstances, deduce improvements of theorem 3.1. We leave it as an exercise for the reader to formulate such theorems.

THE ELEMENTARY METHOD AND ITS REFINEMENTS

In 1949, Erdös [1] and Selberg [2] independently published elementary proof of the Prime Number Theorem. In both cases their starting point was the relationship

$$\sum_{p \leqslant x} (\operatorname{Log} p)^2 + \sum_{p_1 p_2 \leqslant x} (\operatorname{Log} p_1)(\operatorname{Log} p_2) = 2\,x \operatorname{Log} x + \mathrm{O}(x) \qquad (3.21)$$

which is known as the Selberg inequality. Using different complicated, but elementary arguments, they deduced that $\theta(x) = \sum_{p \leqslant x} \operatorname{Log} p \sim x$. As we know this is equivalent to the Prime Number Theorem.

By Abel summation one can write (3.21) as

$$\theta(x) \operatorname{Log} x - \int_1^x \frac{\theta(u)}{u}\,du + \sum_{p \leqslant x} \theta\left(\frac{x}{p}\right) \operatorname{Log} p = 2\,x \operatorname{Log} x + \mathrm{O}(x)$$

and from Tchebycheff's theorem we know that the integral is $\mathrm{O}(x)$, so we have

$$\theta(x) \operatorname{Log} x + \sum_{p \leqslant x} \theta\left(\frac{x}{p}\right) \operatorname{Log} p = 2\,x \operatorname{Log} x + \mathrm{O}(x)\,.$$

The above inequality is obviously equivalent to

$$\psi(x) \operatorname{Log} x + \sum_{n \leqslant x} \psi\left(\frac{x}{n}\right) \Lambda(n) = 2\, x \operatorname{Log} x + O(x)$$

which in turn is equivalent to the inequality

$$\sum_{n \leqslant x} \Lambda(n) \operatorname{Log} n + \sum_{nm \leqslant x} \Lambda(n)\, \Lambda(m) = 2\, x \operatorname{Log} x + O(x)\,. \tag{3.22}$$

As we have seen in § 4 the above inequality could be considered more fundamental than (3.21).

Reasoning from (3.22) instead of from (3.21) one can deduce directly that

$$\psi(x) = \sum_{n \leqslant x} \Lambda(n) \sim x\,.$$

Many later writers chose to work with $\psi(x)$ rather $\theta(x)$ and so we shall state the improvements in the size of the error term which have been obtained by elementary methods as theorems about $\psi(x)$.

After the original publication in 1949 there was a lull for several years before " elementary " improvements were found. In 1955 van der Corput [2] proved that

$$\psi(x) = x + O\left\{\frac{x}{(\operatorname{Log} x)^{\Delta}}\right\} \tag{3.23}$$

with $\Delta \simeq \frac{1}{200}$, this was closely followed by Kuhn [1] who obtained (3.23) with $\Delta \simeq \frac{1}{10}$, then in 1960 Breusch [3] gave a proof of (3.23) with $\Delta \simeq \frac{1}{6}$. A dramatic improvement came in 1960 when Bombieri [1] established (3.23) for any $\Delta > 0$. Two years later Wirsing [1] obtained Bombieri's result by a new method. This technique was elaborated upon by Diamond and Steinig [3], when they proved that

$$|\, \psi(x) - x \,| \leqslant x \exp\{ -(\operatorname{Log} x)^{1/7} (\operatorname{Log} \operatorname{Log} x)^{-2} \}\,. \tag{3.24}$$

for all $x \geqslant \exp\exp 98$. This estimate is the best that has been obtained by elementary methods at the time of writing.

However this latter result is still weaker than estimates which can be obtained by non-elementary methods. For example, in chapter 4 we shall prove that

$$\psi(x) - x = O\{ x \exp(-\alpha \sqrt{\operatorname{Log} x}) \}$$

and in chapter 11 the refinement

$$\psi(x) - x = O(x \exp\{ -\beta(\operatorname{Log} x)^{3/5} (\operatorname{Log} \operatorname{Log} x)^{-1/5} \})\,.$$

The significance of elementary proofs of estimates such as (3.24) is as follows. We have seen in the introduction to chapter 2 that " non-elementary " proofs of the

Prime Number Theorem depend on knowing zero free regions of ζ. The precise relationship between a zero free region of ζ and the order of magnitude of $\psi(x)-x$ will be given in chapter 4. However a converse relationship does exist, that is, a knowledge of the order of magnitude of $\psi(x)-x$ can be used to deduce a zero free region of ζ. Thus, from the elementary proof of an estimate such as (3.24) one can reverse the standard procedure and deduce zero free regions from a knowledge of $|\psi(x)-x|$. It remains to be seen whether or not the elementary methods are capable of proving all the results which can be obtained by complex variable techniques; it certainly does not seem as unlikely as it once did.

After the proof of the Prime Number Theorem by elementary methods it was natural to produce an analogous proof of the Prime Number Theorem for Arithmetic Progressions, namely to prove that if k, l are coprime integers, then

$$\pi(x; k, l) = \sum_{\substack{p \leqslant x \\ p \equiv l \,(\mathrm{mod}\, k)}} 1 \sim \frac{1}{\varphi(k)} \frac{x}{\operatorname{Log} x}$$

and its equivalent formulation

$$\psi(x; k, l) = \sum_{\substack{n \leqslant x \\ n \equiv l \,(\mathrm{mod}\, k)}} \Lambda(n) \sim \frac{1}{\varphi(k)} x .$$

This was done by Selberg [3] in 1950. The pattern of improvements of this result was very similar to the sequence of improvements in the elementary proof of the Prime Number Theorem. For example Levin [1] used the method of Kuhn to show that

$$\pi(x; k, l) = \frac{1}{\varphi(k)} \frac{x}{\operatorname{Log} x} + O\left\{\frac{x}{(\operatorname{Log} x)^{1.14}}\right\}$$

and Wirsing [1] proved that for any $\Delta > 1$ one has

$$\psi(x; k, l) = \frac{1}{\varphi(k)} x + O\left\{\frac{x}{(\operatorname{Log} x)^{\Delta}}\right\}$$

In chapter 8 we shall prove, by rather complicated analytic methods, that the above estimate and that the implied " O " constant does not depend on k or l. It is not cleár whether or not Wirsing's argument can be modified to prove this refinement.

Of course after 1949 there followed a great many other elementary proof of the Prime Number Theorem, which simplified the original proofs of Erdös and Selberg.

Many of these later proofs used the language of " Dirichlet convolutions ". We shall give a very brief account of this topic later and discuss some of the algebraic and topological properties of the associated algebraic structure.

From the variations in the elementary proofs it became clear that the techniques were applicable to a fairly wide class of sequences and not just to the primes. For example, all our theorems so far are applicable, with only trivial modifications to systems of generalised integers whose counting function satisfies

$$N(x;Z) = Ax + O(x^\theta),$$

where $0 \leqslant \theta < 1$. Thus, in particular we can derive an elementary proof of the Prime Ideal Theorem for algebraic number fields. A generalisation in another direction was found by Shapiro [2], who proved by elementary methods a conjecture of Erdös, namely :

Let $\{ a_n \}$ be a sequence of real numbers which satisfies $a_n \geqslant 1$, a_n tends to infinity with n,

$$\sum_{a_n \leqslant x} (\text{Log } a_n)^2 + \sum_{a_r a_s \leqslant x} (\text{Log } a_r)(\text{Log } a_s) \sim 2\, x \text{ Log } x$$

Then we have

$$\sum_{a_n \leqslant x} \text{Log } a_n \sim x .$$

THE DIRICHLET RING OF ARITHMETIC FUNCTIONS

Working within the confines of " elementary number theory " one usually manipulates sequences $\{ a_n \}$ which arise because the a_n are coefficients of Dirichlet series such as $\zeta(s)$, $\zeta^{-1}(s)$, $Z(s)$, etc. The associated Dirichlet series are all absolutely uniformly convergent on all compact sets in the half plane $\mathcal{R}(s) > 1$. Thus all natural analytic prossesses, such as differentiation, multiplication and the rearrangement of terms in the series are justified. These operations permit us to deduce algebraic relationships between the coefficients of various Dirichlet series. For example, if $\mathcal{R}(s) > 1$ we have

$$\zeta(s) = \sum_{n=1}^{\infty} \frac{1}{n^s} \quad \text{and} \quad \zeta^{-1}(s) = \sum_{n=1}^{\infty} \frac{\mu(n)}{n^s}$$

and by multiplication and rearrangement of terms it follows that

$$1 = \zeta(s)\,\zeta^{-1}(s) = \sum_{n=1}^{\infty} \frac{\delta_n}{n^s}$$

where $c_n = \sum_{r \mid n} \mu(r)$. Thus we can now conclude that

$$\sum_{r \mid n} \mu(r) = \begin{cases} 1 & \text{if } n = 1 \\ 0 & \text{if } n > 1 . \end{cases}$$

With such an analytic structure, leading to purely algebraic relationships one quite naturally considers the concept of " formal Dirichlet series ". That is, expressions of the type

$$\sum_{n=1}^{\infty} \frac{a_n}{n^s} \quad \text{with} \quad a_n \in \mathbb{C}$$

and to define, in a purely formal way, the above analytic operations. Once one is concerned only with the formal relationships between the coefficients of Dirichlet series it is natural to make the correspondence

$$\sum_{n=1}^{\infty} \frac{a_n}{n^s} \leftrightarrow \{ a_n \} \in \mathbb{C}^{\mathbb{N}}$$

and then to study the algebraic structure induced on the space of complex sequences.

Thus, the addition and multiplication of Dirichlet series leads us to define an addition $\oplus$ and a multiplication $*$ on $\mathbb{C}^{\mathbb{N}}$ as follows.

If $\mathbf{a} = \{ a_n \}$, $\mathbf{b} = \{ b_n \}$, then

(i) $\mathbf{a} \oplus \mathbf{b} = \mathbf{c}$, where $\mathbf{c} = \{ a_n + b_n \}$

(ii) $\mathbf{a} * \mathbf{b} = \mathbf{c}$, where $\mathbf{c} = \left\{ \sum_{rs=n} a_r b_s \right\}$.

It is not difficult to verify that $\{ \mathbb{C}^{\mathbb{N}}, *, \oplus \}$ is an associative, commutative ring with identity $\mathbf{1} = \{ 1, 0, 0, ... \}$. This ring, which we denote by $\mathfrak{D}$, is usually called the *Dirichlet ring of arithmetic functions.*

As far as " elementary number theory " is concerned the ring $\mathfrak{D}$ is usually employed only as a convenient shorthand for writing out identities between arithmetic functions. The simplest example is as follows. The Riemann zeta function corresponds to the sequence $\mathbf{z} = \{ 1, 1, ... \}$ and $1/\zeta(s)$ to the sequence $\mathbf{m} = \{ \mu(n) \}$, and we have the relationship

$$\mathbf{1} = \mathbf{z} * \mathbf{m} .$$

Thus, $\mathbf{m}$ is the inverse of $\mathbf{z}$ in $\mathfrak{D}$. The Möbius inversion formula admits a similar intuitive interpretation. The pair $(\mathbf{a}, \mathbf{b})$ is a Möbius pair of sequences if $\mathbf{b} = \mathbf{z} * \mathbf{a}$ and then

$$\mathbf{b} = \mathbf{z} * \mathbf{a} \Leftrightarrow \mathbf{a} = \mathbf{z}^{-1} * \mathbf{b}$$

We leave it as an exercise for the reader to translate our lemmas on arithmetic functions into statements about $\mathfrak{D}$ and its endomorphism ring.

As an algebraic object the ring $\mathfrak{D}$ is extremely interesting. It is an easy exercise to show that the units of $\mathfrak{D}$ are precisely the sequences $\{ a_n \}$ with $a_1 \neq 0$ and that the set

$$\{ \mathbf{a} \in \mathfrak{D} \mid a_1 = 0 \}$$

forms a maximal prime ideal. A non-trivial fact is that $\mathfrak{D}$ is a unique factorisation domain. This had been conjectured for a long time before it was proved by Cashwell and Everett [1] in 1959.

The formal differentiation of a Dirichlet series :

$$\frac{d}{ds}\left(\sum_{n=1}^{\infty} \frac{a_n}{n^s}\right) = -\sum_{n=1}^{\infty} \frac{a_n \operatorname{Log} n}{n^s}.$$

implies that the endomorphism ring of $\mathfrak{D}$ contains derivations. The natural derivation on $\mathfrak{D}$ being defined by

$$\mathbf{a}' = \{ -a_n \operatorname{Log} n \}$$

An additional structure which one can put on $\mathfrak{D}$ is that of a non-archimedian valuation ring. For example, if we define $\| \, . \, \|$ by

$$\| \mathbf{0} \| = 0 \quad \text{and} \quad \| \mathbf{a} \| = N^{-1}$$

where $\mathbf{a} = \{ a_n \}$ and a_N is the first non-zero term in the sequence, then it is not difficult to verify that $\| \, . \, \|$ is a non-archimedian valuation. A study of some of the interaction between the differential and valuation structures has been given by Shapiro [3].

The ring $\mathfrak{D}$ can be used to prove purely analytical facts. For example, a well known theorem of Hilbert asserts that $\zeta(s)$ does not satisfy an algebraic differential equation. Popken [1] proves the following refinement of Hilbert's theorem. Let $f \in \mathbb{C}[x_0, \ldots, x_r]$ have total degree g, then

$$\| f(\mathbf{z}, \mathbf{z}', \ldots, \mathbf{z}^{(r)}) \| \geqslant \begin{cases} (r+2)^{-1} & \text{si} \quad g = 1 \\ \{ cg \operatorname{Log} \{ (r+2) g \} . \operatorname{Log}^2 (r+2) \}^{-g} & \text{if} \quad g > 1 \end{cases}$$

where c is an absolute constant.

Exercises to chapter 3

3.1 (i) Prove that $\left| \sum_{n \leqslant x} \frac{\mu(n)}{n} \right| \leqslant 1$ for all $x \geqslant 1$ by starting with the relation $[x] = \sum_{n \leqslant x} 1$ and using the Möbius inversion formula.

(ii) Deduce from (i) that

$$\left| \int_1^x \frac{M(u)}{u^2} \, du \right| \leqslant 2 \, .$$

3.2 Use theorem 3.1 to prove that

(i) $\sum_{n=1}^{\infty} \frac{\mu(n)}{n} = 0$ and $\sum_{n=1}^{\infty} \frac{\Lambda(n) - 1}{n} = - 2 \gamma$

where γ is Euler's constant.

3.3 (i) If the functions F and G, which are defined for all $x \geqslant 1$, are related by

$$G(x) = \operatorname{Log} x \sum_{n \leqslant x} F\left(\frac{x}{n}\right) .$$

prove that

$$F(x) \operatorname{Log} x + \sum_{n \leqslant x} F\left(\frac{x}{n}\right) \Lambda(n) = \sum_{n \leqslant x} \mu(n) \, G\left(\frac{x}{n}\right) .$$

(ii) Prove that if $x \geqslant 1$, then

$$\psi(x) \operatorname{Log} x + \sum_{n \leqslant x} \psi\left(\frac{x}{n}\right) \Lambda(n) = 2 \, x \operatorname{Log} x + \mathrm{O}(x)$$

and

$$M(x) \operatorname{Log} x + \sum_{n \leqslant x} M\left(\frac{x}{n}\right) \Lambda(n) = \mathrm{O}(x) \, .$$

(iii) Can you use the above formulae to give " elementary " proofs that $\psi(x) = x + \mathrm{o}(x)$ and $M(x) = \mathrm{o}(x)$?

3.4 Let (f, F) be a Möbius pair of sequences.

(i) Suppose that

$$\lim_{x\to\infty} \frac{1}{x} \sum_{n\leqslant x} f(n) = 0 \quad \text{et} \quad \limsup_{x\to\infty} \frac{1}{x} \sum_{n\leqslant x} |F(n)| < \infty$$

Prove that

$$\lim_{x\to\infty} \left\{ \frac{1}{x} \sum_{n\leqslant x} F(n) - \sum_{n\leqslant x} \frac{f(n)}{n} \right\} = 0 .$$

(ii) Suppose that the following limit exists :

$$\lim_{x\to\infty} \sum_{n=1}^{\infty} \frac{f(n)}{n}$$

Prove that

$$\lim_{x\to\infty} \frac{1}{x} \sum_{n\leqslant x} F(n) = \sum_{n=1}^{\infty} \frac{f(n)}{n} .$$

3.5 Let $\{ a_n \}$ be a sequence of real numbers and F a function defined for all x. Suppose that

(i) $F(x)$ is of bounded variation in every finite interval,

(ii) $\sum_{n\leqslant x} a_n = o(x)$ et $\sum_{n\leqslant x} | a_n | = O(x)$,

(iii) $F(x) = O(x)$.

Prove that

$$\sum_{n\leqslant x} a_n F\left(\frac{x}{n}\right) = o(x) .$$

Replace condition (iii) above by :

(iv) $F(x) = O(\omega(x))$, where $\omega(x)$ is a non-decreasing function such that the following integral exists

$$\int_1^{\infty} \frac{\omega(t)}{t^2}\, dt$$

and deduce that

$$\sum_{n\leqslant x} a_n F\left(\frac{x}{n}\right) = o(x)$$

3.6 By using techniques from the theory of divergent series (see Hardy [2]), prove the following theorem of Rubel [1].

(i) Let $(g(n), G(n))$ be a Möbius pair of sequences. If

$$\lim_{n\to\infty} G(n) = l$$

exists and is equal to l, then

$$\sum_{n=1}^{\infty} \frac{g(n)}{n} = l.$$

(ii) Show, by example, that the convergence of the series

$$\sum_{n=1}^{\infty} \frac{g(n)}{n}$$

is not a consequence of the existence of the limit

$$\lim_{n\to\infty} \frac{1}{n} \sum_{k\leqslant n} G(k) = L.$$

(iii) Prove that if the above series does converge to l and the above limit exists and is equal to L, then $l = L$.

3.7 Prove that

$$\int_1^{\infty} \frac{\psi(x) - x}{x^2}\, dx = -(1+\gamma).$$

Write $g(x) = \sum_{n\leqslant x} \frac{\mu(n)}{n}$ and for any $\sigma \geqslant 1$ show that

$$\int_1^{\infty} \left\{ \sum_{n\leqslant x} \frac{\mu(n)}{n} \right\} \frac{dx}{x^{\sigma}} = \frac{1}{(\sigma-1)\,\zeta(\sigma)}$$

3.8 Let (f, F) be a Möbius pair of functions.

(i) Prove that

$$\int_1^{x} \frac{F(u)}{u}\, du = \int_1^{x} \frac{f(u)}{u} \left[\frac{x}{u}\right] du;$$

(ii) If $F(x) = Ax \operatorname{Log} x + Bx + o(x)$, then show that

$$\frac{1}{x}\int_1^{x} \frac{f(y) - Ay}{y} \left[\frac{x}{y}\right] dy = (B - A\gamma) + o(1).$$

(iii) If $f(x) = Ax + o(x)$, prove that

$$\int_1^\infty \frac{f(y) - Ay}{y^2}\,dy = B - A\gamma\,.$$

3.9 Let f be a real valued function defined for all $x \geqslant 1$ and $\{ a_n \}$ a sequence of positive numbers such that the series $\sum_{n=1}^{\infty} a_n^{-1}$ is convergent to A. Define the function F by

$$F(x) = \sum_{a_n \leqslant x} f\left(\frac{x}{a_n}\right).$$

and prove the following variant of theorem 1.

$$A \liminf_{x\to\infty} \frac{f(x)}{x} \leqslant \liminf_{x\to\infty} \frac{F(x)}{x} \leqslant \limsup_{x\to\infty} \frac{F(x)}{x} \leqslant A \limsup_{x\to\infty} \frac{f(x)}{x}\,.$$

3.10 Let $\{ a_n \}$ be a sequence of real numbers satisfying

(i) $a_n \geqslant 0$ for $n \geqslant 1$,

(ii) $F(x) = \sum_{n\leqslant x} a_n \left[\frac{x}{n}\right] = x \operatorname{Log} x + O(x)\,.$

Define the function S by $S(x) = \sum_{n \leqslant x} a_n$. Prove that there exist real numbers α and β such that

$$0 < \alpha \leqslant \liminf_{x\to\infty} \frac{S(x)}{x} \leqslant \limsup_{x\to\infty} \frac{S(x)}{x} \leqslant \beta < \infty\,.$$

[Hint : (i) Note that if $z \geqslant 0$, then $[z] - 2[z/2] \geqslant 0$.
(ii) Show that $O(x) = F(x) - 2\,F(x/2) \geqslant S(x) - S(x/2)$.
(iii) Write $G(x) = \sum_{n\leqslant x} a_n n^{-1}$ and prove that $|\,G(x) - \operatorname{Log} x\,| < K_2$.
Put $\alpha = e^{-2K_2 - 1}$ and consider $G(x) - G(x/\alpha)$.]

3.11 Modify the discussion in § 4 to give a proof that $\zeta(s)$ does not vanish on the line $\mathfrak{R}(s) = 1$. (This is essentially Hadamards original argument.)

3.12 (i) Let h be an arithmetic function. Prove that if

$$\mathbf{h}(\mathbf{f} * \mathbf{g}) = (\mathbf{hf}) * (\mathbf{hg})$$

for all arithmetic functions f and g, then h is completely multiplicative.

(ii) If f is a multiplicative function and

$$\mathbf{h}(\boldsymbol{\mu} * \mathbf{z}) = (\mathbf{h}.\boldsymbol{\mu}) * (\mathbf{h}.\mathbf{z}) .$$

prove that f is completely multiplicative.

3.13 Prove that the endomorphism d_n of $\mathfrak{D}$ defined by

$$d_h(\mathbf{a}) = \{ a_n h(n) \} .$$

is a derivation if and only if h is completely additive, that is $h(mn) = h(m) + h(n)$ for all positive integers m and n.

3.14 Let h be any real valued, non-negative, completely additive arithmetic function. If $\mathbf{a} \in \mathfrak{D}$, then we define $v(\mathbf{a}, h)$ as the largest integer v such that $h(n) < v \Rightarrow a_n = 0$: next we define $\| \cdot \|_h$ by

$$\| \mathbf{0} \|_h = 0 \quad \text{and} \quad \| \mathbf{a} \|_h = e^{v(\mathbf{a},h)} .$$

Prove that $\| \cdot \|_h$ is a non-archemedian valuation on $\mathfrak{D}$. Show that two valuations $\| \cdot \|_h$ and $\| \cdot \|_g$ induce the same topology on $\mathfrak{D}$ if and only if h and g are proportional.

CHAPTER 4

The remainder term in the Prime Number Theorem

§ 1 INTRODUCTION

In § 2 of chapter 2 we saw that there was a connection between the distribution of prime numbers and the location of zeros of $\zeta(s)$. This relationship will be made quite explicit in this chapter. We shall :

(*a*) demonstrate the existence of a zero free region for $\zeta(s)$ strictly to the left of the line $\Re(s) = 1$,

(*b*) deduce an estimate for the error term in the Prime Number Theorem from a knowledge of a zero free region for $\zeta(s)$,

(*c*) end the chapter by discussing how one can reverse the implication from (*a*) to (*b*) and deduce the existence of a zero free region for $\zeta(s)$ from a non-trivial estimate for $M(x)$.

In our discussion of (*a*) we shall divide the argument into two parts. First, as theorem 4.4, we show that an upper bound for $|\zeta(s)|$ to the left of the line $\Re(s) = 1$ can be used to deduce that $\zeta(s) \neq 0$ in a certain region of the half plane $\Re(s) \leqslant 1$. Then, as a particular application of theorem 4.4, we shall use the upper bound for $|\zeta(s)|$ provided by theorem 2.3 to prove that $\zeta(s) \neq 0$ in the region :

$$R_c = \left\{ s = \sigma + it \mid 1 - \frac{c}{\operatorname{Log} |t|} \leqslant \sigma \leqslant 1, |t| \geqslant t_0 \right\}$$

for all sufficiently large values of $|t|$.

The philosophy behind our treatment is that any future improvement of theorem 2.3 will automatically translate into an improvement of the zero free region. Such an improvement will be given in chapter 11.

Similarly, in our discussion of (*b*) we shall take a "general" zero free region R and then, as theorem 4.6, deduce an estimate for the remainder term in the Prime Number Theorem as a function of R. As a particular application of this theorem we shall use the region R_c to deduce that

$$\psi(x) - x = O\{ x \exp(-a\sqrt{\operatorname{Log} x}) \}.$$

This estimate for $R(x)$ is better than any we have so far obtained and the refinement in our approximation to $\psi(x)$ naturally leads to a better approximation to $\pi(x)$.

The formula which we will obtain for $\pi(x)$ is rather different from what one might expect. It is

$$\pi(x) = \int_2^x \frac{du}{\operatorname{Log} u} + O\{ x \exp(-a\sqrt{\operatorname{Log} x}) \}.$$

We shall often write

$$\operatorname{Li}(x) = \int_2^x \frac{du}{\operatorname{Log} u}.$$

An integration by parts shows that $\operatorname{Li}(x) \sim x/\operatorname{Log} x$, so the above formula for $\pi(x)$ contains the simple form of the Prime Number Theorem proved earlier. The function $\operatorname{Li}(x)$ is quite definitely a better approximation to $\pi(x)$ than $x/\operatorname{Log} x$ in the sense that

$$\limsup_{x\to\infty} \frac{|\pi(x) - \operatorname{Li}(x)|}{x \exp(-a\sqrt{\operatorname{Log} x})} < \infty \quad \text{and} \quad \limsup_{x\to\infty} \frac{|\pi(x) - x/\operatorname{Log} x|}{x \exp(-a\sqrt{\operatorname{Log} x})} = \infty.$$

Our principal tools in this chapter are three general theorems from complex function theory. We shall devote § 2 to their proof.

§ 2 THREE THEOREMS FROM COMPLEX FUNCTION THEORY

2.1 Our first theorem is due to Borel; it will only be used to prove the second theorem. This latter theorem will be of crucial importance in this chapter and in chapters 5 and 8, when we study $\zeta(s)$ and Dirichlet's L-functions. The final theorem in this section is Hadamard's "three circles theorem".

Theorem 4.1 *Let* U, r, R *be real numbers with* $0 < r < R$ *and* f *a holomorphic function in the disc* $|z - z_0| < R$. *Suppose also that* f *satisfies*

$$\Re\{ f(z) \} \leqslant \mathrm{U}.$$

within the disc. If the Taylor series expansion of f about z_0 is

$$f(z) = \sum_{n=0}^{\infty} c_n(z - z_0)^n ,$$

then the following inequalities hold :

(a) $| c_n | \leqslant \dfrac{2}{R^n}(\mathrm{U} - \Re(f(z_0)))$ *for* $n = 1, 2, \ldots,$

(b) *if* $| z - z_0 | \leqslant r < R$, *then*

$$| f(z) - f(z_0) | \leqslant \frac{2r}{R - r}(\mathrm{U} - \Re \{ f(z_0) \}) .$$

(c) *if* $| z - z_0 | \leqslant r < R$, *then for all* $\nu \geqslant 1$

$$\left| \frac{f^{(\nu)}(z)}{\nu !} \right| \leqslant \frac{2R}{(R - r)^{\nu+1}}(\mathrm{U} - \Re \{ f(z_0) \}) .$$

Proof. Without loss of generality we can suppose that $z_0 = 0$. For simplicity we shall define the function φ by

$$\varphi(z) = \mathrm{U} - f(z)$$

The power series expansion of φ about $z = 0$ is thus

$$\varphi(z) = \mathrm{U} - c_0 - \sum_{n=1}^{\infty} c_n z^n$$

$$= \sum_{n=0}^{\infty} b_n z^n$$

We shall denote by P and Q respectively the real and complex parts of $\varphi(z)$ and by Γ the circle $| z | = r < R$.

If $n \geqslant 1$ the function $\varphi(z) z^{n-1}$ is holomorphic in the disc $| z | < R$ and Cauchy's theorem gives

$$0 = \int_{\Gamma} \varphi(z) z^{n-1} \, dz = \int_{-\pi}^{+\pi} (\mathrm{P} + iQ) \, e^{in\theta} \, d\theta$$

hence by taking complex conjugates

$$0 = \int_{-\pi}^{+\pi} (\mathrm{P} - iQ) \, e^{-in\theta} \, d\theta = 0 \qquad (n \geqslant 1) .$$

Also, from Cauchy's theorem we know, for $n \geqslant 0$

$$b_n = \frac{1}{2 i\pi} \int_{\Gamma} \frac{\varphi(z)}{z^{n+1}} \, dz = \frac{1}{2 \pi r^n} \int_{-\pi}^{+\pi} (\mathrm{P} + iQ) \, e^{-in\theta} \, d\theta , \tag{4.2}$$

From (4.1) and (4.2) we conclude

$$b_n = \frac{1}{\pi r^n}\int_{-\pi}^{+\pi} \mathrm{P}\, e^{-in\theta}\, d\theta \qquad (n \geqslant 1)\,.$$

Thus, for each integral $n \geqslant 1$ we certainly have

$$|\, b_n \,| \leqslant \frac{1}{\pi r^n}\int_{-\pi}^{+\pi} |\, \mathrm{P} \,|\, d\theta\,.$$

By hypothesis, for $|\, z \,| < R$,

$$\mathrm{P} = \Re\,\{\, \varphi(z) \,\} = \mathrm{U} - \Re\,\{\, f(z) \,\} \geqslant 0$$

and consequently $\mathrm{P} \geqslant 0$, thus $|\, \mathrm{P} \,| = \mathrm{P}$. Upon taking $n = 0$ in (4.2) we see

$$\Re(b_0) = \frac{1}{2\,\pi}\int_{-\pi}^{+\pi} \mathrm{P}\, d\theta$$

and hence

$$|\, b_n \,| \leqslant \frac{2}{r^n}\,\Re(b_0) = \frac{2}{r^n}(\mathrm{U} - \Re\,\{\, f(0) \,\}) \qquad (n \geqslant 1)\,.$$

Now, letting r increase to R, we deduce

$$|\, b_n \,| \leqslant \frac{2}{R^n}(\mathrm{U} - \Re\,\{\, f(0) \,\}) \qquad (n \geqslant 1)$$

This proves part (*a*) of the theorem.

(*b*) The following inequalities are trivial using part (*a*) :

$$\begin{aligned}
|\, \varphi(z) - \varphi(0) \,| &= \left| \sum_{n=1}^{\infty} b_n\, z^n \right| \leqslant \sum_{n=1}^{\infty} |\, b_n \,|.|\, z \,|^n \\
&\leqslant 2\,\Re\,\{\, \varphi(0) \,\} \sum_{n=1}^{\infty} \left(\frac{r}{R}\right)^n \\
&= 2\,\Re\,\{\, \varphi(0) \,\} \frac{r}{R - r}\,.
\end{aligned}$$

(*c*) By differentiating φ and using (*a*) we see that

$$\begin{aligned}
|\, \varphi^{\nu}(z) \,| &= \left| \sum_{n=\nu}^{\infty} n(n-1)\cdots(n-\nu+1)\, b_n\, z^{n-\nu} \right| \\
&\leqslant 2\,\Re\,\{\, \varphi(0) \,\} \sum_{n=\nu}^{\infty} n(n-1)\cdots(n-\nu+1)\frac{r^{n-\nu}}{R^n} \\
&= 2\,\Re\,\{\, \varphi(0) \,\} \frac{d^{\nu}}{dr^{\nu}} \left\{ \sum_{n=0}^{\infty} \left(\frac{r}{R}\right)^n \right\} \\
&= 2\,\Re\,\{\, \varphi(0) \,\} \frac{R\nu\,!}{(R - r)^{\nu+1}}\,.
\end{aligned}$$

This completes the proof of the theorem. ∎

Our next theorem will be of fundamental importance to our deeper study of Prime Number Theory.

Theorem 4.2 *Let $R > 0$, $q > 1$ and $\mathrm{U} \geqslant 0$ be fixed real numbers. Suppose that f is holomorphic on the disc $| s - s_0 | \leqslant qR$ and satisfies :*

(i) $f(s_0) \neq 0$,

(ii) $\left| \dfrac{f(s)}{f(s_0)} \right| \leqslant e^{\mathrm{U}}$.

Denote by N the number of zeros $\rho = \beta + i\gamma$ of f in the disc $| s - s_0 | \leqslant R$. The following inequalities hold :

(a) $0 \leqslant N \leqslant \dfrac{\mathrm{U}}{\operatorname{Log} q} \leqslant \dfrac{q\mathrm{U}}{q-1}$;

(b) *If* $| s - s_0 | \leqslant r < R$, *then*

$$\left| \frac{f'(s)}{f(s)} - \sum_{\rho} \frac{1}{s - \rho} \right| \leqslant \frac{QR\mathrm{U}}{(R - r)^2},$$

where $Q = \max \{ 2, (q^2 \operatorname{Log} q)^{-1} \}$.

Suppose now that f has no zeros in the region

$$\{ s = \sigma + it \mid \sigma > \sigma_0 - d, \ | s - s_0 | \leqslant R \}$$

where $0 < d < R$.

(c) *If s satisfies* $| s - s_0 | \leqslant r \leqslant d$, *then*

$$\sum_{\{\rho\}} \Re\left(\frac{1}{s - \rho}\right) - \Re\left\{\frac{f'(s)}{f(s)}\right\} \leqslant \frac{QR\mathrm{U}}{(R - r)^2}$$

where the summation is over any subset (ρ) of the zeros of f contained in the disc

$$| s - s_0 | \leqslant R.$$

(d) *If $r < d$, then for s satisfying* $| s - s_0 | \leqslant r$ *we have*

$$\left| \frac{f'(s)}{f(s)} \right| \leqslant \frac{d + r}{d - r} \left\{ \left| \frac{f'(s_0)}{f(s_0)} \right| + \frac{QR\mathrm{U}}{(R - d)^2} \right\}.$$

Proof. Without loss of generality we can suppose that $s_0 = 0$. For technical convenience we shall write f in a " normalised " form. First observe that for $| s | \leqslant Rq$ and ρ any zero of $f(s)$ in the disc $| s | \leqslant R$ we have $(q^2 R^2 - \bar{\rho} s) \neq 0$. Hence we can write

$$\frac{f(s)}{f(0)} = g(s) \prod_{\rho} \frac{qR(s - \rho)}{q^2 R^2 - \bar{\rho} s} \tag{4.3}$$

where g is holomorphic on the disc $|s| \leqslant qR$ and non-zero on the disc $|s| \leqslant R$. Observe that for $|s| = qR$ we have

$$\left| \frac{qR(s-\rho)}{q^2 R^2 - \bar{\rho} s} \right| = \frac{qR\,|s-\rho|}{|s| . |\bar{s} - \bar{\rho}|} = 1 ,$$

Our hypothesis on f implies that for $|s| = qR$

$$|g(s)| \leqslant e^{\mathrm{U}}$$

Whence, by the maximum modulus principle, we deduce that

$$|g(s)| \leqslant e^{\mathrm{U}}$$

for $|s| \leqslant qR$.

Putting $s = 0$ in (4.3) we obtain

$$1 = g(0) \prod_{\rho} \left(\frac{-\rho}{qR} \right)$$

and since $R \geqslant |\rho|$ we certainly have

$$|g(0)| = \prod_{\rho} \frac{qR}{|\rho|} \geqslant q^N,$$

and consequently

$$N \leqslant \frac{\mathrm{U}}{\operatorname{Log} q} .$$

This proves part (*a*) of the theorem.

(*b*) Because g is holomorphic on the disc $|s| \leqslant qR$ and non-zero on the disc $|s| \leqslant R$ we can write, for $|s| \leqslant R$,

$$g(s) = g(0)\, e^{h(s)}$$

where h is holomorphic on the disc $|s| \leqslant R$ and $h(0) = 0$.

We are now going to apply theorem 4.1(*c*) to h. First we must check that the hypothesis of theorem 4.1 is satisfied. For $|s| \leqslant R$ we have

$$\left| \frac{g(s)}{g(0)} \right| \leqslant \frac{e^{\mathrm{U}}}{|g(0)|} = \exp \{ \mathrm{U} - \operatorname{Log} |g(0)| \}$$

Thus, for $|s| \leqslant R$ it follows that

$$\Re \{ h(s) \} \leqslant \mathrm{U} - \operatorname{Log} |g(0)| \leqslant \mathrm{U} - N \operatorname{Log} q .$$

Theorem 4.1(*c*) permits us to conclude that for $|s| \leqslant r < R$

$$|h'(s)| \leqslant \frac{2R}{(R-r)^2} (\mathrm{U} - N \operatorname{Log} q) . \qquad (4.4)$$

From (4.3) it follows, by logarithmic differentiation, that

$$h'(s) + \sum_\rho \frac{\bar\rho}{q^2 R^2 - \bar\rho s} = \frac{f'(s)}{f(s)} - \sum_\rho \frac{1}{s - \rho}. \tag{4.5}$$

By trivial majorisations we see that if $|s| \leqslant r$, then

$$\left| \sum_\rho \frac{\bar\rho}{q^2 R^2 - \bar\rho s} \right| \leqslant \frac{RN}{q^2 R^2 - Rr} \leqslant \frac{RN}{q^2(R - r)^2} \tag{4.6}$$

Combining inequalities (4.4), (4.5) and (4.6) we obtain

$$\left| \frac{f'(s)}{f(s)} - \sum_\rho \frac{1}{s - \rho} \right| \leqslant \frac{R}{(R - r)^2} \left\{ 2\,\mathrm{U} - 2N \operatorname{Log} q + \frac{N}{q^2} \right\}.$$

The right hand side of the above inequality is a linear function of N and so, by (*a*), lies between

$$\frac{R\mathrm{U}.2}{(R - r)^2} \quad \text{and} \quad \frac{R\mathrm{U}}{(R - r)^2} \cdot \frac{1}{q^2 \operatorname{Log} q}.$$

Thus for $|s| \leqslant r$ we certainly have

$$\left| \frac{f'(s)}{f(s)} - \sum_\rho \frac{1}{s - \rho} \right| \leqslant \frac{QR\mathrm{U}}{(R - r)^2}.$$

(*c*) For any complex numbers z and w we always have

$$\Re(z) - \Re(w) \leqslant |z - w|.$$

Hence, from (*b*) it follows that

$$\sum_\rho \Re\left(\frac{1}{s - \rho}\right) - \Re\left\{\frac{f'(s)}{f(s)}\right\} \leqslant \frac{QR\mathrm{U}}{(R - r)^2}.$$

Now for any s satisfying $|s| \leqslant r \leqslant d$ and $\rho = \beta + i\gamma$ a zero of $f(s)$,

$$\Re\left(\frac{1}{s - \rho}\right) = \frac{\sigma - \beta}{|s - \rho|^2} \geqslant 0$$

since $\sigma \geqslant -d \geqslant \beta$. Thus we can omit any number of terms from the summation over ρ and (*c*) follows.

(*d*) Let us write

$$F(s) = -\frac{f'(s)}{f(s)} \quad \text{and} \quad \mathrm{U}_d = \frac{QR\mathrm{U}}{(R - d)^2}$$

Then F is holomorphic on the disc $|s| \leqslant d$ and by (*c*) with $(\rho) = \varnothing$ and $r = d$ we see

$$\Re(F(s)) \leqslant \mathrm{U}_d.$$

We now apply theorem 4.1(*b*) to deduce that for $|s| \leqslant r < d$ the following inequality holds :

$$|F(s) - F(0)| \leqslant \frac{2r}{d-r}(\mathrm{U}_d - \Re\{F(0)\}).$$

with the consequence that

$$\begin{aligned} |F(s)| &\leqslant |F(0)| + \frac{2r}{d-r}(\mathrm{U}_d + |F(0)|) \\ &\leqslant \left(1 + \frac{2r}{d-r}\right)(\mathrm{U}_d + |F(0)|). \end{aligned}$$

This completes the proof of the theorem. ∎

The final theorem of this section is the classical " three circles " theorem of Hadamard.

Theorem 4.3 *Let f be holomorphic on the annulus $0 < r_1 \leqslant |z| \leqslant r_2$. Denote by $M(r)$ the maximum value of $|f(z)|$ on the circle $|z| = r$. Then* Log $M(r)$ *is a convex function of* Log r, *i.e. if $r_1 \leqslant r \leqslant r_2$, then*

$$\mathrm{Log}\left(\frac{r_2}{r_1}\right) \mathrm{Log}\, M(r) \leqslant \mathrm{Log}\left(\frac{r_2}{r}\right) \mathrm{Log}\, M(r_1) + \mathrm{Log}\left(\frac{r}{r_1}\right) \mathrm{Log}\, M(r_2).$$

Proof. Let μ be the real number defined by

$$r_1^{\mu} M(r_1) = r_2^{\mu} M(r_2). \tag{4.7}$$

and denote by T the common value of these two numbers. The function $z^{\mu} f(z)$ is holomorphic at each point of the annulus $0 < r_1 \leqslant |z| \leqslant r_2$ and its absolute value is single valued. On the frontier of the annulus we have

$$\max |z^{\mu} f(z)| = T;$$

Thus, by the maximum modulus principle, we have for all z satisfying $|z| = r \in [r_1, r_2]$ the inequality

$$r^{\mu} M(r) \leqslant T. \tag{4.8}$$

By taking logarithms in (4.7), (4.8) and then eliminating μ and T we obtain the stated relationship between $M(r_1)$, $M(r_2)$ and $M(r)$. ∎

§ 3 ZERO FREE REGIONS FOR $\zeta(s)$

3.1 We are now in a position to prove one of the principal results of the chapter, namely the existence of a zero free region for $\zeta(s)$ in the half plane $\Re(s) < 1$. There are several ways in which one can obtain such zero free regions. We have chosen a method which does not use any deep properties of $\zeta(s)$ and which is capable of generalisation to other functions. In the *Notes* to this chapter several other techniques are briefly described.

Before stating our main theorem we shall introduce some notation. Let T and ξ be real numbers satisfying

$$T > 0 \quad \text{and} \quad 0 < \xi < \min \left\{1, \frac{1}{2} T\right\} ,$$

Suppose that $M = M(T, \xi)$ is an upper bound for $|\zeta(s)|$ in the pair of regions :

$$\{ s = \sigma + it \mid \sigma \geqslant 1 - \xi , \mid t - T \mid \leqslant 3 \xi \}$$

and

$$\{ s = \sigma + it \mid \sigma \geqslant 1 - \xi , \mid t - 2T \mid \leqslant 3 \xi \} .$$

then the following theorem is true.

Theorem 4.4 *There exist absolute constants A, K, K_1, K_2 such that if η is defined by*

$$\frac{1}{\eta} = \frac{A}{\xi} \operatorname{Log} \frac{MK}{\xi} ,$$

then for $s = \sigma + iT$ with $\sigma > 1 - \eta$ we have :

(a) $\zeta(s) \neq 0$,

(b) $|Z(s)| < K_1 \eta^{-1}$,

(c) $|\zeta(s)|^{\pm 1} < K_2 \eta^{-1}$.

Possible values for the constants are :

$$A = 216 , \quad K = 75 , \quad K_1 = 83 , \quad K_2 = \frac{3}{2} e^{166} .$$

Proof. The basic ideas behind the proof are quite simple. Unfortunately the technical details will tend to obscure this fact. The reason being that we shall have a rather bewildering array of parameters at our disposal and they will only be explicitly chosen at the end of the proof. From a strictly logical point of view we could dispense with the parameters and work with completely explicit functions. This would make the proof look like a series of arithmetical accidents and completely obscure the reasoning behind the choice of parameters. We have chosen a middle way. Whenever parameters are introduced we shall always state their ultimate numerical value. Thus the reader who wishes to get the feel of the argument with completely explicit functions can do so.

(*a*) First we shall outline the argument. Take $\lambda \in (0, a\xi)$, where a is an absolute constant and λ is a function of M and ξ which is to be chosen explicitly later. Assume that $\zeta(s)$ has a zero $\rho = \beta + i\gamma$ which satisfies

$$1 - \lambda < \beta < 1 , \qquad | \gamma - T | \leqslant \frac{1}{2} \xi . \tag{4.9}$$

We shall use theorem 4.2(*c*) to show that the above assumption implies that $\lambda > F(M, \xi)$, where as we shall see, a possible choice for $F(M, \xi)$ is

$$F(M, \xi) = \frac{7 \xi}{1\,440} \left(\operatorname{Log} \frac{150\, M}{\xi} \right)^{-1} \Big) .$$

Thus, if we choose $\lambda = F(M, \xi)$, then the assumption (4.9) implies a contradiction, namely $\lambda > F(M, \xi)$. Thus, with our choice of λ we do *not* have a zero of $\zeta(s)$ which satisfies (4.9). In particular $\zeta(s) \neq 0$ if $s = \sigma + iT$ with $\sigma \geqslant 1 - F(M, \xi)$.

Now we begin the details of the proof. We shall apply theorem 4.2 to the function f defined by

$$f(s) = \zeta^4(s + i\gamma)\, \zeta(s + 2\, i\gamma)$$

where $\rho = \beta + i\gamma$ is a zero of $\zeta(s)$ satisfying (4.9).

By our initial assumption f has a zero of order at least 4 at $s = \beta$. The only singularities of f in the half plane $\Re(s) > 0$ are poles at $s = 1 - i\gamma$ and $s = 1 - i\, 2\, \gamma$. The parameters of theorem 4.2 are to be chosen as follows :

$$s_0 = 1 + b\xi , \quad R = (a + b)\, \xi , \quad qR = (1 + b)\, \xi , \quad d = r = b\xi$$

where a, b are absolute constants to be chosen later. For the moment we need only suppose that $0 < a < 1$, $0 < b < \frac{1}{4}$. $\left(\text{Later we shall choose } a = \frac{7}{10},\ b = \frac{1}{50}.\right)$

First we must verify that the hypotheses of theorem 4.2 are satisfied. The function f is holomorphic on the disc $| s - s_0 | \leqslant qR$, since the only singularities of f are poles at $s = 1 - i\gamma$ and $s = 1 - i\, 2\, \gamma$, which are outside the disc. We know that $\zeta(s) \neq 0$ in the half plane $\Re(s) > 1$, hence $f(s_0) \neq 0$.

It remains to obtain an upper bound for $| f(s)/f(s_0) |$ on the disc $| s - s_0 | \leqslant qR$. The upper bound M for $| \zeta(s) |$ in the two regions

$$\{ s = \sigma + it \mid 1 - \xi \leqslant \sigma , \ | t - T | \leqslant 3\, \xi \}$$
$$\{ s = \sigma + it \mid 1 - \xi \leqslant \sigma , \ | t - 2\, T | \leqslant 3\, \xi \} .$$

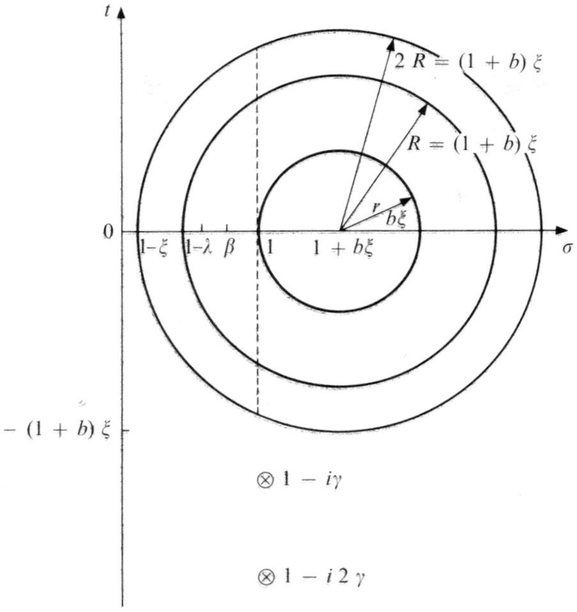

Fig. 5

yields the upper bound

$$|f(s)| \leqslant M^5 \quad \text{for} \quad |s - s_0| \leqslant qR = (1 + b)\,\xi\,.$$

For $\sigma > 1$ we trivially have

$$|\zeta(\sigma + it)|^{-1} = \left| \sum_{n=1}^{\infty} \frac{\mu(n)}{n^{\sigma+it}} \right| \leqslant \sum_{n=1}^{\infty} \frac{1}{n^{\sigma}} = \zeta(\sigma)$$

with the consequence that

$$|f(s_0)|^{-1} \leqslant \zeta^5(\sigma_0)\,.$$

Thus for s satisfying $|s - s_0| \leqslant qR$

$$\left|\frac{f(s)}{f(s_0)}\right| \leqslant (\zeta(\sigma_0)\,M)^5\,,$$

which we write as

$$\left|\frac{f(s)}{f(s_0)}\right| \leqslant \left\{\frac{(\sigma_0 - 1)\,\zeta(\sigma_0)\,M}{\sigma_0 - 1}\right\}^5 = \left\{(\sigma_0 - 1)\,\zeta(\sigma_0)\,\frac{M}{b\xi}\right\}^5.$$

If K_3 is an upper bound for $(\sigma - 1)\zeta(\sigma)$ when $\sigma \in \left[1, \frac{3}{2}\right] \supseteq [1, 1 + b\xi]$, then we have

$$\left| \frac{f(s)}{f(s_0)} \right| \leqslant \left(K_3 \frac{M}{b\xi}\right)^5 = \exp\left\{5 \operatorname{Log}\left(\frac{K_3 M}{b\xi}\right)\right\}.$$

for $|s - s_0| \leqslant R$. $\left(\text{A possible choice for } K_3 \text{ is } \frac{3}{2} \text{ see exercise 4.3.}\right)$ Thus, all the hypotheses of theorem 4.2 are satisfied and we can take $U = 5 \operatorname{Log}(K_3 M/b\xi)$ and $d = \sigma_0 - 1$.

We know that $\zeta(s)$ and hence $f(s)$ has no zeros in the region

$$\sigma \geqslant 1 = s_0 - d, \quad |s - s_0| \leqslant R.$$

Thus part (*c*) of theorem 4.2 is applicable. Since $q = \dfrac{1 + b}{a + b}$ it follows that if a and b are suitably small we have $Q = \max\{2, (q^2 \operatorname{Log} q)^{-1}\} = 2$. We shall take the set (ρ) to be the quadruple zero at $s = \beta$ and conclude that for $\sigma > 1$

$$\frac{4}{\sigma - \beta} \leqslant \Re\left\{\frac{f'(\sigma)}{f(\sigma)}\right\} + \frac{10(a + b)}{a^2 \xi} \operatorname{Log}\left(\frac{K_3 M}{b\xi}\right).$$

We next obtain an upper bound for the right hand side of the above expression. First we show that for $\sigma > 1$

$$\Re\left\{\frac{f'(\sigma)}{f(\sigma)}\right\} \leqslant 3 Z(\sigma).$$

The proof is quite straightforward. From the definition of f we have

$$f(s) = \zeta^4(s + i\gamma)\,\zeta(s + i\,2\,\gamma)$$

and logarithmic differentiation gives

$$\frac{f'(\sigma)}{f(\sigma)} = 4\frac{\zeta'(\sigma + i\gamma)}{\zeta(\sigma + i\gamma)} + \frac{\zeta'(\sigma + 2\,i\gamma)}{\zeta(\sigma + 2\,i\gamma)}$$

Consequently it follows that

$$\frac{f'(\sigma)}{f(\sigma)} - 3\,Z(\sigma) = -\{3\,Z(\sigma) + 4\,Z(\sigma + i\gamma) + Z(\sigma + 2\,i\gamma)\}$$

For $\sigma > 1$ we use the Dirichlet series representation of $Z(\sigma)$ to deduce

$$\Re\left\{\frac{f'(\sigma)}{f(\sigma)}\right\} - 3\,Z(\sigma) = -\sum_{n=1}^{\infty} \frac{\Lambda(n)}{n^\sigma}\{3 + 4\cos(\gamma \operatorname{Log} n) + \cos(2\,\gamma \operatorname{Log} n)\}$$

$$= -\sum_{n=1}^{\infty} 2\frac{\Lambda(n)}{n^\sigma}\{1 + \cos(\gamma \operatorname{Log} n)\}^2 \leqslant 0.$$

Thus we have shown

$$\frac{4}{\sigma - \beta} \leqslant 3\,Z(\sigma) + \frac{10(a+b)}{a^2\,\xi}\,\mathrm{Log}\left(\frac{K_3\,M}{b\xi}\right).$$

Let $K_4 \geqslant 0$ be an upper bound for $Z(\sigma) - (\sigma - 1)^{-1}$ when $\sigma \in \left[1, \frac{3}{2}\right] \supseteq [1, 1 + 2\,b\xi]$. (A possible choice for K_4 is 0, see *Notes* to chapter 5.) We can now conclude that

$$\begin{aligned}\frac{4}{\sigma - \beta} - \frac{3}{\sigma - 1} &\leqslant 3\,K_4 + \frac{10(a+b)}{a^2\,\xi}\,\mathrm{Log}\left(\frac{K_3\,M}{b\xi}\right)\\ &= \frac{10(a+b)}{a^2\,\xi}\,\mathrm{Log}\left\{\frac{K_3\,M}{b\xi}\exp\left(\frac{3\,K_4\,a^2\,\xi}{10(a+b)}\right)\right\}\\ &\leqslant \frac{10(a+b)}{a^2\,\xi}\,\mathrm{Log}\,\frac{K_5\,M}{b\xi}\end{aligned}$$

where

$$K_5 \geqslant K_3 \exp\left\{\frac{3\,K_4\,a^2}{10(a+b)}\right\}.$$

If we write $\sigma = 1 + \mu$, then by (4.9) we have $\lambda + \mu > \sigma - \beta$ and so from the above inequality we obtain

$$\frac{4}{\lambda + \mu} - \frac{3}{\mu} \leqslant \frac{10(a+b)}{a^2\,\xi}\,\mathrm{Log}\left(\frac{K_5\,M}{b\xi}\right).$$

Now we make a specific choice for μ, namely $\mu = 6\,\lambda$, and conclude that

$$\frac{1}{14\,\lambda} < \frac{10(a+b)}{a^2\,\xi}\,\mathrm{Log}\left(\frac{K_5\,M}{b\xi}\right)$$

This is the lower bound for λ which the existence of the zero $\rho = \beta + i\gamma$ forces upon us. Hence if we define λ by

$$\lambda > \frac{a^2\,\xi}{140(a+b)}\left\{\mathrm{Log}\left(\frac{K_5\,M}{b\xi}\right)\right\}^{-1}.$$

then we can conclude that there are *no* zeros $\rho = \beta + i\gamma$ of $\zeta(s)$ which satisfy

$$1 - \lambda < \beta \leqslant 1, \qquad |\gamma - T| \leqslant \frac{1}{2}\xi;$$

Although $\zeta(s)$ has no zeros which satisfy the above inequalities we shall not take $\eta = \lambda$. Instead we will choose $\eta = \theta\lambda$, where $0 < \theta < 1$ is an absolute constant $\left(\text{the value of which will be } \theta = \frac{20}{21}\right)$. The reason for doing this is that in order to prove parts (*b*) and (*c*) of the theorem we can only use the zero free region $1 - \lambda \leqslant \sigma \leqslant 1$, $|t - T| \leqslant \frac{1}{2}\xi$ to deduce upper bounds for $|Z(s)|$ and $|\zeta(s)|^{\pm 1}$ in the strictly smaller region :

$$1 - \lambda < \sigma < 1\,, \qquad |t - T| \leqslant \frac{1}{2}\xi$$

(*b*) To obtain the upper bound for $|Z(s)|$ we are going to apply part (*d*) of theorem 4.2, taking $f(s) = \zeta(s)$. The parameters are taken as follows :

$$s_0 = 1 + \lambda + iT\,, \quad R = \frac{1}{2}\xi\,, \quad qR = \xi\,, \quad d = 2\lambda\,, \quad r = (1 + \theta)\lambda\,,$$

The hypotheses of theorem 4.2 are easily verified, for trivially $\zeta(s_0) \neq 0$ and since the only singularity of $\zeta(s)$ in the half plane $\mathcal{R}(s) > 0$ is a simple pole at $s = 1$ we see that $\zeta(s)$ is holomorphic in the disc $|s - s_0| \leqslant \xi$. Furthermore, on the disc $|s - s_0| \leqslant qR = \xi$ we have

$$\left|\frac{\zeta(s)}{\zeta(s_0)}\right| \leqslant \left|(\sigma_0 - 1)\,\zeta(\sigma_0)\,\frac{\zeta(s)}{\sigma_0 - 1}\right| \leqslant \frac{K_3 M}{\lambda}$$

Moreover, since $\zeta(s)$ has no zeros in the region $\sigma > \sigma_0 - d = 1 - \lambda$, $|s - s_0| \leqslant R$ and $r < d$, part (*d*) of theorem 4.2 is applicable and we can conclude that on the disc $|s - s_0| \leqslant r$

$$|Z(s)| \leqslant \frac{3 + \theta}{1 - \theta}\left\{|Z(s_0)| + \frac{\xi}{\left(\frac{1}{2}\xi - 2\lambda\right)^2}\operatorname{Log}\left(\frac{K_3 M}{\lambda}\right)\right\}. \qquad (4.10)$$

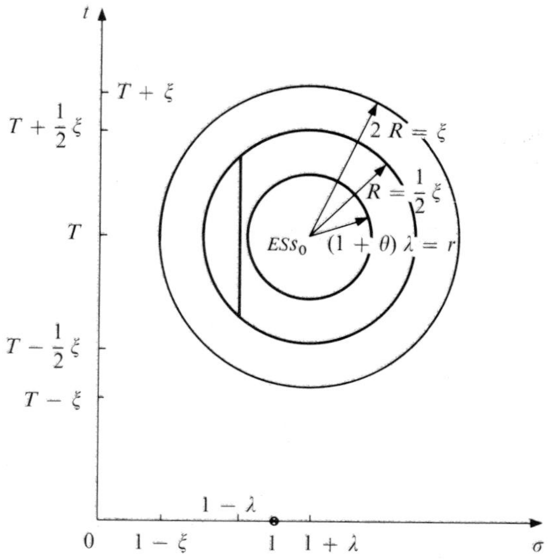

Fig. 6

Now we are going to obtain a " tidy " upper bound for the right hand side of the above inequality. Observe that

$$3 + \theta < 4, \quad |Z(s_0)| = \left| \sum_{n=1}^{\infty} \frac{\Lambda(n)}{n^{s_0}} \right| \leqslant \sum_{n=1}^{\infty} \frac{\Lambda(n)}{n^{1+\lambda}} = Z(1 + \lambda)$$

and

$$\frac{\lambda}{\xi} = \frac{a^2}{140(a + b) \operatorname{Log}\left(\frac{K_5 M}{b\xi}\right)} < \frac{a^2}{140(a + b)} < \frac{a}{140}$$

$$\leqslant \frac{1}{200}$$

of a is chosen small enough. From the last observation we deduce that

$$\frac{\xi}{\left(\frac{1}{2}\xi - 2\lambda\right)^2} \leqslant \frac{5}{\xi}.$$

Thus, from the above remarks and (4.10) we have, for $|s - s_0| \leqslant r$:

$$|Z(s)| \leqslant \frac{4}{1 - \theta} \left\{ Z(1 + \lambda) + \frac{5}{\xi} \operatorname{Log}\left(\frac{K_3 M}{\lambda}\right) \right\},$$

The above expression can now be simplified upon noting that

$$\begin{aligned} \operatorname{Log}\left(\frac{K_3 M}{\lambda}\right) &= \operatorname{Log}\left(\frac{K_3 M}{b\xi}\right) + \operatorname{Log}\left(\frac{b\xi}{\lambda}\right) \\ &\leqslant \frac{a^2 \xi}{140(a + b)\lambda} + \frac{b\xi}{\lambda} \\ &\leqslant \left\{ \frac{a^2}{140(a + b)} + b \right\} \frac{1}{\lambda} \leqslant \frac{K_6}{\lambda} \end{aligned}$$

$\left(\text{a possible choice for } K_6 \text{ is } \frac{1}{40}\right)$ and that

$$Z(1 + \lambda) \leqslant \frac{1}{\lambda} + K_4$$

Thus we certainly have, for $|s - s_0| \leqslant r$:

$$|Z(s)| < \frac{4}{(1 - \theta)\lambda} (1 + \lambda K_4 + K_6) \leqslant \frac{K_7}{\lambda} \tag{4.11}$$

where $K_7 \geqslant \frac{4}{1 - \theta}(1 + K_4 + K_6)$. (A possible value for K_7 being 87.)

If we now take $\eta = \theta\lambda$, then we have proved part (*b*) of the theorem, with $K_1 = \theta K_7$.

(*c*) The final part of the theorem is an easy consequence of (4.11). Quite trivially we have

$$\int_{\sigma+iT}^{1+\lambda+iT} Z(s)\, ds = -\int_{\sigma+iT}^{1+\lambda+iT} \frac{\zeta'(s)}{\zeta(s)}\, ds = \operatorname{Log} \zeta(\sigma + iT) - \operatorname{Log} \zeta(1 + \lambda + iT)$$

for $1 - \theta\lambda \leqslant \sigma \leqslant 1 + \lambda$, and consequently, using (4.11), we see that in the disc $| s - s_0 | \leqslant r$

$$\left| \operatorname{Log} \frac{\zeta(s)}{\zeta(s_0)} \right| \leqslant 2\, K_7\,.$$

Now for any non zero complex number z

$$| z |^{\pm 1} = \exp \{ \pm \Re(\operatorname{Log} z) \} \leqslant \exp | \operatorname{Log} z |\,.$$

and on taking $z = \dfrac{\zeta(s)}{\zeta(s_0)}$ and $K_8 = e^{2K_7}(> 1)$, we deduce that

$$\begin{aligned} | \zeta(s) |^{\pm 1} &\leqslant K_8\, | \zeta(s_0) | \\ &\leqslant K_8\, \zeta(1 + \lambda) \\ &\leqslant \frac{K_8\, K_3}{\lambda}\,. \end{aligned}$$

Thus if we choose $K_2 = \theta K_8\, K_3$ we have proved the final part of the theorem, with $\eta = \theta\lambda$.

To sum up, the theorem has been proved if we take

$$\eta = \theta\lambda = \frac{\theta a^2\, \xi}{140(a + b)} \left\{ \operatorname{Log} \left(\frac{KM}{\xi} \right) \right\}^{-1}$$

where $K = K_5/b$, $K_1 = \theta K_7$ and $K_2 = K_8\, K_3$ and the parameters a, b, θ are chosen to be

$$a = \frac{7}{10}, \quad b = \frac{1}{50}, \quad \theta = \frac{20}{21}.$$

All the hypotheses which we have imposed upon the parameters are seen to be satisfied. The corresponding numerical values of the constants are

$$A = \frac{140(a + b)}{\theta a^2} = 216\,, \quad K = 50\, K_5 = 75\,, \quad K_1 = 83\,, \quad K_2 = \frac{3}{2} e^{166}\,. \qquad \blacksquare$$

3.2 We shall now use theorem 2.3 to obtain an explicit upper bound for $M(\xi, T)$ and then apply theorem 4.4 to deduce a zero-free region for $\zeta(s)$. In chapter 11 we will obtain a refinement of theorem 2.3 which, when combined with theorem 4.4, will lead to a new zero-free region. A result of the following type was first obtained by de la Vallée Poussin [2].

Theorem 4.5 *There exist absolute constants c and $t_0(c)$ such that in the region*

$$\left\{ s = \sigma + it \mid \sigma > 1 - \frac{c}{\operatorname{Log} |t|}, |t| > t_0 \right\};$$

we have :

(a) $\zeta(s) \neq 0$,

(b) $Z(s) = O(\operatorname{Log} |t|)$ *uniformly in* σ *as* $|t| \to \infty$,

(c) $\zeta(s)$ *and* $1/\zeta(s)$ *are both* $O(\operatorname{Log} |t|)$ *uniformly in* σ *as* $|t| \to \infty$.

The constant c can be any number in the interval $(0, A^{-1})$, where A is the absolute constant in theorem 4.4, a possible value of which is 216. The implied " O " constants and $t_0(c)$ are effective functions of c.

Proof. Let ξ be any fixed real number in $(0, 1)$ and let T be a " large " real number (the precise meaning of " large " will be made quite explicit later). We now apply theorem 2.3 with $\theta = 1 - \xi$ to deduce upper bounds for $|\zeta(s)|$ in the two regions

$$\{ s = \sigma + it \mid \sigma \geqslant 1 - \xi, |t - T| \leqslant 3\xi \}$$

and

$$\{ s = \sigma + it \mid \sigma \geqslant 1 - \xi, |t - 2T| \leqslant 3\xi \}.$$

In the first region we have

$$|\zeta(s)| \leqslant \frac{7}{4\xi(1-\xi)}(T + 3\xi)^{\xi},$$

and in the second

$$|\zeta(s)| \leqslant \frac{7}{4\xi(1-\xi)}(2T + 3\xi)^{\xi}.$$

Since $\xi < 1$ we can conclude

$$M = M(\xi, T) < \frac{7}{4\xi(1-\xi)}(2T + 3)^{\xi}.$$

With the notation of theorem 4.4 we write

$$\frac{1}{\eta} = \frac{A}{\xi} \operatorname{Log}\left(\frac{KM}{\xi}\right).$$

Now for any fixed $k > 1$ and all sufficiently large T, say $T > t_0(\xi, k)$,

$$\frac{KM}{\xi} \leqslant \frac{7(2\,T + 3)^{\xi}}{4\,\xi^2(1 - \xi)} \leqslant T^{k\xi}$$

and consequently

$$\frac{1}{\eta} = \frac{A}{\xi}\,\mathrm{Log}\left(\frac{KM}{\xi}\right) \leqslant Ak\,\mathrm{Log}\,T\,.$$

If we take $c = (Ak)^{-1}$ the result follows by appealing to theorem 4.4 for each $T > t_0(c) = t_0(\xi, k)$. ■

§ 4 THE ERROR TERM IN THE PRIME NUMBER THEOREM

4.1 In this section we shall show how a knowledge of a zero free region for $\zeta(s)$ can be used to deduce a non-trivial estimate for the error term in the Prime Number Theorem. As usual we shall divide our discussion into two parts. First we shall suppose that we have a " sufficiently regular " zero free region R for $\zeta(s)$. Then, as theorem 4.6, we shall find an upper bound for the remainder term in the Prime Number Theorem as a function of R. Afterwards, as an application, we will use theorem 4.5 to deduce that

$$\pi(x) - \int_2^x \frac{du}{\mathrm{Log}\,u} = \mathrm{O}\,\{\,x\,\exp(-\,a\,\sqrt{\mathrm{Log}\,x})\,\}\,.$$

In chapter 11, after we have improved upon theorem 4.5 we shall deduce the corresponding improvement for the estimate of the above error term.

4.2 Now we define what is meant by " sufficiently regular " zero free region. Let η be a continuous, real valued function defined for all $t \geqslant 0$. Suppose also that $\eta'(t)$ is continuous, except possibly at a finite number of points where $\eta'(t + 0)$ and $\eta'(t - 0)$ both exist and are unequal. In addition we assume that for $t \geqslant 0$ we have

$$0 < \eta(t) < 1 \qquad \text{and} \qquad -1 < \eta'(t) \leqslant 0 \tag{4.12}$$

Let $R(\eta)$ be the region of the complex plane defined by

$$R = \{\,s = \sigma + it \mid 1 - \eta(|\,t\,|) \leqslant \sigma\,\}\,.$$

We say that R is " sufficiently regular " if the following two conditions hold for $s \in R$:

(i) $\zeta(s) \neq 0$,

(ii) there exists ε satisfying $0 < \varepsilon < 1$ such that as $|t| \to \infty$

$$Z(s) = O(|t|^{\varepsilon}), \quad \zeta(s) = O(|t|^{\varepsilon}), \quad \zeta^{-1}(s) = O(|t|^{\varepsilon})$$

uniformly in σ.

Before stating the principal theorem of this section we need to introduce some notation. Let λ and ε satisfy $0 < \varepsilon < 1$ and $0 \leqslant \lambda < 1 - \varepsilon$. Define the function $H_{\lambda,\varepsilon}$ by

$$H(x) = H_{\lambda,\varepsilon}(x) = \inf_{t \geqslant 0} \left\{ \frac{\eta(t) \operatorname{Log} x + \lambda \operatorname{Log} t}{1 + \lambda + \varepsilon} \right\}.$$

Our theorem is as follows.

Theorem 4.6. *Suppose that $R(\eta)$ is a region with the above properties. Then, as x tends to infinity the following asymptotic relations hold :*

$$\psi(x) - x = O(x\, e^{-H(x)}),$$
$$M(x) = O(x\, e^{-H(x)}),$$
$$\Pi(x) - \operatorname{Li}(x) = O(x\, e^{-H(x)}),$$

and

$$\theta(x) - x = O(x\, e^{-H(x)}) + O(x^{1/2})$$
$$\pi(x) - \operatorname{Li}(x) = O(x\, e^{-H(x)}) + O(x^{1/2}).$$

The implied " O " constants can all be determined effectively.

Proof. Our proof will follow the pattern of theorem 2.8. Let us denote by L the contour :

$$L = \{ s = \sigma + it : \sigma = 1 - \eta(|t|) \}.$$

Then, as in the proof of theorem 2.8 we have

$$\psi_1(x) = \frac{1}{2} x^2 + \frac{1}{2\, i\pi} \int_L \frac{Z(s)}{s(s+1)} x^{s+1}\, ds. \quad (4.12)$$

Let h be a real valued function, to be chosen later, which is defined for $x \geqslant 2$ and satisfies

$$0 < h(x) < x.$$

It now follows that

$$\frac{\psi_1(x-h) - \psi_1(x)}{-h} \leqslant \psi(x)$$

$$\psi(x) \leqslant \frac{\psi_1(x+h) - \psi_1(x)}{h}. \quad (4.13)$$

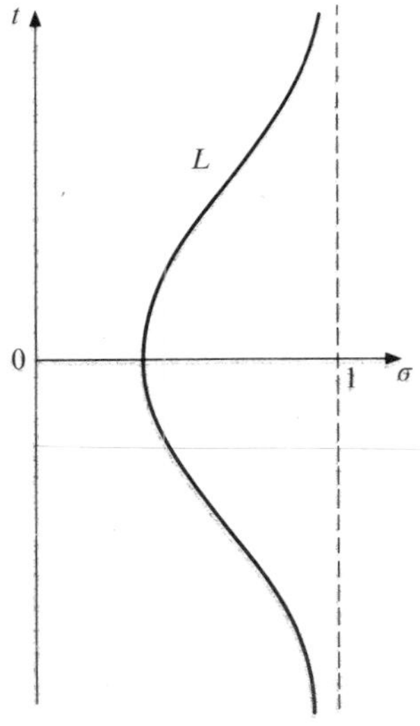

Fig. 7

From equation (4.12) we have

$$\frac{\psi_1(x \pm h) - \psi_1(x)}{\pm h} = x \pm \frac{1}{2} h + \frac{1}{2\, i\pi} \int_L \frac{Z(s)}{s(s+1)} \frac{(x \pm h)^{s+1} - x^{s+1}}{\pm h} \, ds$$

which, when combined with (4.13) yields

$$\psi(x) = x + \mathrm{O}(h) + \mathrm{O}\left\{ \int_L \frac{Z(s)}{s} \frac{(x \pm h)^{s+1} - x^{s+1}}{\pm (s+1)\, h} \, ds \right\}.$$

All that remains now is to estimate the above integral and then choose the function h.

Quite trivially we have

$$\int_L \frac{Z(s)}{s} \frac{(x \pm h)^{s+1} - x^{s+1}}{\pm (s+1)\, h} \, ds = \mathrm{O}\left\{ \int_L \left| \frac{Z(s)}{s} \right| \cdot \left| \frac{(x \pm h)^{s+1} - x^{s+1}}{\pm (s+1)\, h} \right| \cdot |\, ds \,| \right\}.$$

By symmetry it will suffice to estimate this last integral on the upper half of the contour L. By our hypothesis about the region R there exist constants K_1 and K_2 such that on the contour L

$$|\, Z(s) \,| \leqslant K_1 (1 + t)^{\varepsilon}; \quad \sigma = 1 - \eta(t);$$

$$\left| \frac{ds}{dt} \right| = |\, -\eta'(t) + i \,| < \sqrt{2}; \quad |\, s \,| \geqslant K_2 (1 + |\, t \,|).$$

Finally we note that since $\sigma < 1$ and $h \leqslant x$

$$\left| \frac{(x \pm h)^{s+1} - x^{s+1}}{\pm (s+1)\, h} \right| = \left| \frac{1}{\pm h} \int_x^{x \pm h} u^s \, du \right| \leqslant \frac{1}{\pm h} \int_x^{x \pm h} u^{\sigma} \, du$$

$$\leqslant (x + h)^{\sigma} \leqslant 2\, x^{\sigma},$$

and

$$\left| \frac{(x \pm h)^{s+1} - x^{s+1}}{\pm (s+1)\, h} \right| \leqslant \frac{(x + h)^{\sigma+1} + x^{\sigma+1}}{\frac{1}{2}(t+1)\, h} \leqslant \frac{10\, x^{\sigma+1}}{(t+1)\, h},$$

Thus we see that

$$\left| \frac{(x \pm h)^{s+1} - x^{s+1}}{\pm (s+1)\, h} \right| \leqslant \min \left\{ 2\, x^{\sigma}, \frac{10\, x^{\sigma+1}}{(t+1)\, h} \right\}.$$

Using the above estimates we can conclude

$$\psi(x) = x + \mathrm{O}(h) + \mathrm{O}\left\{ \int_0^{\infty} (1 + t)^{\varepsilon - 1} \min \left\{ x^{\sigma}, \frac{x^{\sigma+1}}{(t+1)\, h} \right\} dt \right\}$$

and consequently, upon writing $\sigma = 1 - \eta(t)$,

$$\frac{\psi(x)}{x} - 1 = \mathrm{O}\left(\frac{h}{x}\right) + \mathrm{O}\left\{\int_0^\infty (1+t)^{\varepsilon-2} x^{-\eta(t)} \min\left\{(1+t), \frac{x}{h}\right\} dt\right\}.$$

In order to simplify the above integral we make the substitution $1 + t = v$, then, upon recalling that $\eta'(u) \leqslant 0$ for all $u > 0$, we have $\eta(t) \geqslant \eta(v)$ and so

$$\begin{aligned}(1+t)^{\varepsilon-2} x^{-\eta(t)} &\leqslant v^{\varepsilon-2} \exp\{-\eta(v) \operatorname{Log} x\} \\ &= v^{\lambda+\varepsilon-2} \exp\{-\eta(v) \operatorname{Log} x - \lambda \operatorname{Log} v\} \\ &= v^{\lambda+\varepsilon-2} \exp\{-\xi(x)\}\end{aligned}$$

where ξ is defined by

$$\xi(x) = \inf_{v \geqslant 1} \{\eta(v) \operatorname{Log} x + \lambda \operatorname{Log} v\}.$$

Thus we can now conclude that

$$\frac{\psi(x)}{x} - 1 = \mathrm{O}\left(\frac{h}{x}\right) + \mathrm{O}\left\{e^{-\xi(x)} \int_1^\infty v^{\lambda+\varepsilon-2} \min\left\{v, \frac{x}{h}\right\} dv\right\}.$$

Make the substitution $\omega = v(h/x)$ in the above integral and observe

$$\int_1^\infty v^{\lambda+\varepsilon-2} \min\left\{v, \frac{x}{h}\right\} dv \leqslant \left(\frac{x}{h}\right)^{\lambda+\varepsilon} \int_0^\infty \omega^{\lambda+\varepsilon-2} \min\{\omega, 1\}\, d\omega =$$

$$= \left(\frac{x}{h}\right)^{\lambda+\varepsilon} \left(\int_0^1 \omega^{\lambda+\varepsilon-1}\, d\omega + \int_1^\infty \omega^{\lambda+\varepsilon-2}\, d\omega\right) = \mathrm{O}\left\{\left(\frac{x}{h}\right)^{\lambda+\varepsilon}\right\}.$$

Hence

$$\frac{\psi(x)}{x} - 1 = \mathrm{O}\left(\frac{h}{x}\right) + \mathrm{O}\left\{\left(\frac{x}{h}\right)^{\lambda+\varepsilon} e^{-\xi(x)}\right\}.$$

and it remains to choose h so that both the above " O " terms have the same order of magnitude, namely

$$h = h(x) = x \exp\left\{-\frac{\xi(x)}{1+\lambda+\varepsilon}\right\} = x \exp\{-H(x)\}.$$

We have proved $\psi(x) = x + \mathrm{O}(xe^{-H(x)})$, which is our first asymptotic formula. To obtain an asymptotic formula for $\prod(x)$ we recall the identity

$$\Pi(x) = \sum_{n \leqslant x} \frac{\Lambda(n)}{\operatorname{Log} n},$$

and by the Abel summation formula we have

$$\Pi(x) = \frac{\psi(x)}{\operatorname{Log} x} + \int_2^x \frac{\psi(u)}{u(\operatorname{Log} u)^2}\, du\,.$$

Upon integrating by parts we see

$$\int_2^x \frac{du}{\operatorname{Log} u} = \frac{x}{\operatorname{Log} x} + \mathrm{O}(1) + \int_2^x \frac{1}{(\operatorname{Log} u)^2}\, du$$

with the consequence

$$\Pi(x) - \int_2^x \frac{du}{\operatorname{Log} u} = \frac{\psi(x) - x}{\operatorname{Log} x} + \int_2^x \frac{\psi(u) - u}{u\,(\operatorname{Log} u)^2}\, du + \mathrm{O}(1)$$

$$= \mathrm{O}\left\{\frac{x\, e^{-H(x)}}{\operatorname{Log} x} + \int_2^x \frac{u\, e^{-H(u)}}{u(\operatorname{Log} u)^2}\, du\right\}.$$

Because the function $\operatorname{Log} u - H(u)$ is non-decreasing it follows that an upper bound for the above error term is

$$\mathrm{O}\left\{\frac{x\, e^{-H(x)}}{\operatorname{Log} x} + x\, e^{-H(x)} \int_2^x \frac{du}{u(\operatorname{Log} u)^2}\right\}$$

The above integral is O(1) and hence we have

$$\Pi(x) - \int_2^x \frac{du}{\operatorname{Log} u} = \mathrm{O}(x\, e^{-H(x)})\,.$$

The asymptotic formulae for $\theta(x)$ and $\pi(x)$ are now very easy to obtain. We recall, from chapter 1, § 4, that

$$\psi(x) - \theta(x) = \sum_{m=2}^{\infty} \theta(x^{1/m}) = \mathrm{O}(x^{1/2})$$

and

$$\Pi(x) - \pi(x) = \sum_{m=2}^{\infty} \frac{1}{m}\, \pi(x^{1/m}) = \mathrm{O}(x^{1/2})\,,$$

Thus, it is trivial that

$$\theta(x) = \psi(x) + \mathrm{O}(x^{1/2}) = x + \mathrm{O}(x\, e^{-H(x)}) + \mathrm{O}(x^{1/2})$$

and

$$\pi(x) = \Pi(x) + \mathrm{O}(x^{1/2}) = \mathrm{Li}\,(x) + \mathrm{O}(x\, e^{-H(x)}) + \mathrm{O}(x^{1/2})\,.$$

The proof of the asymptotic formula for $M(x)$ is left as an exercise for the reader.

■

4.3 As an application of the previous theorem we shall use theorem 4.5, which gave us a " sufficiently regular " zero free region for $\zeta(s)$, to deduce explicit error terms. We shall only state the following theorem for the functions $\psi(x)$ and $\pi(x)$ Of course, the result holds for all the functions mentioned in theorem 4.6.

Theorem 4.7 *There exists an absolute constant* $\alpha > 0$ *such that*

$$\psi(x) = x + O(xe^{-\alpha\sqrt{\operatorname{Log} x}}) \quad \text{and} \quad \pi(x) = \int_2^x \frac{du}{\operatorname{Log} u} + O(xe^{-\alpha\sqrt{\operatorname{Log} x}})$$

as x tends to infinity. A possible choice for α *is* $\frac{1}{15}$. *The implied " O " constants can be determined effectively.*

Proof. Let c and t_0 be the constants of theorem 4.5. Fint recall the conditions of § 4.2. Let $t_1 \geqslant t_0$ be chosen so that $c/t_1(\operatorname{Log} t_1)^2 < 1$ and so that $\zeta(s)$ has no zeros in the rectangle

$$\left\{ s = \sigma + it \mid 1 - \frac{c}{\operatorname{Log} t_1} \leqslant \sigma \leqslant 1 \, ; \, |t| \leqslant t_1 \right\} .$$

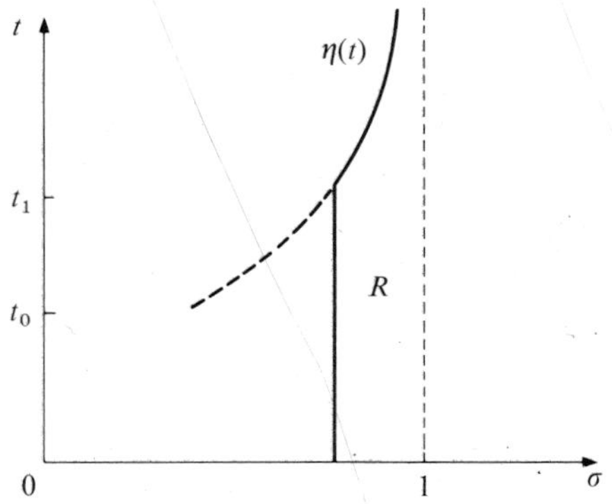

Fig. 8

Define η by

$$\eta(t) = \begin{cases} \dfrac{c}{\operatorname{Log} t_1} & \text{if } \ |t| \leqslant t_1 \\[2ex] \dfrac{c}{\operatorname{Log} |t|} & \text{if } \ t_1 \leqslant |t| \end{cases}$$

and the region R by

$$R = \{ s = \sigma + it \mid 1 - \eta(t) \leqslant \sigma \} .$$

From theorem 4.5 we know that, within R, $\zeta(s) \neq 0$ and $|Z(s)|$, $|\zeta(s)|^{\pm 1}$ are $O(\text{Log}\,|t|)$. Also, for *any* $0 < \varepsilon < 1$, we have $\text{Log}\,|t| = O(|t|^{\varepsilon})$. Hence the region R satisfies all the regularity conditions of theorem 4.6 and all that remains is to compute the function $H(x)$.

Thus, theorem 4.6 is applicable; all that remains is to compute the function $H(x)$. By the definition of R and H we have

$$H(x) = \inf_{t \geqslant 1} \left\{ \frac{\eta(t)\,\text{Log}\,x + \lambda\,\text{Log}\,t}{1 + \lambda + \varepsilon} \right\}$$

$$= \min \left\{ \inf_{t \geqslant t_1} \frac{c\dfrac{\text{Log}\,x}{\text{Log}\,t} + \lambda\,\text{Log}\,t}{1 + \lambda + \varepsilon},\ \inf_{1 \leqslant t \leqslant t_1} \frac{\dfrac{c}{\text{Log}\,t_1}\text{Log}\,x + \lambda\,\text{Log}\,t}{1 + \lambda + \varepsilon} \right\}.$$

It is trivial that

$$\inf_{1 \leqslant t \leqslant t_1} \frac{\dfrac{c}{\text{Log}\,t_1}\text{Log}\,x + \lambda\,\text{Log}\,t}{1 + \lambda + \varepsilon} = \frac{c}{(1 + \lambda + \varepsilon)\,\text{Log}\,t_1}\,\text{Log}\,x\,.$$

and it follows from the arithmetic-geometric mean inequality that

$$c\frac{\text{Log}\,x}{\text{Log}\,t} + \lambda\,\text{Log}\,t \geqslant 2\sqrt{\lambda c\,\text{Log}\,x}\,,$$

hence

$$H(x) \geqslant \min \left\{ \frac{2\sqrt{\lambda c}}{1 + \lambda + \varepsilon}\sqrt{\text{Log}\,x},\ \frac{c}{(1 + \lambda + \varepsilon)\,\text{Log}\,t_1}\,\text{Log}\,x \right\}.$$

Thus for all sufficiently large x

$$H(x) \geqslant \frac{2\sqrt{\lambda c}}{1 + \lambda + \varepsilon}\sqrt{\text{Log}\,x}\,.$$

If we now take any $\alpha \in (0, \sqrt{c})$, then by choosing $\lambda = 1 - 2\,\varepsilon$, with ε sufficiently small we see that

$$H(x) \geqslant a\sqrt{\text{Log}\,x}$$

■

§ 5 ZERO FREE REGIONS DEDUCED FROM ERROR TERMS

5.1 We have just seen that the knowledge of a sufficiently regular zero free region for $\zeta(s)$ can be used to deduce a non-trivial error term in the Prime Number Theorem. Remarkably enough a converse relationship exists. Namely, if we assume that either

$$\psi(x) = x + O\{x\,e^{-\varphi(x)}\} \quad \text{ou} \quad M(x) = O\{x\,e^{-\varphi(x)}\}$$

where φ satisfies some very mild regularity conditions, then one can deduce the existence of a zero free region $R(\varphi)$ for $\zeta(s)$. The region $R(\varphi)$ which can be deduced is the best that one could hope for, in the sense that from the region $R(\varphi)$ one can use theorem 4.6 to deduce an error term $O(x\,e^{-\varphi(x)})$.

The deduction of a zero free region from the estimate $\psi(x) = x + O(x\,e^{-\varphi(x)})$ is rather delicate. It seems to depend essentially on theorems of Turán which give lower estimates for

$$\max_{N \leqslant \nu \leqslant N+M} \left| \sum_{n=A}^{B} a_n z_n^{\nu} \right|$$

and on an explicit formula for $\psi(x)$ in terms of the complex zeros of $\zeta(s)$. Unfortunately lack of space will prevent us from giving an account of Turán's work and its application to prime number theory. The reader is refered to Turán [2], Stas [2] and Wiertelak [1] for an account of the derivation.

We shall content ourselves with solving the simpler problem of deducing a zero free region for $\zeta(s)$ from an estimate of the type $M(x) = O(x \exp\{-\varphi(x)\})$. The argument in this case is due to Allison [1].

5.2 The precise conditions on $\varphi(x)$ could be varied quite a lot, the main point being that φ and its inverse both increase at a " moderate " rate. Our particular set of conditions is chosen so as to simplify some of the calculations in the proof of theorem 4.8.

Let x_0, t_0 be fixed positive real numbers. For all $x > x_0$ and $t > t_0$ we suppose :

(i) $\varphi(x)$ is real valued, positive and continuous,

(ii) $(\operatorname{Log} x)^2\, e^{-\varphi(x)}$ decreases monotonically,

(iii) the function $X(t)$, defined implicitly by

$$\varphi(X(t)) = 20 \operatorname{Log} t\,,$$

satisfies the condition $4 \operatorname{Log} t \leqslant \operatorname{Log} X(t)$.

It is clear from the above conditions that φ and X both tend monotonically to infinity.

Throughout the proof of theorem 4.8 we will do various calculations which hold only if x and t are " large " enough. We assume once and for all that x_0 and t_0 are sufficiently large to cover all the situations which are going to arise.

Theorem 4.8 *Suppose that* $|M(x)| < Kx\,e^{-\varphi(x)}$ *for all* $x \geqslant 1$, *where K is an absolute constant and* φ *satisfies the above conditions. Then* $\zeta(s)$ *does not vanish in the region*

$$\{\, s = \sigma + it \mid 1 - \eta(|\,t\,|) \leqslant \sigma,\ |\,t\,| > t_0 \,\}$$

where t_0 *is an effectively computable constant and the function* η *is defined by*

$$\eta(t) = \frac{\operatorname{Log} t}{\operatorname{Log} X(t)}.$$

Proof. The idea behind the proof is very simple. One considers the function f defined by

$$f(s) = 1 - \zeta(s) \sum_{n \leqslant x} \frac{\mu(n)}{n^s}$$

and a circle with centre $2 + it$ and radius $1 + \eta(t)$. Then one proves that for $t > t_0$ we have $|f(s)| < 1$ on this circle. By the maximum modulus principle we can then conclude that $\zeta(s) \neq 0$ on or within the circle.

To obtain an upper bound for $|f(s)|$ on the circle we shall employ theorem 4.3 with $r_1 = 1$, $r = 1 + \eta(t)$, $r_2 = \frac{3}{2}$. Denoting by M_1, M, M_2 the maximum value of $|f(s)|$ on the circles of radii r_1, r, r_2 respectively we have, by theorem 4.3 :

$$\operatorname{Log}\frac{3}{2}.\operatorname{Log} M \leqslant \operatorname{Log}\frac{3}{2(1+\eta)} \operatorname{Log} M_1 + \operatorname{Log}(1+\eta) \operatorname{Log} M_2 . \tag{4.14}$$

By finding sufficiently good upper bounds for M_1 and M_2 we will show that for all $t > t_0$ we have $\operatorname{Log} M < 1$, which implies the theorem. In order to obtain upper bounds for M_1 and M_2 we need the following result about $\zeta(s)$, which follows from theorem 2.3 upon taking $\theta = \frac{1}{2}$:

For $\frac{1}{2} \leqslant \sigma$ *and* $|t| \geqslant 1$ *we have* $\zeta(\sigma + it) = O(|t|^{1/2})$. (4.15)

(a) *An upper bound for* M_1. We shall find an upper bound for

$$\left| \zeta^{-1}(s) - \sum_{n \leqslant x} \mu(n)\, n^{-s} \right|$$

in the region $1 < \sigma \leqslant 4$, $t > 0$. This upper bound, when combined with (4.15) will then yield an upper bound for $|f(s)|$ on the circle $|s - 2 - it| = r_1 = 1$.

By the Abel summation formula, for $x > 1$

$$\sum_{n \leqslant x} \mu(n)\, n^{-s} = \frac{M(x)}{x^s} + s \int_1^x \frac{M(u)}{u^{s+1}}\, du \tag{4.16}$$

and if $\sigma > 1$ we have

$$\frac{1}{\zeta(s)} = \sum_{n=1}^{\infty} \mu(n)\, n^{-s} = s \int_1^{\infty} \frac{M(u)}{u^{s+1}}\, du ,$$

Consequently

$$\frac{1}{\zeta(s)} - \sum_{n \leqslant x} \mu(n)\, n^{-s} = s \int_x^{\infty} \frac{M(u)}{u^{s+1}}\, du - \frac{M(x)}{x^s}.$$

The two terms on the right hand side will now be estimated separately. By our hypothesis on $M(x)$ we have

$$\left| \frac{M(x)}{x^s} \right| \leqslant \frac{Kx\, e^{-\varphi(x)}}{x^{\sigma}} \leqslant K\, e^{-\varphi(x)}.$$

If $1 < \sigma \leqslant 4$, $t > 0$ and $x > x_0$, then

$$\left| s \int_x^{\infty} \frac{M(u)}{u^{s+1}}\, du \right| \leqslant |\, s\, | \int_x^{\infty} \frac{|\, M(u)\, |}{u^{\sigma+1}}\, du \leqslant K(t+4) \int_x^{\infty} \frac{e^{-\varphi(u)}}{u^{\sigma}}\, du$$

$$\leqslant K(t+4) \int_x^{\infty} \frac{e^{-\varphi(u)} \operatorname{Log}^2 u}{u \operatorname{Log}^2 u}\, du.$$

By condition (ii) the above integral is less than

$$e^{-\varphi(x)} \operatorname{Log}^2 x \int_x^{\infty} \frac{du}{u \operatorname{Log}^2 u} = e^{-\varphi(x)} \operatorname{Log} x.$$

Thus, for $1 < \sigma \leqslant 4$, $t > 0$, $x > x_0$ we certainly have

$$\left| \frac{1}{\zeta(s)} - \sum_{n \leqslant x} \frac{\mu(n)}{n^s} \right| \leqslant K(t+5)\, e^{-\varphi(x)} \operatorname{Log} x.$$

From the above and (4.15)

$$|\, f(s)\, | = \left| 1 - \zeta(s) \sum_{n \leqslant x} \mu(n)\, n^{-s} \right| \leqslant K_1\, t^{3/2}\, e^{-\varphi(x)} \operatorname{Log} x \tag{4.17}$$

for $1 < \sigma \leqslant 4$, $t > 1$ and $x > x_0$. By the continuity of f the above upper bound also holds for $1 \leqslant \sigma \leqslant 4$, $t > 1$, $x > x_0$. Hence, for $t > 1$, $x > x_0$

$$M_1 \leqslant K_1\, t^{3/2}\, e^{-\varphi(x)} \operatorname{Log} x.$$

(b) *An upper bound for* M_2. To find an upper bound for M_2 we shall consider two cases : first we obtain an upper bound for $|\, f(s)\, |$ on the segment

$$\left\{ s : |\, s - 2 - it\, | \leqslant \frac{3}{2},\ \sigma > 1 \right\}$$

and then an upper bound on the segment

$$\left\{ s : |\, s - 2 - it\, | \leqslant \frac{3}{2}, \frac{1}{2} \leqslant \sigma \leqslant 1 \right\}.$$

The first case is easily dealt with, since (4.17) gives such an upper bound. The second case is almost as easy. From (4.16) we have, for $\frac{1}{2} \leqslant \sigma \leqslant 1$, $t \geqslant 1$ and $x > 1$

$$\left| \sum_{n \leqslant x} \frac{\mu(n)}{n^s} \right| \leqslant (t+1) \int_1^x \frac{du}{\sqrt{u}} + \sqrt{x} \leqslant K_2 \, t \sqrt{x} \, .$$

By using (4.15) we see that for $\frac{1}{2} \leqslant \sigma \leqslant 1$, $t \geqslant 1$

$$| f(s) | = \left| 1 - \zeta(s) \sum_{n \leqslant x} \frac{\mu(n)}{n^s} \right| \leqslant K_3 \, t^{3/2} \, x^{1/2} \, ,$$

Thus

$$M_2 \leqslant \max \{ K_3 \, t^{3/2} \, x^{1/2}, K_1 \, t^{3/2} \, e^{-\varphi(x)} \operatorname{Log} x \} \, .$$

By condition (ii) it follows that $(\operatorname{Log} x) \, e^{-\varphi(x)}$ tends to zero, hence for $x > x_0$

$$M_2 \leqslant K_3 \, t^{3/2} \, x^{1/2} \, .$$

Substitute the upper bounds for M_1 and M_2 into (4.14) to obtain

$$\operatorname{Log}\left(\frac{3}{2}\right) \operatorname{Log} M \leqslant \operatorname{Log}\left(\frac{3}{2(1+\eta)}\right) \operatorname{Log} (K_1 \, t^{3/2} \, e^{-\varphi(x)} \operatorname{Log} x) + \operatorname{Log} (1+\eta) \operatorname{Log} (K_4 \, t^{3/2} \, x^{1/2}) \, . \quad (4.18)$$

Put $x = \mathrm{X}(t)$ in the above expression, where $\mathrm{X}(t)$ is the function defined in (iii). We are now going to obtain a " tidy " upper bound for the right hand side and show that for all $t > t_0$ we have $M < 1$.

The first step is to prove that for all $t > t_0$

$$K_1 \, t^{3/2} \operatorname{Log} \mathrm{X} \, e^{-\varphi(\mathrm{X})} < 1 \, .$$

By (ii) there exists K_5 such that for all $x > x_0$

$$e^{-\varphi(x)} \operatorname{Log}^2 x \leqslant K_5$$

hence writing $x = \mathrm{X}(t)$ and recalling that $\varphi(\mathrm{X}(t)) = 20 \operatorname{Log} t$ we have

$$\operatorname{Log} \mathrm{X}(t) \leqslant \sqrt{K_5} \, t^{10} \, .$$

Thus

$$K_1 \, t^{3/2} \, e^{-\varphi(\mathrm{X})} \operatorname{Log} \mathrm{X} = K_1 \, t^{3/2} \, t^{-20} \operatorname{Log} \mathrm{X} \leqslant K_1 \sqrt{K_5} \, t^{-17/2} < 1$$

for all $t > t_0$.

In addition to the above inequality we also note that (ii) implies that

$$\frac{3}{2(1+\eta)} \geqslant \frac{6}{5} \, .$$

Hence

$$\operatorname{Log}\left(\frac{3}{2(1+\eta)}\right) \operatorname{Log}\,(K_1\, t^{3/2}\, e^{-\varphi(X)} \operatorname{Log} X) \leqslant \operatorname{Log}\left(\frac{6}{5}\right) \operatorname{Log}\, (K_1 \sqrt{K_5}\, t^{-17/2})\,.$$

Finally, we note that for $\eta > 0$ we have $\operatorname{Log}(1+\eta) \leqslant \eta$.

Using the above inequalities in (4.18) we obtain

$$\begin{aligned}\operatorname{Log}\left(\frac{3}{2}\right) \operatorname{Log} M &\leqslant \operatorname{Log}\left(\frac{6}{5}\right) \operatorname{Log}\,(K_1 \sqrt{K_5}\, t^{-17/2}) + \eta \operatorname{Log}\,(K_4\, t^{3/2}\, X^{1/2}) \\ &\leqslant K_6 - \frac{17}{2} \operatorname{Log}\left(\frac{6}{5}\right) \operatorname{Log} t + \eta \operatorname{Log} K_4 + \\ &\qquad\qquad \frac{3}{2}\eta \operatorname{Log} t + \frac{1}{2}\eta \operatorname{Log} X\,.\end{aligned}$$

Since $\eta \leqslant \frac{1}{4}$ the above upper bound is

$$K_7 + \left\{\frac{3}{8} - \frac{17}{2}\operatorname{Log}\left(\frac{6}{5}\right)\right\} \operatorname{Log} t + \frac{1}{2}\eta \operatorname{Log} X\,.$$

Writing $\eta = \dfrac{\operatorname{Log} t}{\operatorname{Log} X}$ we have shown that for $t > t_0$

$$\operatorname{Log}\left(\frac{3}{2}\right) \operatorname{Log} M \leqslant K_7 + \left\{\frac{7}{8} - \frac{17}{2}\operatorname{Log}\left(\frac{6}{5}\right)\right\} \operatorname{Log} t\,.$$

The right hand side of the above inequality is negative for all $t > t_0$, consequently $M < 1$ for all $t > t_0$ and the theorem is proved. ■

5.3 As was mentioned earlier the above result is very precise. For example, if one takes

$$\varphi(x) = A(\operatorname{Log} x)^{\alpha}$$

with $0 < \alpha < 1$ and $A > 0$ $\left(\text{the classical case being } \alpha = \frac{1}{2}\right)$, then we find that

$$\eta(t) = \frac{\operatorname{Log} t}{\operatorname{Log} X(t)} = A_1(\operatorname{Log} t)^{(\alpha-1)/\alpha}\,.$$

Conversely, from this we can use theorem 4.6 to deduce the error term

$$O\,\{\, x \exp(-\, c(\operatorname{Log} x)^{\alpha})\,\}\,.$$

Notes to chapter 4

ZERO FREE REGIONS FOR $\zeta(s)$

In our deduction of the existence of a zero free region for $\zeta(s)$ we did not use any detailed information about the analytic character of the Riemann zeta function. Our motive for doing this was that we wished our theorem and its proof to be valid for the " zeta " function associated with generalised number systems whose counting function satisfies $N(x) = Ax + O(x^{\theta})$. For such systems the associated zeta function is only defined in the half plane $\mathfrak{R}(s) > \theta$ and one does not possess very much information about its analytic properties.

The original proof of de la Vallée Poussin [2], in contrast used very detailed information about $\zeta(s)$ and the theory of integral functions. An account of this approach is given by Davenport [4]. However, the original proof still makes interesting reading as some effort is made to produce explicit and reasonable numerical constants. For example, de la Vallée Poussin proved the following version of theorem 4.5 :

$\zeta(s)$ has no zeros in the regions

$$\left\{ s = \sigma + it \mid |t| \geqslant 705, \quad \sigma > 1 - \frac{0.034\,666}{\operatorname{Log}(|t|/47.886)} \right\}$$

and

$$\left\{ s = \sigma + it \mid |t| < 705, \quad \sigma > 1 - \frac{0.034\,666}{\operatorname{Log}(705/47.886)} \right\}.$$

A rather different approach to finding zero free regions has been given by Montgomery [1]. His technique, which uses more facts about $\zeta(s)$ and $Z(s)$ than we have yet proved, is roughly as follows. Suppose that $\zeta(s)$ has a zero $\rho_0 = \beta_0 + i\gamma_0$ with β_0 " near " to 1, then one can prove that the existence of ρ_0 implies that there exist other zeros of $\zeta(s)$ in the neighbourhood of ρ_0. Moreover, the closer β_0 is to 1 the more zeros one has. This will lead to a contradiction if β_0 is very close to 1, because one can deduce an upper bound for the number of zeros in the neighbourhood of ρ_0 by using an upper bound for $|\zeta(s)|$ in conjunction with theorem 4.2(*a*).

More precisely, let us denote by $n(T, \omega, h)$ the number of zeros of $\zeta(s)$ in the rectangle

$$\left\{ s = \sigma + it \mid 1 - \omega \leqslant \sigma \leqslant 1, \quad T - \frac{1}{2}h \leqslant t \leqslant T + \frac{1}{2}h \right\}.$$

If $\rho_0 = \beta_0 + i\gamma_0$ is a zero of $\zeta(s)$ which satisfies $\gamma_0 > 0$ and $\beta_0 > 1 - (\text{Log Log } \gamma_0)^{-1}$, then Montgomery conjectures that there is an absolute constant K_1 such that

$$n(\gamma_0, \delta, \delta) > K_1\, \delta(1 - \beta_0)^{-1}$$

uniformly for $1 - \beta_0 \leqslant \delta \leqslant (\text{Log Log } \gamma_0)^{-1}$. Unfortunately all that has been proved is that a suitable average of

$$n(\gamma_0, \omega, h) + n(2\,\gamma_0, \omega, h)$$

is as large as one would expect. Montgomery's principal result is as follows.

If $\zeta(\beta_0 + i\gamma_0) = 0$ *with* $\beta_0 > \frac{1}{2}$ *and* $\gamma_0 > 0$, *then there exists an absolute constant* K_2 *such that for* $1 - \beta_0 \leqslant \delta \leqslant 1$ *we have*

$$\delta^2 \int_0^1 \int_0^\infty \{\, n(\gamma_0, \omega, h) + n(2\,\gamma_0, \omega, h) \,\}\, (h + \delta)^{-5}\, e^{-\omega/\delta}\, dh\, d\omega \geqslant K_2(1 - \beta_0)^{-1}.$$

One can still use upper bounds for $n(\gamma_0, \omega, h)$ and $n(2\,\gamma_0, \omega, h)$ derived from upper bounds for $|\,\zeta(s)\,|$ to deduce an upper bound for $(1 - \beta_0)^{-1}$ in terms of γ_0. The zero free regions which can be deduced in this way seem to be exactly the same as the regions which can be deduced from theorem 4.4.

Exercises to chapter 4

4.1 Show that for all sufficiently large values of n there are more prime numbers in the interval $[1, n]$ than in the interval $[n, 2n]$.

4.2 From tables of prime numbers Legendre conjectured that $x/(\text{Log}\, x - 1.08366...)$ was a better asymptotic approximation to $\pi(x)$ than $x/\text{Log}\, x$. Is Legendre's conjecture true ?

4.3 Show that

$$\max_{1 \leqslant \sigma \leqslant 3/2} (\sigma - 1)\, \zeta(\sigma) \leqslant \frac{3}{2}.$$

[Hint : Use the analytic continuation given by theorem 2.1.]

4.4* Suppose that one has an estimate for *one* of

$$\psi(x) - x, \quad \theta(x) - x, \quad \Pi(x) - \text{Li}\,(x), \quad M(x)$$

of the form $O(x \exp(-\varphi(x)))$, where φ satisfies some conditions of regularity such as those imposed in § 5. Is it possible to deduce, by elementary methods, that the estimate holds for each of the functions ?

CHAPTER 5

The Riemann Zeta function

§ 1 INTRODUCTION

The Riemann zeta function will now be studied in some detail. The pure theory of $\zeta(s)$ has been the subject of very extensive investigations. However, we shall confine our attention to the proof of results which will be needed in chapter 6. For a comprehensive account of the general theory up to about 1950 see Titchmarsh [2]. Some of the more recent developments and their implications for Prime Number Theory are discussed by Montgomery [1].

The contents of the chapter are as follows. Section 2 is a miscellany. First we shall prove a formula which will be a vital tool in our proof of the relationship between $\zeta(s)$ and $\zeta(1 - s)$. (This relationship, known as the functional equation for $\zeta(s)$ lies at the heart of the deeper study of the zeta function) Then we will state, without proof, several well known properties of the Γ function. These latter results will be used frequently later on in the chapter. In § 3 we will give a proof of the functional equation and several important consequences. We denote § 4 to the study of the general distribution of the zeros of $\zeta(s)$. The connection between the zeros of $\zeta(s)$ and prime number theory will then be made quite explicit in § 5.

§ 2 USEFUL FORMULAE

2.1 The following result is, in fact, a very particular case of a modular transformation of a theta function. However, we shall give a simple direct proof of the formula and not assume any knowledge of modular functions.

Theorem 5.1 (a) *If a is a positive real number and z an arbitrary complex number, then*

$$\sum_{n=-\infty}^{+\infty} e^{-\pi(n+z)^2 a} = \frac{1}{\sqrt{a}} \sum_{n=-\infty}^{+\infty} e^{-\pi n^2/a + 2i\pi nz} .$$

(b) *Upon writing* $\omega(a) = \sum_{n=1}^{\infty} e^{-\pi n^2 a}$ *we have*

$$1 + 2\,\omega(a) = \frac{1}{\sqrt{a}}\left(1 + 2\,\omega\left(\frac{1}{a}\right)\right)$$

and

$$0 < \omega(a) < \frac{1}{2\sqrt{a}}.$$

Proof. (*a*) This result is essentially trivial. One shows that the left hand side is an entire function which is periodic with period 1. Thus it can be expanded as a Fourier series. The right hand side is simply this Fourier expansion. Now for the details.

Let R be a fixed positive number. If $z = x + iy$ and $|z| \leqslant R$, then for each integer n satisfying $|n| > 3R$ we have $|x|, |y| \leqslant n/3$ and

$$|e^{-\pi(n+z)^2 a}| = e^{-\pi(n+x)^2 a + \pi y^2 a} < e^{-\frac{1}{3}\pi n^2 a} < e^{-\frac{1}{3}\pi|n|a},$$

Consequently the series of holomorphic functions :

$$\sum_{n=-\infty}^{+\infty} e^{-\pi(n+z)^2 a} \tag{5.1}$$

is absolutely uniformly convergent on the disc $|z| \leqslant R$. Thus the series (5.1) defines an entire function $g_a(z)$. Clearly $g_a(z)$ is periodic with period 1. Thus $g_a(z)$ can be expanded as a Fourier series :

$$g_a(z) = \sum_{m=-\infty}^{+\infty} c_m\, e^{2i\pi mz}$$

where

$$c_m = \int_{\Delta} g_a(z)\, e^{-2i\pi mz}\, dz\,,$$

and Δ denotes any segment of length 1 parallel to the real axis.

By the definition of g_a we have

$$c_m = \int_{\Delta} \left\{ \sum_{n=-\infty}^{+\infty} \exp(-\pi(n+z)^2 a - 2i\pi mz) \right\} dz$$

$$= \sum_{n=-\infty}^{+\infty} \int_{\Delta} \exp(-\pi(n+z)^2 a - 2i\pi mz)\, dz\,, \tag{5.2}$$

The justification for the interchange of summation and integration being that the series is absolutely, uniformly convergent on all compact sets. Now make the transformation $\omega = z + n$ in (5.2). The result is

$$c_m = \sum_{n=-\infty}^{+\infty} \int_{n+\Delta} \exp(-\pi z^2 a - 2i\pi m z)\,dz$$

$$= \int_{-\infty+ib}^{+\infty+ib} \exp(-\pi z^2 a - 2i\pi m z)\,dz\,,$$

where b is the ordinate of the segment Δ. Finally we make the substitution

$$u = \omega\sqrt{a} + i\frac{m}{\sqrt{a}}$$

and choose $b = -m/a$ to obtain

$$c_m = \frac{e^{-\pi m^2/a}}{\sqrt{a}} \int_{-\infty}^{+\infty} e^{-\pi u^2}\,du = \frac{e^{-\pi m^2/a}}{\sqrt{a}} J\,.$$

Thus we now have

$$g_a(z) = \frac{J}{\sqrt{a}} \sum_{m=-\infty}^{+\infty} \exp\left(-\pi\frac{m^2}{a} + 2i\pi m z\right). \tag{5.3}$$

By taking $a = 1$ and $z = 0$ in the above equation we deduce that $J = 1$, since $g_1(0) \neq 0$. This proves part (*a*) of the theorem. Part (*b*) follows upon taking $z = 0$ in (5.3). The final part of the theorem now follows, since

$$\omega(a) = \sum_{n=1}^{\infty} e^{-\pi a n^2} < \sum_{n=1}^{\infty} \int_{n-1}^{n} e^{-\pi a x^2}\,dx$$

$$= \int_0^{\infty} e^{-\pi a x^2}\,dx = \frac{J}{2\sqrt{a}} = \frac{1}{2\sqrt{a}}.$$

■

2.2 The gamma function. The properties of the Γ function which we shall need are summarised below. All the results are classical, for their proofs we refer the reader to Titchmarsh [2] or to Valiron [1].

(*a*) If $s = \sigma + it$ and $\sigma > 0$, then

$$\Gamma(s) = \int_0^{\infty} e^{-u} u^{s-1}\,du\,.$$

(*b*) For all values of s we have the representation

$$\frac{1}{s\Gamma(s)} = e^{\gamma s} \prod_{n=1}^{\infty} \left(1 + \frac{s}{n}\right) e^{-s/n}$$

where γ is Euler's constant.

(*c*) The Γ function does not have any zeros.

(*d*) The only singularities of the Γ function are simple poles at the points $s = -n$, where $n = 0, 1, 2, \ldots$

(*e*) The Γ function satisfies the following functional relations; valid for all s :

$$\Gamma(s + 1) = s\Gamma(s) ,$$

$$\Gamma(s)\, \Gamma(1 - s) = \frac{\pi}{\sin \pi s} ,$$

$$\Gamma(s)\, \Gamma\left(s + \frac{1}{2}\right) = \frac{\sqrt{\pi}}{2^{2s-1}}\, \Gamma(2\, s) .$$

(*f*) The following two generalisations of Stirling's formula will be needed :

(i) As $|z|$ tends to infinity

$$\operatorname{Log} \Gamma(z + \alpha) = \left(z + \alpha - \frac{1}{2}\right) \operatorname{Log} z - z + \frac{1}{2} \operatorname{Log} 2\,\pi + \mathrm{O}(|z|^{-1}) ,$$

the expansion is valid uniformly in any fixed sector $|\arg z| \leqslant \pi - \delta < \pi$ and for any bounded range of α.

(ii) As $|t|$ tends to infinity

$$\Gamma(\sigma + it) = t^{\sigma + it - \frac{1}{2}} \exp\left\{-\frac{1}{2}\pi t - it - i\frac{\pi}{2}\left(\sigma - \frac{1}{2}\right)\right\} \sqrt{2\,\pi}\left(1 + \mathrm{O}\left(\frac{1}{|t|}\right)\right)$$

the expansion is valid uniformly in all vertical strips $-\infty < a \leqslant \sigma \leqslant b < \infty$.

§ 3 THE FUNCTIONAL EQUATION

3.1 The fact that there is a relationship between $\zeta(s)$ and $\zeta(1 - s)$ was first proved by Riemann [1] in his famous (and only !) paper on prime number theory. This relationship is called the functional equation of $\zeta(s)$. It will be proved in theorem 5.2, then we shall use it to deduce some order of magnitude estimates for $\zeta(s)$ and $Z(s)$. These latter results will be needed in § 4 and § 5.

Theorem 5.2 (a) *The function $\zeta(s)$ has an analytic continuation over the whole complex plane. Its only singularity is a simple pole at $s = 1$ with residue* 1.

(b) *If we define* ξ *by*

$$\xi(s) = \frac{1}{2} s(s-1)\, \pi^{-s/2}\, \Gamma\left(\frac{s}{2}\right) \zeta(s)$$

then ξ *is an integral function and satisfies* $\xi(1-s) = \xi(s)$.

(c) *The Riemann zeta function satisfies the functional equation*

$$\zeta(1-s) = \frac{2}{(2\pi)^s} \cos\left(\frac{\pi s}{2}\right) \Gamma(s)\, \zeta(s) .$$

Proof. If $s = \sigma + it$ and $\sigma > 0$, then for each positive integer n we have, upon making the substitution $u = \pi n^2 x$ in (*a*) of § 2.2 :

$$\Gamma\left(\frac{s}{2}\right) (\pi n^2)^{-\frac{s}{2}} = \int_0^\infty x^{\frac{s}{2}-1} e^{-\pi n^2 x}\, dx .$$

Upon supposing that $\sigma > 1$ we can sum both sides of the above equation over all positive integers n to obtain

$$\Gamma\left(\frac{s}{2}\right) \pi^{-\frac{s}{2}} \zeta(s) = \sum_{n=1}^\infty \int_0^\infty x^{\frac{s}{2}-1} e^{-\pi n^2 x}\, dx . \tag{5.4}$$

Because $\Re(s) > 1$, we have

$$\sum_{n=1}^\infty \int_0^\infty | x^{\frac{s}{2}-1} e^{-\pi n^2 x} |\, dx = \Gamma\left(\frac{\sigma}{2}\right) \pi^{-\frac{\sigma}{2}} \zeta(\sigma)$$

Thus, an appeal to the theorem of dominated convergence allows us to interchange the signs of integration and summation in (5.4). Hence for $\Re(s) > 1$ we have shown that

$$\Gamma\left(\frac{s}{2}\right) \pi^{-\frac{s}{2}} \zeta(s) = \int_0^\infty x^{\frac{s}{2}-1} \omega(x)\, dx , \tag{5.5}$$

where $\omega(x)$ is the function defined in theorem 5.1.

Now we consider the right hand side of (5.5). First we divide the range of integration into two parts : $(0, 1]$ and $[1, \infty)$.

In the first interval we make the substitution $x \mapsto x^{-1}$ and so (5.5) now becomes :

$$\Gamma\left(\frac{s}{2}\right) \pi^{-\frac{s}{2}} \zeta(s) = \int_1^\infty x^{\frac{s}{2}-1} \omega(x)\, dx + \int_1^\infty x^{-\frac{s}{2}-1} \frac{\sqrt{x}}{2} \left\{1 - \frac{1}{\sqrt{x}} + \omega(x)\right\} dx .$$

From theorem 5.1 we know that for $x > 0$ we have

$$\omega\left(\frac{1}{x}\right) = -\frac{1}{2} + \frac{1}{2}x^{1/2} + x^{1/2}\,\omega(x),$$

with the consequence that

$$\Gamma\left(\frac{s}{2}\right)\pi^{-\frac{s}{2}}\zeta(s) = \frac{1}{s(s-1)} + \int_1^\infty x^{\frac{s}{2}-1}\,\omega(x)\,dx + \int_1^\infty x^{-\frac{s+1}{2}}\,\omega(x)\,dx$$

Upon multiplying by $\frac{1}{2}s(s-1)$ and recalling the definition of $\xi(s)$ we see that for $\mathfrak{R}(s) > 1$ we have proved

$$\xi(s) = \frac{1}{2} + \frac{1}{2}s(s-1)\int_1^\infty\left(x^{\frac{s}{2}} + x^{\frac{1-s}{2}}\right)\omega(x)\frac{dx}{x}. \tag{5.6}$$

since $\omega(x) = O(e^{-\pi x})$ the above integral represents an entire function of s, consequently the above formula provides us with an analytic continuation of $\xi(s)$ over the whole complex plane so $\xi(s)$ is an entire function. From the definition of $\xi(s)$ we have

$$\zeta(s) = \frac{2\,\xi(s)\,\pi^{s/2}}{\Gamma(1+s/2)\,(s-1)}.$$

By the known properties of the Γ function it now follows that $\zeta(s)$ is defined as a holomorphic function over the whole complex plane, except for a simple pole at $s = 1$ with residue 1.

(*b*) It is an immediate consequence of (5.6) that for all s we have

$$\xi(1-s) = \xi(s).$$

(*c*) From part (*b*) we can now deduce that

$$\frac{\zeta(1-s)}{\zeta(s)} = \frac{\pi^{-s/2}\,\Gamma(s/2)}{\pi^{(s-1)/2}\,\Gamma\left(\frac{1-s}{2}\right)} = \pi^{\frac{1}{2}-s}\frac{\Gamma\left(\frac{s}{2}\right)\Gamma\left(\frac{s+1}{2}\right)}{\Gamma\left(\frac{1-s}{2}\right)\Gamma\left(\frac{1+s}{2}\right)}$$
$$= 2(2\pi)^{-s}\,\Gamma(s)\cos\left(\frac{\pi s}{2}\right).$$

This completes the proof of the theorem. ■

3.2 We now use the functional equation to obtain upper bounds for $|\zeta(s)|$ and $|Z(s)|$ in some special regions of the complex plane. The estimates which we shall obtain can be improved upon by using more complex function theory, see for example Titchmarsh [2]. Our results will be used to prove the convergence of certain integrals which occur in § 4 and § 5 and to do this comparatively weak upper bounds will suffice.

Theorem 5.3 *Let* $\varepsilon > 0$ *be a fixed real number. In the half plane* $\mathcal{R}(s) \geqslant -\varepsilon$ *the following estimate holds uniformly as* $|t|$ *tends to infinity* :

$$|\zeta(s)| \leqslant A_1(\varepsilon)\, |t|^{\frac{1}{2}+\varepsilon}.$$

where the implied " O " *constant is an effectively computable function of* ε.

Proof. Let δ be a fixed real number which satisfies

$$0 < 2\,\delta < \min\left(\frac{1}{2}, \varepsilon\right).$$

We divide the half plane $\mathcal{R}(s) \geqslant -\varepsilon$ into the four parts :

(a) $\mathcal{R}(s) \geqslant 1 + \delta$; (b) $\frac{1}{2} \leqslant \mathcal{R}(s) \leqslant 1 + \delta$;

(c) $-\varepsilon \leqslant \mathcal{R}(s) \leqslant -\delta$; (d) $-\delta \leqslant \mathcal{R}(s) \leqslant \frac{1}{2}$;

and then estimate $|\zeta(s)|$ in each part.

(*a*) For $\mathcal{R}(s) \geqslant 1 + \delta$ we have

$$|\zeta(s)| \leqslant \zeta(1 + \delta) = \mathrm{O}(1).$$

thus in this half plane $\zeta(s) = \mathrm{O}(|t|^0)$.

(*b*) If s lies in the vertical strip $\frac{1}{2} \leqslant \mathcal{R}(s) \leqslant 1 + \delta$, then we employ theorem 2.3 with $\theta = \sigma - 2\,\delta$ to deduce that

$$|\zeta(s)| \leqslant \frac{7}{4}\frac{|t|^{1-\sigma+2\delta}}{(\sigma - 2\,\delta)(1 - \sigma + 2\,\delta)} \leqslant \frac{7}{4}\frac{|t|^{1-\sigma+2\delta}}{\delta\left(\frac{1}{2} - 2\,\delta\right)} = \mathrm{O}\left(|t|^{\frac{1}{2}+2\delta}\right).$$

(*c*) Now we suppose that s satisfies $-\varepsilon \leqslant \mathcal{R}(s) \leqslant -\delta$. From the functional equation we have

$$\zeta(s) = \frac{(2\,\pi)^s}{2}\frac{\zeta(1 - s)}{\Gamma(s)\cos(\pi s/2)}. \tag{5.7}$$

It is clear that $\zeta(1 - s) = O(1)$ if $-\varepsilon \leqslant \Re(s) \leqslant -\delta$. Upon using the asymptotic expansion of $\Gamma(\sigma + it)$ which is valid in a vertical strip and noting that as t tends to infinity $\cos\left(\frac{\pi s}{2}\right) = O\left(e^{\frac{1}{2}\pi t}\right)$ we see that

$$\frac{(2\pi)^s}{2\,\Gamma(s)\cos(\pi s/2)} = O\left(|t|^{\frac{1}{2}-\sigma}\right)$$

Hence we can conclude that in this case

$$\zeta(s) = O\left(|t|^{\frac{1}{2}+\varepsilon}\right).$$

(*d*) Finally, in the vertical strip $-\delta \leqslant \Re(s) \leqslant \frac{1}{2}$ we have

$$\zeta(1 - s) = O(|t|^{\sigma + 2\delta}) \quad \text{and} \quad \frac{(2\pi)^s}{2\,\Gamma(s)\cos(\pi s/2)} = O\left(|t|^{\frac{1}{2}-\sigma}\right),$$

Thus, from (5.7),

$$\zeta(s) = O\left(|t|^{\frac{1}{2}+2\delta}\right).$$

in the strip $-\delta \leqslant \Re(s) \leqslant \frac{1}{2}$.

The theorem now follows after we observe that the exponent of $|t|$ in each of the four cases is less than $\frac{1}{2} + \varepsilon$. ■

We now turn to estimating $Z(s)$ in various regions of the complex plane, described as follows.

Let δ be a fixed real number satisfying $0 < \delta < 1$. Denote by $R(\delta)$ the following region of the complex plane :

$$R(\delta) = \{ s \mid \Re(s) \leqslant -\delta, |s + 2r| \geqslant \delta \text{ pour } r = 1, 2, 3, \ldots \}.$$

We are going to estimate $|Z(s)|$ when $s \in R(\delta)$. The reason why we are considering this " perforated half plane " is that $Z(s)$ has simple poles at the points $s = -2r$ for $r = 1, 2, \ldots$, a fact to be proved later. When we come to study integrals involving $Z(s)$ we must avoid these singularities. This will be done by confining our paths of integration to lie in a region $R(\delta)$. The following theorem will then be used to prove that the integrals which arise are, in fact convergent.

Theorem 5.4 *Using the above notation we suppose that $s \in R(\delta)$. The following estimate then holds :*

$$|Z(s)| < A_2(\delta)(\operatorname{Log}|s| + 1).$$

where $A(\delta)$ is an explicitly computable function of δ.

Proof. From the functional equation for $\zeta(s)$ we have

$$\zeta(s) = 2(2\,\pi)^{s-1} \sin\left(\frac{\pi s}{2}\right) \Gamma(1 - s) . \zeta(1 - s) .$$

If we take logarithms and then differentiate with respect to s, then we find

$$- \mathrm{Z}(s) = \operatorname{Log}(2\,\pi) + \frac{\pi}{2} \operatorname{cotg}\left(\frac{\pi s}{2}\right) - \frac{\Gamma'(1 - s)}{\Gamma(1 - s)} + \mathrm{Z}(1 - s)$$

with the consequence that

$$| \mathrm{Z}(s) | \leqslant \operatorname{Log}(2\,\pi) + \frac{\pi}{2} \left| \operatorname{cotg}\left(\frac{\pi s}{2}\right) \right| + \left| \frac{\Gamma'(1 - s)}{\Gamma(1 - s)} \right| + | \mathrm{Z}(1 - s) | .$$

For $s \in R(\delta)$ it is trivial that $| \mathrm{Z}(1 - s) | = \mathrm{O}(1)$ and that $\left| \cot\left(\frac{\pi s}{2}\right) \right| = \mathrm{O}(1)$.
From the asymptotic formula for $\operatorname{Log} \Gamma(z)$ we easily deduce that there exists a constant $A_1(\delta)$ such that

$$\left| \frac{\Gamma'(1 - s)}{\Gamma(1 - s)} \right| \leqslant \operatorname{Log} | 1 - s | + A_3(\delta) .$$

Upon combining the above estimates we deduce that if $s \in R(\delta)$, then for suitable constants $A_2(\delta)$ and $A(\delta)$

$$| \mathrm{Z}(s) | \leqslant A_4(\delta) + \operatorname{Log}(1 - s) \leqslant A_2(\delta) (\operatorname{Log} | s | + 1) .$$

This completes the proof of the theorem. ∎

§ 4 THE ZEROS OF $\zeta(s)$

4.1 As we shall see in § 5 the distribution of the zeros of $\zeta(s)$ is intimately connected with the distribution of prime numbers. In this section we shall discuss some elementary facts concerning the zeros of $\zeta(s)$. Unfortunately the precise localisation of the zeros is unknown. Riemann advanced the hypothesis that all the complex zeros of $\zeta(s)$ actually lie on the line $\mathfrak{R}(s) = \frac{1}{2}$. There is some numerical support for this conjecture. For example, Rosser, Schoenfeld and Yohe [4] have verified that the first $3{,}5 \times 10^6$ complex zeros, ordered by increasing imaginary part, all lie on the line $\mathfrak{R}(s) = \frac{1}{2}$ and that all of them are simple. However, this is not a very weighty argument for believing the truth of the Riemann Hypothesis, because as we shall see in chapter 6, conjectures supported only by copious numerical verification are often completely false. On the positive side, it has been rigourously proved that a positive proportion of the complex zeros actually lie on the line $\mathfrak{R}(s) = \frac{1}{2}$ and that *if* the Riemann Hypothesis is true, then a positive proportion of the complex zeros are simple. We shall return briefly to this topic in the *notes* at the end of the chapter.

4.2 Our first theorem is a comparative triviality. However it does bring the "critical strip" into prominence. (The "critical strip" is the popular name for the region $0 < \mathfrak{R}(s) < 1$, which is, for number theorists, heavily veiled in mystery.) We use the notation of theorem 5.2.

Theorem 5.5 (a) *The function* ξ *is real valued on the lines* $\mathfrak{Im}(s) = 0$ *and* $\mathfrak{R}(s) = \frac{1}{2}$.

(b) *The function* ξ *has no zeros in the half planes* $\mathfrak{R}(s) \geqslant 1$ *and* $\mathfrak{R}(s) \leqslant 0$ *or on the real axis. If any zeros do exist, then they lie symmetrically about the two lines* $\mathfrak{R}(s) = \frac{1}{2}$ *and* $\mathfrak{Im}(s) = 0$.

(c) *The zeros of* $\zeta(s)$ *are the same as the zeros of* $\xi(s)$, *together with simple zeros at the points* $s = -2r$, $r = 1, 2, \ldots$

Proof. (*a*) From (5.6) and the functional equation for $\xi(s)$ we see that

$$\overline{\xi(\sigma + it)} = \xi(\sigma - it) = \xi(1 - \sigma + it) \tag{5.8}$$

It now follows that $\overline{\xi(\sigma)} = \xi(\sigma)$ and $\overline{\xi\left(\frac{1}{2} + it\right)} = \xi\left(\frac{1}{2} + it\right)$, hence $\xi(s)$ is real valued on the two lines $\mathfrak{R}(s) = \frac{1}{2}$ and $\mathfrak{Im}(s) = 0$.

(*b*) Neither of the functions $(s - 1)\,\zeta(s)$ and $\Gamma(s/2)$ vanishes in the half plane $\mathfrak{R}(s) \geqslant 1$, consequently the function

$$\xi(s) = \frac{1}{2} s\Gamma\left(\frac{s}{2}\right) \pi^{-s/2} (s - 1)\, \zeta(s)$$

is also non zero in this half plane. The functional equation for $\xi(s)$ then implies that $\xi(s) \neq 0$ in the half plane $\mathfrak{R}(s) \leqslant 0$. If s is a real number satisfying $0 < s < 1$, then it follows from theorem 5.1(*b*) and (5.6) that $\xi(s) > 0$. Thus $\xi(s)$ has no real zeros.

The symmetry of the zeros is an immediate consequence of (5.8) for if $\beta + i\gamma$ is a zero of $\xi(s)$, then so are $\beta - i\gamma$ and $1 - \beta \pm i\gamma$.

(*c*) We know that

$$\zeta(s) = \frac{\pi^{s/2}}{\Gamma\left(1 + \frac{s}{2}\right)} \cdot \frac{\xi(s)}{s - 1}$$

where $\Gamma\left(\frac{s}{2} + 1\right)^{-1}$ is an integral function whose only zeros are simple ones at the points $s = -2r$, $r = 1, 2, \ldots$ Since $\xi(s)$ has no poles part (*c*) is proved. ■

4.3 We shall write a typical zero ρ of $\xi(s)$ as $\rho = \beta + i\gamma$ and denote by $\mathcal{N}(T)$ the number of zeros whose imaginary parts satisfy $0 < \gamma \leqslant T$. It is clear that $\mathcal{N}(T)$ is well defined, since the possible zeros of $\xi(s)$ are confined to the strip $0 < \Re(s) < 1$, so only a finite number can have imaginary parts satisfying $0 < \gamma \leqslant T$. The next theorem provides us with some estimates which will be needed for the proof of theorem 5.7, in § 5 and in chapter 6.

Theorem 5.6 (a) *As T tends to infinity we have*

$$\mathcal{N}(T + 1) - \mathcal{N}(T) = \mathrm{O}(\operatorname{Log} T) .$$

(b) *Let η be a fixed real number which satisfies $0 < \eta \leqslant 1$. If $s = \sigma + iT$ where $-1 \leqslant \sigma \leqslant 2$, then*

$$\mathrm{Z}(s) + \sum_{|\gamma - T| < \eta} \frac{1}{s - \rho} = \mathrm{O}\left(\frac{\operatorname{Log} T}{\eta}\right) .$$

(c) *As T tends to infinity the following two estimates hold :*

$$\sum_{0 < \gamma \leqslant T} \frac{1}{\gamma} = \mathrm{O}(\operatorname{Log}^2 T)$$

$$\sum_{\gamma > T} \frac{1}{\gamma^2} = \mathrm{O}\left(\frac{\operatorname{Log} T}{T}\right) .$$

In particular, the series $\sum \gamma^{-2}$ is convergent.

(d) *There exists an infinite sequence of real numbers $\{ T_n \}$ satisfying $n < T_n < n + 1$ and such that if $s = \sigma + iT_n$, with $-1 \leqslant \sigma \leqslant 2$, then*

$$\mathrm{Z}(s_n) = \mathrm{O}\{ \operatorname{Log}^2 n \} .$$

The implied " O " constants in the above estimates are all absolute and all can be effectively determined.

Proof. In order to prove parts (*a*) and (*b*) of the theorem we will apply theorem 4.2 to $\zeta(s)$ with the following choice of parameters :

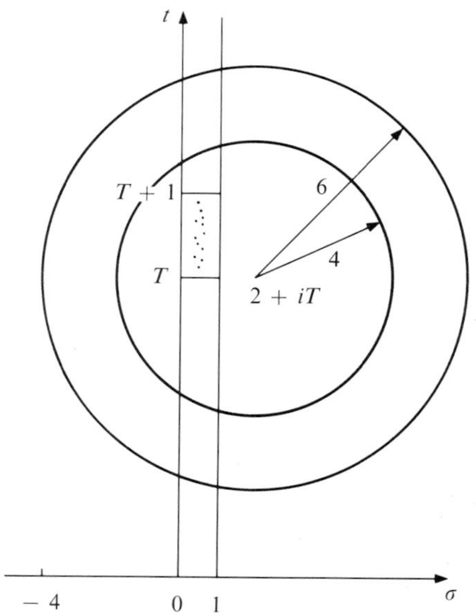

Fig. 9

$$s_0 = 2 + iT\,, \quad R = 4\,, \quad q = \frac{3}{2}\,, \quad r = 3\,.$$

First we verify that the hypotheses of theorem 4.2 are satisfied by $\zeta(s)$ for all sufficiently large values of T. If $T > 6$, then the simple pole at $s = 1$ lies outside the disc $|s - s_0| \leqslant b = qR$ and $\zeta(s)$ is holomorphic on the disc. It is clear that $\zeta(s_0) = \zeta(2 + iT) \neq 0$. All that remains now is to find an upper bound for $|\zeta(s)/\zeta(s_0)|$ on the disc $|s - s_0| \leqslant 6$. Such a bound is given by theorem 5.3 with $\varepsilon = 4$, the result is

$$\left|\frac{\zeta(s_0)}{\zeta(S_0)}\right| \leqslant \zeta(2)\,|\zeta(s)| < e^{5\,\mathrm{Log}\,T}$$

provided that T is sufficiently large. Thus we can take $\mathrm{U} = 5\,\mathrm{Log}\,T$.

(*a*) If we denote by $\mathcal{M}(T)$ the number of zeros of $\zeta(s)$ which satisfy $|\rho - s_0| \leqslant 4$, then

$$\mathcal{N}(T + 1) - \mathcal{N}(T) \leqslant \mathcal{M}(T)\,.$$

and theorem 4.2(*a*) yields $\mathcal{M}(T) \leqslant 15\,\mathrm{Log}\,T$. This proves part (*a*).

(*b*) From theorem 4.2(*b*) we have, for $|s - s_0| \leqslant 3$

$$\left| -\mathrm{Z}(s) - \sum_{|\rho - s_0| \leqslant 4} \frac{1}{s - \rho} \right| \leqslant \frac{2.4}{(4 - 3)^2}\, 5\,\mathrm{Log}\,T$$

We clearly have

$$\left| \sum_{\substack{|\gamma - t| \geqslant \eta \\ |\rho - s_0| \leqslant 4}} \frac{1}{s - \rho} \right| \leqslant \sum_{\substack{|\gamma - t| \geqslant \eta \\ |\rho - s_0| \leqslant 4}} \frac{1}{\eta} \leqslant \frac{\mathcal{M}(t)}{\eta} \leqslant \frac{15\,\mathrm{Log}\,t}{\eta}$$

from which we conclude

$$\left| \mathrm{Z}(s) + \sum_{|\gamma - t| < \eta} \frac{1}{s - \rho} \right| \leqslant 40\,\mathrm{Log}\,t + \frac{15\,\mathrm{Log}\,t}{\eta} = \mathrm{O}\left(\frac{\mathrm{Log}\,t}{\eta}\right).$$

The last term in the above expression contains at most $\mathcal{M}(T)$ terms, each in absolute value at most η^{-1}. Thus an upper bound for the sum is

$$\frac{\mathcal{M}(T)}{\eta} \leqslant \frac{15}{\eta} \operatorname{Log} T$$

and we have

$$Z(s) + \sum_{|\gamma - T| \leqslant \eta} (s - \rho)^{-1} = O\left(\frac{\operatorname{Log} T}{\eta}\right).$$

(*c*) If we group the terms of the two series into blocks which satisfy

$$n < \gamma \leqslant n + 1 \qquad (n = 0, 1, 2, \ldots) .$$

then by (*a*) we have

$$S_1(T) = O\left(1 + \sum_{n=2}^{[T]} \frac{\operatorname{Log} n}{n}\right) = O\{(\operatorname{Log} T)^2\}$$

and

$$S_2(T) = O\left(\sum_{n \geqslant T} \frac{\operatorname{Log} n}{n^2}\right) = O\left(\frac{\operatorname{Log} T}{T}\right).$$

(*d*) Divide the interval $[n, n + 1]$ into $\nu = \mathcal{N}(n + 1) - \mathcal{N}(n) + 1$ equal subintervals. Some subinterval does not contain a γ. We define T_n to be the midpoint of such an interval. To see that the sequence $\{T_n\}$ defined in this way has all the required properties apply part (*b*) with $T = T_n$ and $\eta = \frac{1}{2}\nu^{-1}$. The sum which occurs in (*b*) is then empty and so for $-1 \leqslant \sigma \leqslant 2$ we deduce

$$Z(s) = O\left(\frac{\operatorname{Log} T_n}{\eta}\right) = O\{(\operatorname{Log} n)^2\}. \qquad \blacksquare$$

4.4 We have not yet proved that $\xi(s)$ ever vanishes. The fact that $\xi(s)$ has an infinity of zeros will now be proved by establishing an asymptotic formula for $\mathcal{N}(T)$. It is also possible to show that $\xi(s)$ does have zeros by appealing to the general theory of integral functions. For an account of this approach we refer the reader to Davenport [4].

Theorem 5.7 (a) *As T tends to infinity*

$$\mathcal{N}(T) = \frac{T}{2\pi} \log\left(\frac{T}{2\pi}\right) - \frac{T}{2\pi} + O(\log T).$$

(b) *The function $\xi(s)$ possesses an infinity of zeros which we write as $\rho_{\pm n} = \beta_n \pm i\gamma_n$ where $0 < \gamma_1 \leqslant \gamma_2 \leqslant \ldots$*

(c) *As n tends to infinity $|\rho_n| \sim \gamma_n \sim \dfrac{2\pi n}{\operatorname{Log} n}$.*

Proof. We first show that (*b*) and (*c*) are trivial consequences of (*a*). It is clear that (*b*) follows directly from (*a*). To deduce (*c*) note

$$\mathcal{N}(\gamma_n - 1) < n \leqslant \mathcal{N}(\gamma_n)$$

whence from (*a*)

$$n \sim \frac{\gamma_n}{2\pi} \operatorname{Log} \gamma_n$$

Consequently we have $\operatorname{Log} n \sim \operatorname{Log} \gamma_n$. Since $|\rho_n| = \gamma_n + O(1)$, part (*c*) follows. All that remains now is to prove (*a*).

Let $\mathcal{C}$ be the rectangular path in figure 10, where T is not equal to any γ_n. The number of zeros of $\xi(s)$ within $\mathcal{C}$ is $2\,\mathcal{N}(T)$.

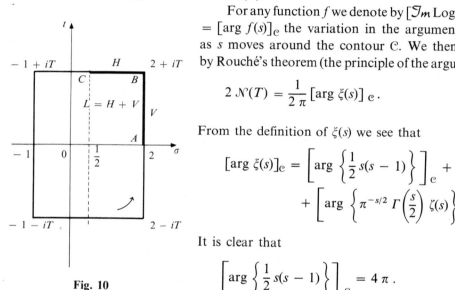

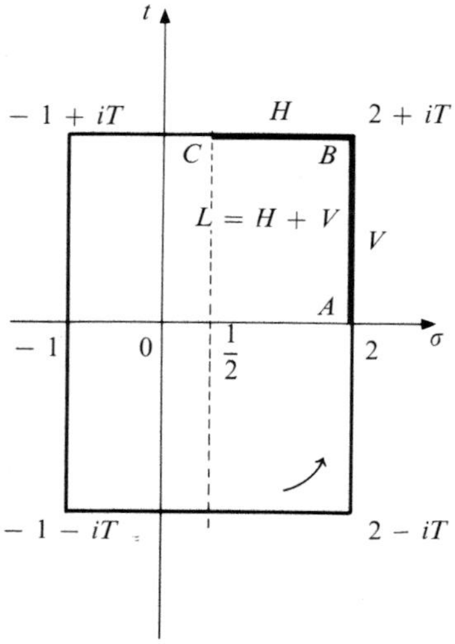

Fig. 10

For any function f we denote by $[\mathcal{Im} \operatorname{Log} f(s)]_{\mathcal{C}} = [\arg f(s)]_{\mathcal{C}}$ the variation in the argument of f as s moves around the contour $\mathcal{C}$. We then have by Rouché's theorem (the principle of the argument)

$$2\,\mathcal{N}(T) = \frac{1}{2\pi} [\arg \xi(s)]_{\mathcal{C}} .$$

From the definition of $\xi(s)$ we see that

$$[\arg \xi(s)]_{\mathcal{C}} = \left[\arg \left\{\frac{1}{2} s(s-1)\right\}\right]_{\mathcal{C}} + \left[\arg \left\{\pi^{-s/2}\,\Gamma\left(\frac{s}{2}\right) \zeta(s)\right\}\right]_{\mathcal{C}},$$

It is clear that

$$\left[\arg \left\{\frac{1}{2} s(s-1)\right\}\right]_{\mathcal{C}} = 4\pi .$$

and since $\pi^{-s/2}\,\Gamma(s/2)\,\zeta(s)$ takes equal values at the points s and $1-s$ and conjugate values at the points $\sigma \pm it$ it follows that

$$\left[\arg \left\{\pi^{-s/2}\,\Gamma\left(\frac{s}{2}\right) \zeta(s)\right\}\right]_{\mathcal{C}} = 4\left[\arg \left\{\pi^{-s/2}\,\Gamma\left(\frac{s}{2}\right) \zeta(s)\right\}\right]_{\mathcal{L}}$$

where $\mathcal{L}$ is the quarter of the rectangle shown in figure 10. So we now have

$$\mathcal{N}(T) = 1 + \frac{1}{\pi}\left[\arg \pi^{-s/2}\right]_{\mathcal{L}} + \frac{1}{\pi}\left[\arg \Gamma\left(\frac{s}{2}\right)\right]_{\mathcal{L}} + \frac{1}{\pi}\left[\arg \zeta(s)\right]_{\mathcal{L}} . \tag{5.9}$$

It is trivial to see that

$$\left[\arg \pi^{-s/2}\right]_{\mathcal{L}} = -\frac{1}{2} T \operatorname{Log} \pi ,$$

In order to evaluate $[\arg \Gamma(s/2)]_{\mathfrak{L}}$ we shall employ the asymptotic expansion for $\operatorname{Log} \Gamma(z + \alpha)$ given in § 2.2, choosing $\alpha = \frac{1}{4}$ and $z = i\frac{T}{2}$.

Clearly

$$\left[\arg \Gamma\left(\frac{s}{2}\right)\right]_{\mathfrak{L}} = \left[\mathfrak{Im}\left\{\operatorname{Log} \Gamma\left(\frac{s}{2}\right)\right\}\right]_{\mathfrak{L}}$$

$$= \mathfrak{Im}\left\{\operatorname{Log} \Gamma\left(\frac{1}{4} + i\frac{T}{2}\right) - \operatorname{Log} \Gamma(1)\right\}$$

$$= \mathfrak{Im}\left\{\left(-\frac{1}{4} + i\frac{T}{2}\right) \operatorname{Log}\left(i\frac{T}{2}\right) - i\frac{T}{2} + \frac{1}{2}\operatorname{Log} 2\pi\right\} + O\left(\frac{1}{T}\right)$$

$$= \frac{1}{2} T \operatorname{Log}\frac{T}{2} - \frac{T}{2} - \frac{\pi}{8} + O\left(\frac{1}{T}\right).$$

and combining the above with (5.9) we now deduce that

$$\mathcal{N}(T) = \frac{T}{2\pi}\operatorname{Log}\left(\frac{T}{2\pi}\right) - \frac{T}{2\pi} + \frac{7}{8} + \frac{1}{\pi}[\arg \zeta(s)]_{\mathfrak{L}} + O\left(\frac{1}{T}\right).$$

It only remains to estimate $[\arg \zeta(s)]_{\mathfrak{L}}$. Upon writing $\mathfrak{L} = V + H$ we see that

$$[\arg \zeta(s)]_{\mathfrak{L}} = [\arg \zeta(s)]_V + [\arg \zeta(s)]_H .$$

On the segment V we certainly have $\mathfrak{R}(\zeta(s)) \geqslant 1 - \sum_{n=2}^{\infty} n^{-2} > 0$ and consequently

$$|[\arg \zeta(s)]_V| < \pi .$$

and we now have

$$\mathcal{N}(T) = \frac{T}{2\pi}\operatorname{Log}\left(\frac{T}{2\pi}\right) - \frac{T}{2\pi} + \frac{1}{\pi}[\arg \zeta(s)]_H + O(1).$$

In order to compute $[\arg \zeta(s)]_H$ we first note that

$$[\arg \zeta(s)]_H = \mathfrak{Im}[\operatorname{Log} \zeta(s)]_H .$$

then we obtain an expression for $\operatorname{Log} \zeta(s)$ as follows. From theorem 5.6(*b*) with $\eta = 1$ we know that

$$\frac{\zeta'}{\zeta}(s) = \sum_{|\gamma - T| < 1} \frac{1}{s - \rho} + O(\operatorname{Log} T).$$

By integrating the above expression along the segment H and then taking the imaginary part we obtain

$$[\arg \zeta(s)]_H = \sum_{|\gamma - T| < 1} [\arg (s - \rho)]_H + O(\operatorname{Log} T)$$
$$= O(\operatorname{Log} T) .$$

since the sum contains $O(\operatorname{Log} T)$ terms, each in absolute value at most π. Thus we have shown that

$$\mathcal{N}(T) = \frac{T}{2\pi} \operatorname{Log} \frac{T}{2\pi} - \frac{T}{2\pi} + O(\operatorname{Log} T) .$$

and the proof of the theorem is complete. ■

§ 5 EXPLICIT FORMULAE

5.1 The asymptotic formulae for the functions $\psi_1(x)$, $\psi(x)$, $\Pi(x)$ and $M(x)$ are intimately related to the zeros of $\zeta(s)$. One can exhibit this relationship quite explicity and write the above expressions as functions of x and the zeros ρ of $\zeta(s)$. We shall give such explicit formulae for the functions $\psi_1(x)$ and $\psi(x)$. The first formula is relatively straightforward to obtain and the result will be utilised several times in this and the next chapter. The formula for $\psi(x)$ involves several technical complications which do not occur in the derivation of the formula for $\psi_1(x)$. We only prove the result to illustrate the basic technique and we shall make no further use of the formula. The analogous formulae for $\Pi(x)$ and $M(x)$ are relegated to the exercises at the end of the chapter.

Theorem 5.8 (a) *If $x \geqslant 1$, then we have*

$$\psi_1(x) = \frac{x^2}{2} + Z(0)\, x - Z(-1) - \sum_{\rho} \frac{x^{\rho+1}}{\rho(\rho+1)} - \sum_{r=1}^{\infty} \frac{x^{1-2r}}{2r(2r-1)} .$$

(b) *If a, b are fixed real numbers which satisfy $1 \leqslant a < b < \infty$ and g is any function with a continuous derivative on the interval $[a, b]$, then*

$$\int_a^b g(x) \left\{ \psi(x) - x - Z(0) + \frac{1}{2} \operatorname{Log}\left(1 - \frac{1}{x^2}\right) \right\} dx = - \sum_{\rho} \int_a^b \frac{g(x)\, x^{\rho}}{\rho}\, dx .$$

(c) *The above series are all absolutely convergent.*

Proof. (*a*) By theorem 2.7 we know

$$\psi_1(x) = \lim_{T \to \infty} \frac{1}{2 i\pi} \int_{2-iT}^{2+iT} \frac{Z(s)}{s(s+1)}\, x^{s+1}\, ds .$$

In order to evaluate this latter integral we consider

$$I_n = \frac{1}{2\,i\pi} \int_{\mathcal{C}_n} \frac{Z(s)}{s(s+1)} x^{s+1}\, ds$$

where the contour $\mathcal{C}_n$ is described as follows. For $n \geqslant 3$ an odd integer and $\{T_n\}$ the sequence of real numbers defined in theorem 5.6 we denote by $\mathcal{C}_n$ the rectangular contour with vertices $2 \pm iT_n$, $-n \pm iT_n$.

We write $I_n = J_n + K_n$, where J_n denotes the integral along the line joining $2 - iT_n$ to $2 + iT_n$ and K_n denotes the integral along the other three sides of the rectangle.

The only singularities of the integrand within the contour $\mathcal{C}_n$ are simple poles at the points $s = 1, 0, -1, -2, -4, \ldots, -2[n/2]$ and at the zeros ρ of $\xi(s)$ which satisfy $|\gamma| < T_n$. Cauchy's residue theorem now allows us to conclude

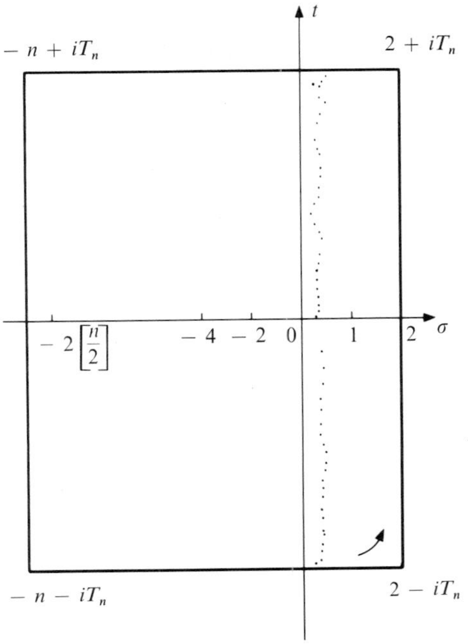

Fig. 11

$$I_n = \frac{x^2}{2} + xZ(0) - Z(-1) - \sum_{2r<n} \frac{x^{1-2r}}{2\,r(1-2\,r)} - \sum_{|\gamma|<T_n} \frac{x^{\rho+1}}{\rho(\rho+1)}\,.$$

In the last sum, if any of the zeros ρ are multiple, then the corresponding term occurs with the multiplicity of ρ.

We already know $\lim J_n = \psi_1(x)$ so we must show $K_n \to 0$ as n tends to infinity. From theorem 5.4 with $\delta = \frac{1}{2}$ and theorem 5.6(d) we easily see that on the remaining sides of the rectangle

$$|Z(s)| < A_1 \operatorname{Log}^2 n\,,$$
$$|s| \geqslant n\,, \qquad |s+1| \geqslant n\,,$$
$$|x^{s+1}| \leqslant x^3\,.$$

Hence we have

$$|K_n| \leqslant \frac{1}{2\,\pi} \int \frac{|Z(s)|}{|s(s+1)|} \cdot |x^{s+1}| \cdot |ds| = O\left\{\frac{\operatorname{Log}^2 n}{n^2}(2n + 4 + 2\,T_n)\,x^3\right\}$$

and for fixed $x > 1$, $K_n \to 0$ as $n \to \infty$.

Examine the two series

$$\sum_{r=1}^{\infty} \frac{x^{1-2r}}{2r(2r-1)} \quad \text{and} \quad \sum_{\rho} \frac{x^{1+\rho}}{\rho(1+\rho)}$$

On any fixed interval [1, X], the absolute values of the general terms are less than r^{-2} and $X^2 |\gamma|^{-2}$ respectively. Hence both series are absolutely, uniformly convergent by appeal to theorem 5.6(*c*).
Thus part (*a*) of the theorem is proved.

(*b*) The uniform convergence of the series $\sum \frac{x\rho}{\rho(\rho+1)}$ and the continuity of g' on the interval $[a, b]$ imply

$$\int_a^b g'(x) f_1(x)\,dx = -\sum_{\rho}\int_a^b g'(x)\frac{x^{\rho+1}}{\rho(\rho+1)}\,dx \tag{5.10}$$

where we have written $f_1(x)$ for the function

$$f_1(x) = \psi_1(x) - \frac{1}{2}x^2 - xZ(0) + Z(-1) + \sum_{r=1}^{\infty}\frac{x^{1-2r}}{2r(2r-1)}.$$

It is clear that $f_1(x)$ is an indefinite integral of the function

$$f(x) = \psi(x) - x - Z(0) + \frac{1}{2}\operatorname{Log}(1 - x^2).$$

and since $\psi(x)$ is discontinuous only at prime powers, $f_1'(x) = f(x)$ whenever x is not equal to a prime power. Thus one can integrate both sides of equation (5.10) by parts. Upon using part (*a*), part (*b*) follows.

The absolute convergence of the series

$$\sum_{\rho}\int_a^b \frac{g(x)\,x^{\rho}}{\rho}\,dx.$$

is a consequence of the absolute convergence of the right hand side of (5.10). ■

We next prove an explicit formula for $\psi(x)$. Unlike $\psi_1(x)$ this function has finite discontinuities whenever x is equal to a prime power. This complication means our discussion has to be rather more delicate than was necessary for the proof of theorem 5.8. In fact, the function which arises is not $\psi(x)$, but the function $\psi_0(x)$ defined by

$$\psi_0(x) = \frac{1}{2}\{\psi(x+0) + \psi(x-0)\}.$$

Of course we see $\psi_0(x) = \psi(x)$ if x is not equal to a prime power and

$$\psi_0(x) = \psi(x) - \frac{1}{2}\Lambda(x)$$

otherwise.

Theorem 5.9 *If $x > 1$, then we have*

$$\psi_0(x) = x + Z(0) - \frac{1}{2}\operatorname{Log}(1 - x^{-2}) - \lim_{T\to\infty} \sum_{|\gamma|<T} \frac{x^\rho}{\rho}.$$

Proof. In its general structure the proof of the theorem is very similar to the proof of theorem 5.8, but the details are a little different. We again use the rectangular contour $\mathcal{C}_n$ (Fig. 11). Consider the integral

$$I_n = \frac{1}{2\,i\pi}\int_{\mathcal{C}_n} \frac{Z(s)}{s} x^s\,ds$$

which we decompose into the two integrals

$$J_n = \frac{1}{2\,i\pi}\int_{2-iT_n}^{2+iT_n} \frac{Z(s)}{s} x^s\,ds$$

and

$$K_n = I_n - J_n\,.$$

The residue theorem gives

$$I_n = J_n + K_n = x + Z(0) + \sum_{2r<n} \frac{x^{-2r}}{2\,r} - \sum_{|\gamma|<T_n} \frac{x^\rho}{\rho}.$$

It is easy to estimate K_n. Theorem 5.4, with $\delta = \frac{1}{2}$, and theorem 5.6(*d*) tell us that if $s \in \mathfrak{D}_n$, then

$$Z(s) = O(\operatorname{Log}^2 n)$$

since $|s| \geqslant n$ if $s \in \mathfrak{D}_n$ we certainly have

$$|K_n| \leqslant \frac{1}{2\,\pi}\int_{\mathfrak{D}} \frac{|Z(s)|}{|s|}\,|x^s\,ds| = O\left\{\frac{\operatorname{Log}^2 n}{n}\int_{\mathfrak{D}} |x^s\,ds|\right\}$$

$$= O\left\{\frac{\operatorname{Log}^2 n}{n}\,\frac{x^2}{\operatorname{Log} x}\right\}.$$

Now we are going to study J_n. Our aim is to show that as n tends to infinity J_n tends to $\psi_0(x)$. In order to do this we shall need the following variant of theorem 2.6 :

If $1 < q \leqslant 2$ *and* $T > 0$, *then*

$$\frac{1}{2\,i\pi}\int_{q-iT}^{q+iT} \frac{y^s}{s}\,ds = I(y) + \Delta(y, T) \qquad (5.11)$$

where $I(y) = 0, \frac{1}{2}, 1$ *according as* $y < 1$, $y = 1$ *or* $y > 1$ *and the quantity* $\Delta(y, T)$ *satisfies*

$$| \Delta(y, T) | < \begin{cases} \dfrac{y^q}{\pi T \mid \operatorname{Log} y \mid} & \text{if} \quad y \neq 1 \\ \dfrac{q}{\pi T} & \text{if} \quad y = 1 . \end{cases} \tag{5.12}$$

A proof of the above result is given in Ingham [1], page 75.

Returning now to our study of J_n, we recall that for $\mathfrak{R}(s) > 1$

$$Z(s) = \sum_{m=1}^{\infty} \frac{\Lambda(m)}{m^s}$$

and so

$$J_n = \frac{1}{2\,i\pi} \int_{2-iT_n}^{2+iT_n} \sum_{m=1}^{\infty} \frac{\Lambda(m)}{s} \left(\frac{x}{m}\right)^s ds .$$

Interchanging summation and integration, then using (5.11) gives us

$$\begin{aligned} J_n &= \psi_0(x) + \sum_{m=1}^{\infty} \Lambda(m)\, \Delta\!\left(\frac{x}{m}, T_n\right) \\ &= \psi_0(x) + R_n . \end{aligned}$$

Finally we estimate the series R_n. Let u be the integer defined by

$$u - \frac{1}{2} < x \leqslant u + \frac{1}{2} .$$

then from (5.12) we have

$$\begin{aligned} | R_n | &\leqslant \sum_{m=1}^{\infty} \Lambda(m) \left| \Delta\!\left(\frac{x}{m}, T_n\right) \right| \\ &\leqslant \sum_{\substack{m=1 \\ m \neq u}}^{\infty} \Lambda(m) \left(\frac{x}{m}\right)^2 \frac{1}{\pi T_n \left| \operatorname{Log}\left(\dfrac{x}{m}\right) \right|} + \Lambda(u) \left| \Delta\!\left(\frac{x}{u}, T_n\right) \right| . \end{aligned}$$

Because $T_n \sim n$ and the series $\sum \Lambda(m)\, m^{-2}$ is convergent,

$$\begin{aligned} | R_n | &= O\left(\frac{x^2}{n \operatorname{Log} x}\right) + O\left\{ \Lambda(u) \left| \Delta\!\left(\frac{x}{u}, T_n\right) \right| \right\} \\ &= O\left(\frac{x^2}{n \operatorname{Log} x}\right) + O\left(\frac{\operatorname{Log} x}{n \delta(x)}\right) \end{aligned}$$

where $\delta(x)=1$ if $x=u$ and $\left|\text{Log}\left(\frac{x}{u}\right)\right|$ if $x \neq u$. Thus, for fixed x we see that as $n \to \infty$, $|R_n| \to 0$. Hence we have shown

$$\psi_0(x) = x + \text{Z}(0) - \frac{1}{2}\text{Log}(1 - x^{-2}) - \lim_{T\to\infty} \sum_{|\gamma|<T} \frac{x^\rho}{\rho}. \qquad \blacksquare$$

For many purposes the explicit formula given by theorem 5.9 is not useful. It is often vital to have the summation over the complex zeros kept as a finite sum and then to have an explicit estimate for the remainder. For example, we have, in effect proved that

$$\psi_0(x) = x + \text{Z}(0) - \frac{1}{2}\text{Log}(1 - x^{-2}) - \sum_{|\gamma|<T_n} \frac{x^\rho}{\rho} + E(x, n)$$

where

$$E(x, n) = \text{O}\left\{\frac{x^2}{\text{Log}\, x}\frac{\text{Log}^2 n}{n} + \frac{x^2}{n\,\text{Log}\, x} + \frac{\text{Log}\, x}{n\delta(x)}\right\}.$$

It is possible to obtain a better estimate for the error term by taking as a contour the rectangle with vertices

$$1 + (\text{Log}\, x)^{-1} \pm iT_n\,, \qquad n \pm iT_n\,.$$

instead of $\mathcal{C}_n$. One then obtains, without difficulty the estimate

$$E(x, n) = \text{O}\left\{\frac{x}{\text{Log}\, x}\frac{\text{Log}^2 n}{n} + \frac{x\,\text{Log}^2 x}{n} + \frac{\langle x \rangle\,\text{Log}\, x}{n}\right\}$$

where the implied " O " constant is absolute and the symbol $\langle x \rangle$ means :

$$\langle x \rangle = \begin{cases} \dfrac{x}{\| x \|} & \text{if } x \notin \mathbb{Z} \\ 1 & \text{if } x = p^r \\ 0 & \text{if } x \in \mathbb{Z},\, x \neq p^r\,. \end{cases}$$

5.2 Now we are going to use the explicit formula for $\psi_1(x)$ to relate the error terms in the asymptote expansions for $\psi(x)$, $\pi(x)$ etc., to the real parts of the zeros of $\zeta(s)$. Let Θ be defined by

$$\Theta = \sup_\rho \beta\,,$$

All that is known about Θ is that following inequalities are satisfied :

$$\frac{1}{2} \leqslant \Theta \leqslant 1\,.$$

Of course, the Riemann Hypothesis is equivalent to $\Theta = \frac{1}{2}$, but this is far from being proved. If $\Theta = 1$, then our next theorem is a triviality, since the " error " terms are larger than the " main " terms. But, if $\Theta < 1$, then the error term is better than any result which has yet been proved.

Theorem 5.10 *As x tends to infinity the following relations are true :*

(a) $\psi_1(x) = \frac{1}{2}x^2 + O(x^{\Theta+1})$,

(b) $\psi(x) = x + O(x^{\Theta} \operatorname{Log}^2 x)$,

(c) $\pi(x) = \operatorname{Li}(x) + O(x^{\Theta} \operatorname{Log} x)$.

Proof. (*a*) From theorem 5.8(*a*) we have

$$\psi_1(x) - \frac{1}{2}x^2 = \sum_{\rho} \frac{x^{1+\rho}}{\rho(1+\rho)} + O(x).$$

Since, by theorem 5.6, the series $\sum_{\rho} \gamma^{-2}$ is convergent, we have

$$\left| \psi_1(x) - \frac{1}{2}x^2 \right| \leqslant \sum_{\rho} \frac{x^{1+\Theta}}{|\rho(\rho+1)|} + O(x)$$
$$\leqslant x^{1+\Theta} \sum_{\rho} |\rho|^{-2} + O(x)$$

(*b*) As usual, the following inequality holds for all $x \geqslant 2$:

$$\psi_1(x) - \psi_1(x-1) \leqslant \psi(x) \leqslant \psi_1(x+1) - \psi_1(x).$$

It is clear, from theorem 5.8(*a*) that

$$\pm \{ \psi_1(x \pm 1) - \psi_1(x) \} = x \pm \frac{1}{2} - \sum_{\rho} \frac{(x \pm 1)^{\rho+1} - x^{\rho+1}}{\pm \rho(\rho+1)} - Z(0) + O\left(\frac{1}{x}\right).$$

Upon denoting by ω_ρ the expression

$$\omega_\rho = \frac{(x \pm 1)^{\rho+1} - x^{\rho+1}}{\pm \rho(\rho+1)}.$$

we easily see

$$|\omega_\rho| \leqslant \frac{(x+1)^{\Theta+1} + x^{\Theta+1}}{\gamma^2} \leqslant 5\frac{x^{\Theta+1}}{\gamma^2}$$

and that

$$|\omega_\rho| = \left|\frac{1}{\rho}\int_x^{x\pm 1} u^\rho\, du\right| \leqslant \frac{1}{|\rho|}(x+1)^\Theta < 2\frac{x^\Theta}{|\gamma|}.$$

Thus, the following estimate is true :

$$\begin{aligned}\left|\sum_\rho \omega_\rho\right| &\leqslant 10\, x^\Theta \sum_{\gamma>0} \min\left\{\frac{x}{\gamma^2}, \frac{1}{\gamma}\right\}\\ &= 10\, x^\Theta \left\{\sum_{0<\gamma<x}\frac{1}{\gamma} + \sum_{\gamma\geqslant x}\frac{x}{\gamma^2}\right\}\\ &= \mathrm{O}\{x^\Theta(\mathrm{Log}^2 x + \mathrm{Log}\, x)\}\\ &= \mathrm{O}\{x^\Theta\, \mathrm{Log}^2 x\}.\end{aligned}$$

Consequently we certainly have

$$\psi(x) = x + \mathrm{O}(x^\Theta\, \mathrm{Log}^2 x).$$

(*c*) This follows from (*b*) by the usual partial summation argument. ■

If the Riemann Hypothesis were true, then it would follow from the above theorem that $\psi_1(x) = \frac{1}{2}x^2 + \mathrm{O}(x^{\frac{3}{2}})$. In the next chapter we shall obtain results in the opposite direction and study how large the " error " terms $\psi_1(x) - \frac{1}{2}x^2$ and $\psi(x) - x$ can become. A simple consequence of theorem 6.1 is that there exists $K > 0$ and infinite sequences of integers $\{x_n\}$, $\{y_n\}$ such that

$$\psi_1(x_n) - \frac{1}{2}x_n^2 > Kx_n^{\frac{3}{2}} \quad \text{and} \quad \psi_1(y_n) - \frac{1}{2}y_n^2 < -Ky_n^{\frac{3}{2}}.$$

Thus, if the Riemann Hypothesis is true, then we know the true order of magnitude of $\psi_1(x) - \frac{1}{2}x^2$. The situation concerning the exact order of magnitude of $\psi(x) - x$ is much more complex and the true order of magnitude is unknown, even on the assumption of the Riemann Hypothesis.

Notes to chapter 5

SOME GENERALISATIONS OF $\zeta(s)$

There are many functions which could be considered as generalisations of the Riemann zeta function. We shall briefly discuss three of them. An important class of zeta functions which will *not* be considered are the zeta functions of algebraic varieties.

1. The Dedekind zeta function. Let K be an algebraic number field. The Dedekind zeta function ζ_K of K is defined in the half plane $\Re(s) > 1$ by

$$\zeta_K(s) = \sum_{\mathfrak{a}} (N\mathfrak{a})^{-s}$$

where the summation is over all the integral ideals $\mathfrak{a}$ in the ring of integers of K and $N\mathfrak{a}$ denotes the absolute norm of the ideal $\mathfrak{a}$. (The absolute norm of $\mathfrak{a}$ is defined to be the index of $\mathfrak{a}$ in the ring of integers of K.) Because the absolute norm is a completely multiplicative function we can appeal to theorem 1.2 and write

$$\zeta_K(s) = \prod_{\mathfrak{p}} \{ 1 - (N\mathfrak{p})^{-s} \}^{-1}$$

the product being over all prime ideals.

It is clear that when $K = \mathbb{Q}$ we have $\zeta_Q(s) = \zeta(s)$ and in general the analytic behaviour of $\zeta_K(s)$ resembles that of $\zeta(s)$. It is not very difficult to show that $\zeta_K(s)$ has a simple pole at $s = 1$ with residue

$$\frac{2^{r_1}(2\pi)^{r_2}}{\omega} \frac{hR}{|d|^{1/2}}$$

where h is the class number, R the regulator and d the discriminant of K, ω denotes the number of roots of unity in K and the integers r_1, $2r_2$ are the number of real and complex embeddings of K in $\mathbb{C}$.

A considerably more difficult task is to show that $\zeta_K(s)$ is holomorphic over the complex plane, except for the pole at $s = 1$, and that $\zeta_K(s)$ satisfies a functional equation. More precisely if we define $\xi_K(s)$ for $\Re(s) > 1$ by

$$\xi_K(s) = \zeta_K(s) \left\{ \Gamma\left(\frac{s}{2}\right) \right\}^{r_1} \{ \Gamma(s) \}^{r_2} \{ |d| \pi^{-r_1} (2\pi)^{-2r_2} \}^{s/2} .$$

then $\xi_K(s)$ satisfies the functional equation

$$\xi_K(1 - s) = \xi_K(s)$$

The facts were proved by Hecke [1]; a modern treatment can be found in Cassels and Frohlich [4]. However, the ideas contained in Hecke's original paper are still of great interest.

As well as giving information about the distribution of prime ideals a knowledge of $\zeta_K(s)$ gives some information about the factorisation of rational primes in K. For an introduction to this vast topic we refer the reader to Brauer [1], [2].

2. The Hurwitz zeta function. Let α satisfy $0 < \alpha \leqslant 1$, then for $\Re(s) > 1$ the Hurwitz zeta function $\zeta(s, \alpha)$ is defined by

$$\zeta(s, \alpha) = \sum_{n=0}^{\infty} (n + \alpha)^{-s}.$$

This function was introduced by Hurwitz in his study of Dirichlet's L-functions, a topic which we shall discuss in chapters 7 and 8. The analytic properties of $\zeta(s, \alpha)$ are in many ways quite different from those of $\zeta(s)$. It is not very difficult to prove that $\zeta(s, \alpha)$ has a holomorphic continuation over the complex plane, except at $s = 1$ and that $\zeta(s, \alpha)$ satisfies the following rather awkward " functional equation " :

For $\Re(s) < 1$ one has

$$\zeta(s, \alpha) = \frac{2\,\Gamma(1 - s)}{(2\,\pi)^{1-s}} \sum_{m=1}^{\infty} \frac{\sin \pi \left(\frac{s}{2} + 2\,m\alpha\right)}{m^{1-s}}.$$

It is clear that $\zeta(s, 1) = \zeta(s)$ and $\zeta\left(s, \frac{1}{2}\right) = (2^r - 1)\,\zeta(s)$. For these two values of α the function $\zeta(s, \alpha)$ has an Euler product representation in the half plane $\Re(s) > 1$, which implies that $\zeta(s, 1)$ and $\zeta\left(s, \frac{1}{2}\right)$ have no zeros in this half plane. For all other values of α the function $\zeta(s, \alpha)$ does not have an Euler product representation. In fact, Cassels [2], completing the work of Davenport and Heilbronn [5], proved that $\zeta(s, \alpha)$ possesses an infinity of zeros in the half plane $\Re(s) > 1$ if $\alpha \neq \frac{1}{2}$ or 1.

3. The Epstein zeta function. Let us denote by $Q(x, y)$ the positive definite quadratic form $ax^2 + bxy + cy^2$ with discriminant $-d = b^2 - 4ac$. The Epstein zeta function $\zeta(s, Q)$ is defined in the half plane $\Re(s) > 1$ by

$$\zeta(s, Q) = \frac{1}{2} \sum_{m,n}{}' \{ Q(m, n) \}^{-s}$$

the summation being over all integral pairs (x, y) except $(0, 0)$. As usual $\zeta(s, Q)$ has a holomorphic continuation over the complex plane, except for a simple pole at $s = 1$, and $\zeta(s, Q)$ satisfies a functional equation.

The situation concerning the zeros of $\zeta(s, Q)$ is quite different from that of $\zeta(s)$. It is true that there are an infinity of zeros on the line $\mathfrak{R}(s) = \frac{1}{2}$, but the analogue of the Riemann Hypothesis is false. For example, Bateman and Grosswald [4] show that if $\sqrt{d}/2\, a > 7.0556$, then $\zeta(s, Q)$ has precisely one real zero in the interval $\left(\frac{1}{2}, 1\right)$. Earlier Davenport and Heilbronn [5] had considered the case when a, b, c are integers. They showed that if the class number $h(d)$ of binary quadratic forms with discriminant $-d$ satisfies $h(d) > 1$, $\zeta(s, Q)$ possesses an infinity of zeros in the half plane $\mathfrak{R}(s) > 1$. When $h(d) = 1$ the Epstein zeta function is non-zero in this half plane. On the other hand, Stark [1] proved that there exists an absolute constant K such that if $\sqrt{d}/2\, a = k$ satisfies $k > K$, then all the zeros of $\zeta(s, Q)$ in the rectangle $\{ s = \sigma + it : -1 < \sigma < 2, -2k < t < 2k \}$ are simple and, with the exception of two real zeros, all lie on the line $\mathfrak{R}(s) = \frac{1}{2}$.

HAMBURGER'S THEOREM

Exercise 5.1 to this chapter shows that if a function $f(s)$ can be represented by a Dirichlet series with monotone coefficients and by an Euler product in the region $\mathfrak{R}(s) > 1$, then for some c, $f(s) = \zeta(s - c)$. Hamburger [1] proved the deeper result that $\zeta(s)$ is characterised by its functional equation and the fact that it has an absolutely convergent Dirichlet series representation in the region $\mathfrak{R}(s) > 1$. More precisely, the situation is as follows.

Let G be an entire function of finite order, P *a polynomial and f defined by* $f(s) = G(s)/\mathrm{P}(s)$.

(i) *Suppose that for* $\mathfrak{R}(s) > 1$ *the function f can be represented by an absolutely convergent Dirichlet series.*

(ii) *In the region* $\mathfrak{R}(s) < 0$ *the function f is given by*

$$\pi^{-\frac{s}{2}} \Gamma\left(\frac{s}{2}\right) f(s) = \pi^{-\frac{1}{2}(1-s)} \Gamma\left(\frac{1-s}{2}\right) g(1-s)$$

where $g(s)$ *is defined in* $\mathfrak{R}(s) > 1$ *by an absolutely convergent Dirichlet series.*

Under the above circumstances Hamburger's theorem asserts that

$$f(s) = g(s) = \alpha . \zeta(s)$$

for some complex number α. Elegant proofs of Hamburger's theorem we later given by Siegel [1] and Hecke [1]. The latters proof was the starting point of a great deal of fascinating work on the study of the relationship between modular functions, Dirichlet series and functional equations. For an account of this we refer the reader to Hecke [1], Weil [2] and Ogg [1].

ZEROS ON THE CRITICAL LINE $\mathfrak{R}(s) = \frac{1}{2}$

The functional equation for $\zeta(s)$ shows that the complex zeros of $\zeta(s)$ are symmetric about the line $\mathfrak{R}(s) = \frac{1}{2}$ and the Riemann Hypothesis asserts that all the complex zeros actually lie on the critical line. If we denote by $\mathcal{N}(T)$ the number of zeros in the rectangle

$$\{ s = \sigma + it \mid 0 \leqslant \sigma \leqslant 1 , 0 \leqslant t \leqslant T \} ,$$

then theorem 5.7 tells us that

$$\mathcal{N}(T) = \frac{T}{2\pi} \operatorname{Log} \frac{T}{2\pi} - \frac{T}{2\pi} + \mathrm{O}(\operatorname{Log} T) .$$

Selberg [1] has shown that there exists a constant $c > 0$ such that for all sufficiently large values of T the number of zeros on the line $\mathfrak{R}(s) = \frac{1}{2}$ with $0 < \gamma < T$ is at least

$$c \frac{T}{2\pi} \operatorname{Log} \frac{T}{2\pi} .$$

This means that a positive proportion of the complex zeros of $\zeta(s)$ actually lie on the critical line.

All the known zeros of $\zeta(s)$ are simple. It is conjectured that all the zeros of $\zeta(s)$ are simple. As a first step towards proving this conjecture Montgomery [2] has shown that if the Riemann Hypothesis is true, then the proportion of simple zeros on the critical line is at least $\frac{2}{3}$.

The proofs of the theorems of Selberg and Montgomery are rather difficult. We shall content ourselves with proving the following theorem, due to Hardy.

Theorem 5.11 *The Riemann zeta function has an infinity of zeros on the critical line.*

Proof. The function $\xi(s)$ has precisely the same complex zeros as $\zeta(s)$ and theorem 5.5 tells us that $\xi\left(\frac{1}{2} + it\right)$ is a real valued function of t. For technical reasons, which will appear later, we shall show that the real valued function f, defined by

$$f(t) = \frac{\xi\left(\frac{1}{2} + it\right) e^{\frac{\pi}{4} t}}{t^2 + \frac{1}{4}}$$

has an infinity of real zeros. $\left(\text{It is clear that } f(t) \text{ and } \xi\left(\frac{1}{2} + it\right) \text{ have precisely the}\right.$ same real zeros.)

Our proof will be by contradiction. If we suppose that f has only a finite number of real zeros, then there exists T_0 such that for all $t > T_0$ the function f is of constant sign. A consequence of this observation is that for all $T > T_0$ we have

$$\left| \int_T^{2T} f(t)\, dt \right| = \int_T^{2T} | f(t) |\, dt . \tag{5.13}$$

We shall obtain our contradiction by showing that there exist constants A_1, A_2, T_1 such that for all $T > T_1$

$$\left| \int_T^{2T} f(t)\, dt \right| < A_1\, T^{1/2} \quad \text{and} \quad \int_T^{2T} | f(t) |\, dt > A_2\, T^{3/4} .$$

Thus, for *all* large T equation (5.13) does not hold.

The starting point for our study of the left hand side of (5.13) is to use either the methods of chapter 2 or the Mellin inversion theorem to " invert " the relationship

$$\Gamma(s) = \int_0^\infty e^{-u} . u^{s-1}\, du$$

and obtain the following formula, valid for $c > 0$, $\mathcal{R}(z) > 0$:

$$e^{-z} = \frac{1}{2\pi i} \int_{c-i\infty}^{c+i\infty} \Gamma(s)\, z^{-s}\, ds . \tag{5.14}$$

Upon replacing s by $s/2$ and z by $\pi n^2 z$ in (5.14) we obtain

$$e^{-\pi z n^2} = \frac{1}{2\pi i} \int_{c-i\infty}^{c+i\infty} \Gamma\left(\frac{s}{2}\right) \pi^{-s/2}\, n^{-s}\, z^{-s/2}\, \frac{ds}{2} .$$

If we choose $c > 1$ and $\mathcal{R}(z) > 0$, then sum both sides of the above equation over n we obtain, after interchanging the signs of summation and integration :

$$\begin{aligned} \omega(z) = \sum_{n=1}^{\infty} e^{-\pi z n^2} &= \frac{1}{2\pi i} \int_{c-i\infty}^{c+i\infty} \Gamma\left(\frac{s}{2}\right) \pi^{-s/2}\, \zeta(s)\, \frac{z^{-s/2}}{2}\, ds \\ &= \frac{1}{2\pi i} \int_{c-i\infty}^{c+i\infty} \frac{\xi(s)}{s(s-1)}\, z^{-s/2}\, ds . \end{aligned}$$

We now move the line of integration left to $\mathfrak{R}(s) = \frac{1}{2}$ passing the simple pole at $s = 1$ with residue $z^{-1/2}$. Thus we now have

$$\omega(z) = \frac{1}{2\pi i}\int_{1/2-i\infty}^{1/2+i\infty} \frac{\xi(s)\, z^{-s/2}}{s(s-1)}\, ds + \frac{1}{\sqrt{z}}$$

$$= \frac{z^{-1/4}}{2\pi}\int_{-\infty}^{+\infty} \frac{\xi\left(\frac{1}{2}+it\right)}{t^2+\frac{1}{4}}\, z^{-it/2}\, dt + \frac{1}{\sqrt{z}}$$

Let us now write $z = e^{i\left(\frac{\pi}{2}-\delta\right)}$, where δ will be chosen explicitly later satisfying $\delta \in \left(0, \frac{1}{2}\pi\right)$. We now have

$$\left(\omega(z) - \frac{1}{\sqrt{z}}\right) z^{1/4} = \frac{1}{2\pi}\int_{-\infty}^{+\infty} \frac{\xi\left(\frac{1}{2}+it\right)}{t^2+\frac{1}{4}}\, e^{\frac{\pi}{4}t} . e^{-\frac{\delta t}{2}}\, dt$$

$$= \frac{1}{2\pi}\int_{0}^{\infty} \frac{\xi\left(\frac{1}{2}+it\right)}{t^2+\frac{1}{4}} \left\{ e^{\left(\frac{\pi}{4}-\frac{\delta}{2}\right)t} - e^{-\left(\frac{\pi}{4}-\frac{\delta}{2}\right)t} \right\} dt$$

$$= \frac{1}{\pi}\int_0^\infty f(t)\, e^{-\frac{\pi}{4}t} \cosh\left(\frac{\pi}{4}t - \frac{\delta t}{2}\right) dt .$$

From theorem 5.1, which is easily seen to be valid for complex values of a, provided that $\mathfrak{R}(a) > 0$, we deduce that

$$| z^{1/4}(\omega(z) - z^{-1/2}) | = O(\delta^{-1/2})$$

and hence

$$\int_0^\infty f(t)\, e^{-\frac{\pi}{4}t} \cosh\left(\frac{\pi}{4}t - \frac{\delta}{2}t\right) dt = O(\delta^{-1/2}) . \qquad (5.15)$$

Let us now return to the integral on the left hand side of (5.13). Without loss of generality we can suppose that $T > T_0$ and that f is positive for all $t > T_0$. If $t \in [T, 2T]$, then there exists an absolute constant A_3 such that

$$1 \leqslant A_3\, e^{-\frac{\pi}{4}t} \cosh\left(\frac{\pi}{4}t - \frac{1}{2}\frac{t}{T}\right).$$

Hence we certainly have

$$\int_T^{2T} f(t)\,dt \leqslant A_3 \int_T^{2T} f(t)\, e^{-\frac{\pi}{4}t} \cosh\left(\frac{\pi}{4}t - \frac{1}{2}\frac{t}{T}\right) dt$$

$$\leqslant A_3 \int_T^{\infty} f(t)\, e^{-\frac{\pi}{4}t} \cosh\left(\frac{\pi}{4}t - \frac{1}{2}\frac{t}{T}\right) dt$$

$$= \mathrm{O}\left(\int_0^{\infty} f(t)\, e^{-\frac{\pi}{4}t} \cosh\left(\frac{\pi}{4}t - \frac{1}{2}\frac{t}{T}\right) dt\right)$$

$$= \mathrm{O}(\sqrt{T}),$$

on taking $\delta = T^{-1}$ in (5.15).

Now we consider the integral which occurs on the right hand side of (5.13). From the definition of $\xi(s)$ we have

$$f(t) = \frac{e^{4t}}{t^2 + \frac{1}{4}} \cdot \frac{1}{2}\left(\frac{1}{2} + it\right)\left(-\frac{1}{2} + it\right)\pi^{-1/4 - it/2}\,\Gamma\left(\frac{1}{4} + i\frac{t}{2}\right)\zeta\left(\frac{1}{2} + it\right).$$

The asymptotic formula for $\Gamma\left(\frac{1}{4} + it\right)$ now yields

$$|f(t)| \sim \frac{1}{2}\pi^{-1/4}\, t^{-1/4}\sqrt{2\pi}\left|\zeta\left(\frac{1}{2} + it\right)\right|.$$

Thus, there exists absolute constants A_4 and T_3 such that if $t > T_3$, then

$$|f(t)| > A_4\, t^{-1/4}\left|\zeta\left(\frac{1}{2} + it\right)\right|$$

and so

$$\int_T^{2T} |f(t)|\,dt > A_4\, T^{-1/4} \int_T^{2T} \left|\zeta\left(\frac{1}{2} + it\right)\right| dt.$$

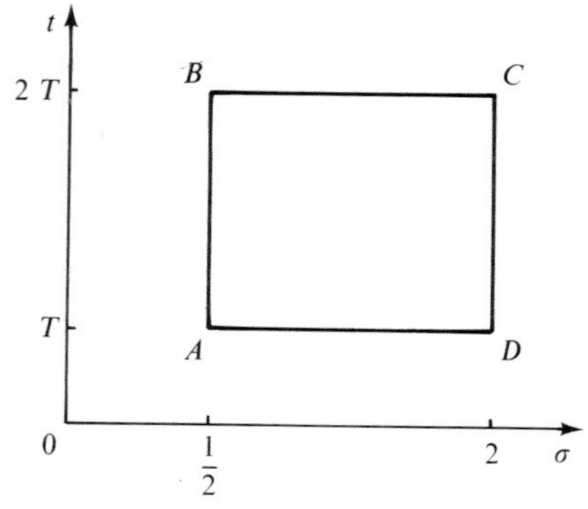

Finally we are going to deduce an asymptotic formula for this last integral.

Let $\mathcal{C}$ be the rectangular contour shown in figure 12. Since $\zeta(s)$ is holomorphic within and on $\mathcal{C}$ it follows that

$$\mathrm{O} = \int_{\mathcal{C}} \zeta(s)\,ds = \int_{1/2+iT}^{1/2+i2T} \zeta(s)\,ds + \int_{BC} \zeta(s)\,ds + \int_{CD} \zeta(s)\,ds + \int_{DA} \zeta(s)\,ds.$$

The last three integrals are very easy to estimate. First of all

$$\int_{CD} \zeta(s)\, ds = -i \int_T^{2T} \zeta(2+it)\, dt = -i \int_T^{2T} \left\{ 1 + \sum_{n=2}^{\infty} n^{-2-it} \right\} dt$$

$$= -iT + O(1).$$

By applying theorem 2.3 with $\theta = \frac{1}{2}$ we easily see that

$$\int_{AD} \zeta(s)\, dt = O(\sqrt{T}) \quad \text{and} \quad \int_{CB} \zeta(s)\, ds = O(\sqrt{T}).$$

Combining the above estimates, we have shown

$$\left| \int_T^{2T} \zeta\left(\frac{1}{2} + it\right) dt \right| = T + O(\sqrt{T})$$

which has the consequence that for all $T > T_4$

$$\int_T^{2T} |f(t)|\, dt > A_2\, T^{3/4}.$$

This completes the proof of the theorem. ■

THE DIFFERENCE Z(s) − (s − 1)$^{-1}$

During the proof of theorem 4.4 we needed an upper bound for $Z(\sigma) - (\sigma - 1)^{-1}$ when $\sigma \in \left[1, \frac{3}{2}\right]$. It was then claimed that 0 was an upper bound. We now make good the claim.

Theorem 5.12 *If* $\sigma \in \left[1, \frac{3}{2}\right]$, *then*

$$Z(\sigma) - \frac{1}{\sigma - 1} < 0.$$

Proof. We will use the following explicit formula for $Z(s)$, the proof of which may be found in Davenport [4], pages 81-84 :

If s is not a zero of $\zeta(s)$, *then*

$$Z(s) - \frac{1}{s-1} = 1 + \frac{1}{2}\gamma - \text{Log}\,(2\pi) + \frac{1}{2} \frac{\Gamma'\left(1 + \frac{s}{2}\right)}{\Gamma\left(1 + \frac{s}{2}\right)} - \sum_{\rho} \left(\frac{1}{s - \rho} + \frac{1}{\rho} \right) \tag{5.16}$$

where γ *is Euler's constant.*

First we show that if $s = \sigma \geqslant 1$, then the sum over the zeros is real and positive. If ρ is a complex zero of $\zeta(s)$, then so is $\bar{\rho}$, hence

$$\sum_\rho \left(\frac{1}{\sigma - \rho} + \frac{1}{\rho}\right) = \frac{1}{2} \sum_\rho \left(\frac{1}{\sigma - \rho} + \frac{1}{\sigma - \bar{\rho}} + \frac{1}{\rho} + \frac{1}{\bar{\rho}}\right)$$
$$= \frac{1}{2} \sum_\rho \left\{ \frac{2(\sigma - \Re(\rho))}{|\sigma - \rho|^2} + \frac{2\,\Re(\rho)}{|\rho|^2} \right\}.$$

The latter sum is clearly real and since all the zeros satisfy $0 < \Re(\rho) < 1$ the sum is positive. Thus, from (5.16) we have

$$Z(\sigma) - \frac{1}{\sigma - 1} \leqslant 1 + \frac{1}{2}\gamma - \operatorname{Log}(2\pi) + \frac{1}{2} \frac{\Gamma'\left(1 + \frac{1}{2}\sigma\right)}{\Gamma\left(1 + \frac{1}{2}\sigma\right)}.$$

The logarithmic derivative of the gamma function will now be studied. Upon taking the logarithmic derivative of (*b*) in § 2.2 and replacing s by $1 + \frac{1}{2}\sigma$ one obtains

$$\frac{1}{2} \frac{\Gamma'\left(1 + \frac{1}{2}\sigma\right)}{\Gamma\left(1 + \frac{1}{2}\sigma\right)} = -\frac{1}{2}\gamma + \sum_{n=1}^{\infty} \left(\frac{1}{2n} - \frac{1}{\sigma + 2n}\right).$$

It is clear from the above formula that for $\sigma \geqslant 1$ the function

$$\frac{\Gamma'\left(1 + \frac{1}{2}\sigma\right)}{\Gamma\left(1 + \frac{1}{2}\sigma\right)}$$

is monotone increasing. In particular for $\sigma \in \left[1, \frac{3}{2}\right]$

$$\frac{1}{2} \frac{\Gamma'\left(1 + \frac{1}{2}\sigma\right)}{\Gamma\left(1 + \frac{1}{2}\sigma\right)} \leqslant \frac{1}{2} \frac{\Gamma'(2)}{\Gamma(2)} = \frac{1}{2} - \frac{1}{2}\gamma\,.$$

which implies that for $\sigma \in \left[1, \frac{3}{2}\right]$

$$Z(\sigma) - \frac{1}{\sigma - 1} \leqslant \frac{3}{2} - \operatorname{Log}(2\pi) < 0\,.$$

■

Exercises to chapter 5

5.1 (i) Let f be a real valued monotonic function which satisfies the functional relation

$$f(mn) = f(m) + f(n).$$

for all pairs of positive integers m and n. Prove that $f(x) = c \operatorname{Log} x$ for some real number c. (See Wirsing [2] for some variations on the hypothesis, but with the same conclusion.) [Hint : Write $g(x) = \exp(f(x))$ and for $a \geqslant 3$ define R_t, S_t by

$$R_t = a^t + \cdots + a + 1, \quad S_t = a^t - \cdots - a - 1.$$

Prove that $g(R_t) \geqslant g(a)^t$ and $g(s_t) \leqslant g(a)^t$. If $n > a$ and $a^r < n \leqslant a^{r+1}$ show that

$$g(a)^{r-1} \leqslant g(n) \leqslant g(a)^{r+2}.$$

Conclude that for any integers a, b, n we have

$$g(a)^{\frac{1}{\operatorname{Log} a} - \frac{2}{\operatorname{Log} n}} \leqslant g(n)^{\frac{1}{\operatorname{Log} n}} \leqslant g(b)^{\frac{1}{\operatorname{Log} b} + \frac{2}{\operatorname{Log} n}}.$$

Now deduce that $g(n)^{\frac{1}{\operatorname{Log} n}}$ is a constant.]

(ii) Let $\{ a_n \}$ be a monotone increasing sequence of real numbers. Suppose there exists a half plane $\Re(s) > \alpha$ in which the function f defined by

$$f(s) = \sum_{n=1}^{\infty} \frac{a_n}{n^s}$$

is holomorphic and possesses an Euler product representation $f(s) = \prod_p (1 - \varepsilon(p)\, p^{-r})^{-1}$.

Prove that $f(s) = \zeta(s - c)$ for some real number c.

5.2 Prove the following results :

(i) $\displaystyle\sum_{n=2}^{\infty} (-1)^n \zeta(n) \frac{x^n}{n!} = x \operatorname{Log} x + (2\gamma - 1)x + \frac{1}{2} + \mathrm{O}\left(\frac{1}{x}\right)$,

(ii) $\displaystyle\sum_{n=2}^{\infty} (-1)^n \zeta(n) \frac{x(x-1)\cdots(x-n+1)}{n!} = x \operatorname{Log} x + (2\gamma - 1)x + \mathrm{O}\left(\frac{1}{x}\right)$.

5.3 Prove that for each positive integer n the number $\pi^{-2n} \zeta(2n)$ is rational.

5.4 For $\mathfrak{R}(s) > 0$ define the function $L(s)$ by

$$L(s) = \sum_{n=0}^{\infty} \frac{(-1)^n}{(2n+1)^s}.$$

show that $L(s)$ has an analytic continuation over the whole complex plane as an holomorphic function and satisfies the functional equation

$$L(1-s) = 2^s \pi^{-s} \sin \frac{\pi s}{2} \Gamma(s) L(s).$$

5.5 Prove the following explicit formula for $\Pi(x)$:

$$\Pi(x) = \mathrm{Li}(x) - \sum_{\gamma > 0} \{ \mathrm{Li}(x^\rho) + \mathrm{Li}(x^{1-\rho}) \} + \int_x^\infty \frac{dt}{(t^2 - 1) \operatorname{Log} t} - \operatorname{Log} 2$$

5.6 Prove that

$$\pi(x) = \sum_{r=1}^{\infty} \mu(r) \Pi(x^{1/2})$$

and so deduce an explicit formula for $\pi(x)$.

5.7 Riemann and Gauss both conjectured that $\pi(x) < \mathrm{Li}(x)$ for all $x > 3$. Use theorem 6.2 to show that this conjecture implies that the Riemann Hypothesis is true.

5.8 Write $\pi(x) - \mathrm{Li}(x) = \mathrm{P}(x)$ and $\psi(x) - x = R(x)$. Prove that

$$\mathrm{P}(x) = -\frac{\sqrt{x}}{\operatorname{Log} x} + \frac{R(x)}{\operatorname{Log} x} + \int_2^x \frac{R(u)\, du}{u \operatorname{Log}^2 u} + \mathrm{o}\left(\frac{\sqrt{x}}{\operatorname{Log} x}\right).$$

Assume the truth of the Riemann Hypothesis and prove that for all $x > x_0$

$$\int_1^x \mathrm{P}(u)\, du < 0.$$

5.9 Ramanujan claimed that

$$\pi(x) = \int_0^\infty \frac{(\operatorname{Log} x)^u \, du}{u\zeta(u+1)\,\Gamma(u+1)} + O(1) .$$

A table given by Hardy [1] shows that the above integral represents $\pi(x)$ with astonishing accuracy over the range $1 \leqslant x \leqslant 10^9$? Prove that

$$\int_0^\infty \frac{(\operatorname{Log} x)^u \, du}{u\zeta(u+1)\,\Gamma(u+1)} = \sum_{m=1}^\infty \frac{\mu(m)}{m} \operatorname{Li}(x^{1/m}) + O(1)$$

and then use theorem 6.3 to deduce that Ramanujan's claim is false.

5.10 Suppose that all the zeros of $\zeta(s)$ are simple. Prove that if x is not an integer, then

$$M(x) = \sum_\rho \frac{x^\rho}{\rho\zeta'(\rho)} - 2 + \sum_{n=1}^\infty (-1)^{n-1} \left(\frac{2\pi}{x}\right)^{2n} \frac{1}{(2n)!\, n\zeta(2n+1)}$$

5.11 Prove that if $x > 1$ is not a prime power and $s \neq 1$, ρ or $-2n$, $n = 1, 2, \ldots$, then

$$\sum_{n \leqslant x} \frac{\Lambda(n)}{n^s} = Z(s) + \frac{x^{1-s}}{1-s} + \sum_{r=1}^\infty \frac{x^{-2r-s}}{2r+s} - \sum_\rho \frac{x^{\rho-s}}{\rho - s} .$$

5.12 Assume the truth of the Riemann Hypothesis and prove that

$$\psi(x) = x + 2\sqrt{x} \sum_{\gamma > 0} \frac{\sin(\gamma \operatorname{Log} x)}{\gamma} + O(\sqrt{x})$$

where the summation is over the complex parts of the zeros of $\zeta(s)$.

5.13 Deduce the Prime Number Theorem from the explicit formula for : (i) $\psi_1(x)$, (ii) $\psi_0(x)$.

5.14 Assume that the Riemann Hypothesis is true. Use the explicit formula for $\psi_0(x)$ to show that the following limit exists :

$$\lim_{x\to\infty} \frac{1}{\operatorname{Log} x} \int_2^x \left\{\frac{\psi(t) - t}{t}\right\}^2 dt$$

and that

$$\int_2^x \left\{\frac{\psi(t) - t}{\sqrt{t}}\right\}^2 dt = O(x) ,$$

5.15 If $\psi(x) = x + O(x^{\alpha+\varepsilon})$ for every $\varepsilon > 0$, then prove that $\psi(x) = x + O(x^\alpha \operatorname{Log}^2 x)$.

5.16 For $\mathfrak{R}(s) > 1$ define the function $P(s)$ by

$$P(s) = \sum_p p^{-s}.$$

(i) Prove that for $\mathfrak{R}(s) > 1$

$$P(s) = \sum_{n=1}^{\infty} \frac{\mu(n)}{n} \operatorname{Log} \zeta(ns) .$$

(ii) Show that $P(s)$ can be analytically continued into the half plane $\mathfrak{R}(s) > 0$ and that the line $\mathfrak{R}(s) = 0$ is a natural boundary.

5.17 Let $s = \sigma + it$, suppose that $\frac{1}{2} \leqslant \sigma \leqslant 1$ and that y is an integer satisfying $\pi y \geqslant |t|$. Prove that

$$\begin{aligned}\zeta(s, \alpha) &= \sum_{n=0}^{y} (n + \alpha)^{-s} + \frac{(y + 1 + \alpha)^{1-s}}{s - 1} + \frac{1}{2}(y + 1 + \alpha)^{-s} \\ &\quad - s\int_{y+1}^{\infty} \frac{u - [u] - \frac{1}{2}}{(u + \alpha)^{s+1}}\, du \\ &= \sum_{n=0}^{y} (n + \alpha)^{-s} - \frac{y^{1-s}}{1 - s} + O(y^{-\sigma}) .\end{aligned}$$

CHAPTER 6

Irregularities in the distribution of the prime numbers

§ 1 INTRODUCTION

So far in this book we have concerned ourselves with the regularity of the sequence of Prime Numbers. In this chapter we shall study some of the irregularities of the sequence. This is a vast topic and lack of space prevents us from giving an adequate discussion of the known results. Consequently we restrict ourselves to the simplest theorems.

The chapter is divided into three independent sections, the only connecting thread being the use of theorem 6.1. In § 2 we shall consider the problems of whether or not the functions $\psi(x)-x$ and $\pi(x)-\mathrm{Li}(x)$ change sign infinitely often and how large can they become. A conjecture of Mertens, which asserts that $|M(x)| < \sqrt{x}$ for all $x > 2$ will be discussed in § 3. Finally, in § 4 we shall use theorem 6.1 to disprove a conjecture of D. Shanks on the distribution of primes in the two arithmetic progressions $\{4n \pm 1\}$.

§ 2 OSCILLATION THEOREMS

2.1 In this section we are going to look at the functions R, Q and P, which are defined by

$$R(x) = \psi(x) - x\,, \quad Q(x) = \Pi(x) - \mathrm{Li}\,(x)\,, \quad \mathrm{P}(x) = \pi(x) - \mathrm{Li}\,(x)\,.$$

and consider the following two questions :

(i) *How large can* $|R(x)|$, $|Q(x)|$ *and* $|\mathrm{P}(x)|$ *become* ?

(ii) *Do* $R(x)$, $Q(x)$ *and* $\mathrm{P}(x)$ *change sign infinitely often as* x *tends to infinity* ?

The first question is a natural one to ask. We have gone to some effort to find upper bounds for $|R(x)|$ etc. and one would like some idea whether or not our upper bounds are close to the true order of magnitude.

One of our reasons for considering the second question, aside from idle curiosity, lies in the fact that *if* $P(x) < 0$ for all $x > 2$, then, as our discussion in 2.5 will show, the Riemann Hypothesis is true. Both Gauss [1] and Riemann [1] conjectured that $P(x) < 0$ for all $x > 2$ and numerical computations, for sporadic values of x up to 10^{13}, support the conjecture. Thus, the problem is of some interest.

2.2 Our first theorem, due originally to Landau, is a result about Dirichlet integrals. However, as we shall see, the theorem has numerous applications to the study of the oscillation of number theoretic functions. Moreover, in chapter 7 we shall use it in the proof of Dirichlet's theorem on primes in arithmetic progressions.

Before stating the theorem we shall introduce some notation. Let A be a real valued function which is integrable on every finite interval $[1, X]$ and satisfies $A(x) \geqslant 0$ for all $x \geqslant 1$. Suppose that the function F is defined for all s satisfying $\mathfrak{R}(s) > a$ by

$$F(s) = \int_1^\infty \frac{A(x)}{x^{s+1}}\, dx$$

Theorem 6.1 *With the above notation, suppose that F can be analytically continued along the segment* $(b, a]$ *of the real axis. Then* :

(a) *the above integral converges for all s with* $\mathfrak{R}(s) > b$ *and so represents a holomorphic function in the half plane* $\mathfrak{R}(s) > b$,

(b) *the function F satisfies* $|F(\sigma + it)| \leqslant F(\sigma)$ *for* $\sigma > b$ *and all t.*

Proof. For any fixed $X \in [1, \infty)$ and all s we define F_X by

$$F_X(s) = \int_1^X \frac{A(x)}{x^{s+1}}\, dx .$$

and in the half plane $\mathfrak{R}(s) > a$ we define R_X by

$$R_X(s) = F(s) - F_X(s) . \tag{6.1}$$

Thus for $\mathfrak{R}(s) > a$, $R_X(s)$ has the representation

$$R_X(s) = \int_X^\infty \frac{A(x)}{x^{s+1}}\, dx = F(s) - F_X(s) .$$

and the uniform convergence of the integral on all compact subsets of the half plane $\mathfrak{R}(s) > a$ allows us to differentiate n times to obtain

$$(-1)^n R_X^{(n)}(s) = \int_X^\infty \frac{A(x)}{x^{s+1}} (\text{Log } x)^n \, dx \tag{6.2}$$

By hypothesis we can analytically continue F along the real axis from $s = \alpha > a$ to $s = \beta > b$, hence from (6.1) it follows that $R_X(s)$ can be continued along this segment and that (6.1) is valid for real s satisfying $b < s \leqslant a$. Thus for any $\alpha > a$ and $\beta > b$ there exists a finite chain of power series $\mathfrak{T}_0, \ldots, \mathfrak{T}_k$, with centres

$$\alpha = \alpha_0 > \alpha_1 > \cdots > \alpha_k = \beta$$

and such that for $1 \leqslant \nu \leqslant k$, α_ν is contained in the circle of convergence of $\mathfrak{T}_{\nu-1}$, which give the analytic continuation of $R_X(s)$.

Consider the series $\mathfrak{T}_0 : \sum_{n=0}^{\infty} a_n(\alpha_0 - s)^n$. We know that

$$a_n = (-1)^n \frac{R_X^{(n)}(\alpha_0)}{n!},$$

and so, by (6.2) it follows that $a_n \geqslant 0$ for $n \geqslant 0$. Next, consider the power series $\mathfrak{T}_1 : \sum_{n=0}^{\infty} b_n(\alpha_1 - s)^n$, where b_n is given by

$$b_n = (-1)^n \frac{R_X^{(n)}(\alpha_1)}{n!}.$$

Upon computing $R_X^{(n)}(\alpha_1)$ from the series $\mathfrak{T}_0$ and noting that $\alpha_1 < \alpha_0$ we see that $b_n \geqslant 0$ for each $n \geqslant 0$. In exactly the same way we deduce that all of the coefficients in the power series $\mathfrak{T}_2, \ldots, \mathfrak{T}_k$ are positive. An immediate consequence of this observation is

$$R_X(\beta) = R_X(\alpha_k) \geqslant 0$$

Since (6.1) is valid for $s = \beta$ the above inequality implies that

$$F_X(\beta) \leqslant F(\beta).$$

Because $A(x) \geqslant 0$ for all $x \geqslant 1$ it follows that $F_X(\beta)$ increases as X tends to infinity and since $F_X(\beta)$ is bounded above it must tend to a finite limit. Thus the integral

$$\int_1^\infty \frac{A(x)}{x^{\beta+1}} \, dx$$

exists. Again using the fact that $A(x) \geqslant 0$ for all $x \geqslant 1$ we can conclude that the integral

$$\int_1^\infty \frac{A(x)}{x^{s+1}}\,dx$$

is absolutely uniformly convergent in the half plane $\Re(s) > \beta$ for any β satisfying $b < \beta \leqslant a$. Parts (*a*) and (*b*) of the theorem now follow. ■

Remark. The above theorem implies that if the half plane of convergence of the Dirichlet series

$$F(s) = \sum_{n=1}^{\infty} \frac{a_n}{n^s}$$

is $\Re(s) > a$ and the coefficients a_n are all real and ultimately all of the same sign, then F has a singularity at $s = a$.

2.3 Before giving applications of theorem 6.1 we shall introduce a very convenient notation. Let f be a real valued function of a real variable and g a strictly positive function of a real variable. We shall write

$$f(x) = \Omega_+ \{ g(x) \} \quad \text{and mean} \quad \limsup_{x\to\infty} \frac{f(x)}{g(x)} > 0\,,$$

$$f(x) = \Omega_- \{ g(x) \} \quad \text{and mean} \quad \liminf_{x\to\infty} \frac{f(x)}{g(x)} < 0\,,$$

$$f(x) = \Omega \{ g(x) \} \quad \text{and mean} \quad \limsup_{x\to\infty} \frac{|f(x)|}{g(x)} > 0\,,$$

$$f(x) = \Omega_\pm \{ g(x) \} \quad \text{and mean} \quad f(x) = \Omega_+ \{ g(x) \} \;\textit{et}\; f(x) = \Omega_- \{ g(x) \}\,.$$

2.4 The following notation, which was introduced in chapter 5, will be used in the statement of theorem 6.2 :

$$\Theta = \sup_\rho \beta\,,$$

the supremum being taken over all the complex zeros $\rho = \beta + i\gamma$ of $\zeta(s)$. Thus we have $\frac{1}{2} \leqslant \Theta \leqslant 1$.

Theorem 6.2 (a) *For any fixed* δ *satisfying* $0 < \delta < \Theta$ *we have*

$$\psi(x) - x = \Omega_\pm(x^{\Theta-\delta})\,, \quad \text{and} \quad \Pi(x) - \mathrm{Li}\,(x) = \Omega_\pm(x^{\Theta-\delta})\,.$$

(b) *If* $\zeta(s)$ *has a zero on the line* $\Re(s) = \Theta$, *then*

$$\psi(x) - x = \Omega_{\pm}(x^{\Theta}), \qquad \text{and} \qquad \Pi(x) - \mathrm{Li}\,(x) = \Omega_{\pm}\left(\frac{x^{\Theta}}{\operatorname{Log} x}\right).$$

Proof. Parts (*a*) and (*b*) can be handled simultaneously if we choose a fixed zero $\rho = \beta + i\gamma$ of $\zeta(s)$ as follows. In case (*a*) choose ρ so that $\gamma > 0$, $\beta > \Theta - \delta$ and β is maximal for our choice of γ. In case (*b*) we choose ρ on the line $\Re(s) = \Theta$ with $\gamma > 0$.

2.5 We shall first prove the assertions about $\psi(x)$. Denote by Λ and λ the following quantities :

$$\Lambda = \limsup_{x \to \infty} \frac{\psi(x) - x}{x^{\beta}},$$

$$\lambda = \liminf_{x \to \infty} \frac{\psi(x) - x}{x^{\beta}}.$$

We will show that

$$\text{(i) } \Lambda \geqslant \frac{m}{|\rho|} \quad \text{and} \quad \text{(ii) } \lambda \leqslant -\frac{m}{|\rho|},$$

where m is the multiplicity of the zero ρ.

Suppose that (i) is *false*, then there exist constants B and C such that

$$\Lambda < B < m\,|\rho|^{-1}$$

and, for all $x \geqslant 1$

$$A(x) = -\psi(x) + x + Bx^{\beta} + \mathrm{C} \geqslant 0\,.$$

Denoting this last expression by $A(x)$ and recalling the integral representation for $Z(s)$ we see that for $\Re(s) > 1$

$$F(s) = \int_1^{\infty} \frac{A(x)}{x^{s+1}}\,dx = -\frac{Z(s)}{s} + \frac{1}{s-1} + \frac{B}{s-\beta} + \frac{\mathrm{C}}{s}. \tag{6.3}$$

Since the poles of $-Z(s)/s$ and $(s-1)^{-1}$ at $s = 1$ cancel, all the conditions of theorem 6.1 are satisfied with $a = 1$ and $b = \beta > 0$. Hence, for all s satisfying $\Re(s) > \beta$ we have

$$|F(\sigma + it)| \leqslant F(\sigma)\,. \tag{6.4}$$

Because $\rho = \beta + i\gamma$ is a zero of order m of $\zeta(s)$, it follows from (6.3) that as $\sigma \to \beta^+$

$$| F(\sigma + i\gamma) | \sim \left| \frac{Z(\sigma + i\gamma)}{\rho} \right| \sim \frac{m}{| \rho | (\sigma - \beta)}$$

and

$$F(\sigma) \sim \frac{B}{\sigma - \beta}$$

Consequently, from (6.4) we conclude that

$$\frac{m}{| \rho |} \leqslant B;$$

which is a contradiction. Hence we must have $\Lambda \geqslant m | \rho |^{-1} > 0$. The fact that $\lambda \leqslant - m | \rho |^{-1}$ is proved in the same manner.

2.6 The proof of the two assertions about $\Pi(x)$ is basically the same as the above. We choose the zero ρ exactly as before. The quantities Λ and λ are defined by

$$\Lambda = \limsup_{x \to \infty} \frac{\Pi(x) - \text{Li}\,(x)}{x^{\beta}/\text{Log}\, x}$$

and

$$\lambda = \liminf_{x \to \infty} \frac{\Pi(x) - \text{Li}\,(x)}{x^{\beta}/\text{Log}\, x}.$$

We shall now prove that

(iii) $\Lambda \geqslant m | \rho |^{-1}$ and (iv) $\lambda \leqslant - m | \rho |^{-1}$.

Suppose that $\Lambda < m | \rho |^{-1}$, then there exist constants B and C such that

$$\Lambda < B < m | \rho |^{-1} \tag{6.3}$$

and, for all $x \geqslant 2$,

$$A(x) = -\Pi(x) + \text{Li}\,(x) + B\frac{x^{\beta}}{\text{Log}\, x} + \text{C} \geqslant 0$$

It now follows that

$$F(s) = \int_2^{\infty} \frac{A(x)}{x^{s+1}}\, dx = \int_2^{\infty} \frac{\Pi(x)}{x^{s+1}}\, dx + s\int_2^{\infty} \frac{\text{Li}\,(x)}{x^{s+1}}\, dx + $$

$$+ \int_2^{\infty} \frac{Bx^{\beta - 1 - s}}{\text{Log}\, x}\, dx + \int_2^{\infty} \frac{\text{C}\, dx}{x^{s+1}}.$$

For $\mathfrak{R}(s) > 1$ we know from chapter 1 that

$$\int_2^\infty \frac{\Pi(x)}{x^{s+1}}\, dx = \frac{1}{s} \operatorname{Log} \zeta(s)$$

By a partial integration and the change of variable $y = (s - 1) \operatorname{Log} x$ we have

$$s \int_2^\infty \frac{\operatorname{Li}(x)}{x^{s+1}}\, dx = \frac{\operatorname{Li}(2)}{2^s} + \int_2^\infty \frac{dx}{x^s \operatorname{Log} x}$$

$$= \frac{\operatorname{Li}(2)}{2^s} + \int_{(s-1)\log 2}^\infty e^{-u} \frac{du}{u}$$

$$= \int_{(s-1)\log 2}^1 \frac{du}{u} + \int_{(s-1)\log 2}^1 \frac{e^{-u} - 1}{u}\, du + \int_1^\infty \frac{e^{-u}}{u}\, du + \frac{\operatorname{Li}(2)}{2^s}$$

$$= -\operatorname{Log}(s - 1) + g(s)$$

where $g_1(s)$ is an entire function. By using an argument similar to the above we see that

$$\int_2^\infty \frac{x^\beta}{\operatorname{Log} x} \frac{dx}{x^{s+1}} = -\operatorname{Log}(s - \beta) + g_1(s)$$

where $g_2(s)$ is an entire function. Thus, for $\mathfrak{R}(s) > 1$ we have shown that

$$F(s) = -\frac{1}{s} \operatorname{Log} \{ (s - 1)\, \zeta(s) \} - B \operatorname{Log}(s - \beta) + h(s)$$

where $g(s)$ is an entire function. It is clear from (2.3), that $\zeta(s) < 0$ on the segment $[0, 1)$, thus $F(s)$ can be analytically continued as a holomorphic function along the segment $s > \beta$ of the real axis. Hence theorem 6.1 is applicable with $a = 1$ and $b = \beta$. We can now conclude that for $\sigma > \beta$ and any real t

$$| F(\sigma + i\gamma) | \leqslant F(\sigma)$$

Because $\rho = \beta + i\gamma$ is a zero of order m of $\zeta(s)$ it follows that as $\sigma \to \beta^+$

$$| F(\sigma + i\gamma) | \sim \left| \frac{1}{\rho} \operatorname{Log} \zeta(\sigma + i\gamma) \right| \sim \frac{m}{|\rho|} | \operatorname{Log}(\sigma - \beta) | .$$

and trivially as $\sigma \to \beta^+$

$$F(\sigma) \sim -B \operatorname{Log}(\sigma - \beta)$$

Consequently we must have

$$\frac{m}{|\rho|} \leqslant B$$

which is a contradiction. Hence we must have $\Lambda \geqslant m|\rho|^{-1} > 0$. The fact that $\lambda \leqslant -m|\rho|$ is proved in exactly the same manner. ■

2.7 It is an immediate consequence of theorem 6.2 that the functions $R(x)$ and $Q(x)$ change sign infinitely often as x tends to infinity. However, we cannot use theorem 6.2 to decide whether or not $\mathrm{P}(x)$ changes sign infinitely often. The reason is as follows. If we write $M = [(\mathrm{Log}\, x)/(\mathrm{Log}\, 2)]$, then we have, by § 4 of chapter 1,

$$\begin{aligned} Q(x) - \mathrm{P}(x) &= \sum_{m=2}^{\infty} \frac{1}{m} \pi(x^{1/m}) \\ &= \frac{1}{2}\pi(\sqrt{x}) + \mathrm{O}\left\{ \left[\frac{\mathrm{Log}\, x}{\mathrm{Log}\, 2}\right] x^{1/3} \right\} \\ &\sim \frac{\sqrt{x}}{\mathrm{Log}\, x} \end{aligned}$$

and consequently

$$\mathrm{P}(x) = \frac{\sqrt{x}}{\mathrm{Log}\, x} \left\{ -1 + Q(x)\frac{\mathrm{Log}\, x}{\sqrt{x}} + \mathrm{o}(1) \right\}. \tag{6.5}$$

Now if $\Theta > \frac{1}{2}$, then theorem 6.2, with $\delta < \Theta - \frac{1}{2}$, implies

$$\liminf_{x \to \infty} Q(x)\frac{\mathrm{Log}\, x}{\sqrt{x}} = -\infty \quad \text{and} \quad \limsup_{x \to \infty} Q(x)\frac{\mathrm{Log}\, x}{\sqrt{x}} = +\infty. \tag{6.6}$$

This latter result when combined with (6.5) implies that $\mathrm{P}(x)$ changes sign infinitely often. But if $\Theta = \frac{1}{2}$, that is if the Riemann Hypothesis is true, then theorem 6.2 only implies that

$$\liminf_{x \to \infty} \frac{Q(x)}{\sqrt{x}/\mathrm{Log}\, x} = -\frac{1}{|\rho|} \quad \text{and} \quad \limsup_{x \to \infty} \frac{Q(x)}{\sqrt{x}/\mathrm{Log}\, x} = +\frac{1}{|\rho|}$$

where ρ is any zero of $\zeta(s)$. Numerical computations show that the largest value of $m/|\rho|$ is 0.07... and so we can deduce nothing about the oscillatory behaviour of $\mathrm{P}(x)$ from (6.5).

2.8 In order to settle the question of the variation of the sign of $P(x)$ we shall prove the following theorem of Littlewood, which implies that (6.6) is true regardless of whether or not the Riemann Hypothesis is true.

Theorem 6.3 *When x tends to infinity we have*

$$\psi(x) - x = \Omega_\pm\{\sqrt{x}\,\mathrm{Log\,Log\,Log}\,x\},$$
$$\theta(x) - x = \Omega_\pm\{\sqrt{x}\,\mathrm{Log\,Log\,Log}\,x\},$$

and

$$\Pi(x) - \mathrm{Li}(x) = \Omega_\pm\left\{\frac{\sqrt{x}}{\mathrm{Log}\,x}\,\mathrm{Log\,Log\,Log}\,x\right\},$$

$$\pi(x) - \mathrm{Li}(x) = \Omega_\pm\left\{\frac{\sqrt{x}}{\mathrm{Log}\,x}\,\mathrm{Log\,Log\,Log}\,x\right\}.$$

Proof. If $\Theta > \frac{1}{2}$, the theorem is a trivial consequence of theorem 6.2. Consequently we shall suppose from now on that $\Theta = \frac{1}{2}$, that is, we shall assume that the Riemann Hypothesis is true.

From § 4 of chapter 1 we know that

$$\psi(x) - \theta(x) = O(\sqrt{x}) \quad \text{et} \quad \Pi(x) - \pi(x) = O\left(\frac{\sqrt{x}}{\mathrm{Log}\,x}\right).$$

Thus it will suffice to prove the first assertion from each of the above pairs.

Our first step is to prove the implication :

$$\psi(x) - x = \Omega_\pm(\sqrt{x}\,\mathrm{Log\,Log\,Log}\,x)$$
$$\Rightarrow \Pi(x) - \mathrm{Li}(x) = \Omega_\pm\left(\frac{\sqrt{x}}{\mathrm{Log}\,x}\,\mathrm{Log\,Log\,Log}\,x\right),$$

then we shall prove the first assertion.

2.9 By the definition of $\Pi(x)$ and the Abel summation formula we have

$$\Pi(x) = \sum_{n\leqslant x}\frac{\Lambda(n)}{\mathrm{Log}\,n} = \frac{\psi(x)}{\mathrm{Log}\,x} + \int_2^x \frac{\psi(u)}{u(\mathrm{Log}\,u)^2}\,du$$
$$= \frac{x}{\mathrm{Log}\,x} + \frac{R(x)}{\mathrm{Log}\,x} + \int_2^x \frac{du}{(\mathrm{Log}\,u)^2} + \int_2^x \frac{R(u)}{u(\mathrm{Log}\,u)^2}\,du$$
$$= \mathrm{Li}(x) + \frac{R(x)}{\mathrm{Log}\,x} + \int_2^x \frac{R(u)}{u(\mathrm{Log}\,u)^2}\,du + \frac{2}{\mathrm{Log}\,2}$$

d'où

$$Q(x) = \Pi(x) - \mathrm{Li}\,(x) = \frac{R(x)}{\mathrm{Log}\, x} + \int_2^x \frac{R(u)}{u(\mathrm{Log}\, u)^2}\, du + \frac{2}{\mathrm{Log}\, 2}\,.$$

which implies

$$Q(x) = \frac{R(x)}{\mathrm{Log}\, x} + \int_2^x \frac{R(u)}{u(\mathrm{Log}\, u)^2}\, du + \mathrm{O}(1)\,.$$

Upon integrating this last expression by parts we obtain

$$Q(x) = \frac{R(x)}{\mathrm{Log}\, x} + \frac{R_1(x)}{x(\mathrm{Log}\, x)^2} - \int_2^x R_1(u) \frac{d}{du}\left\{\frac{1}{u(\mathrm{Log}\, u)^2}\right\} du + \mathrm{O}(1)\,,$$

where we have written $R_1(x)$ for $\psi_1(x) - \frac{1}{2}x^2 = \int_0^x R(u)\, du$. From theorem 5.10 we know that $R_1(x) = \mathrm{O}(x^{\frac{3}{2}})$. Using this result in the above expression and then integrating by parts we find

$$\begin{aligned} Q(x) - \frac{R(x)}{\mathrm{Log}\, x} &= \mathrm{O}\left\{\frac{\sqrt{x}}{(\mathrm{Log}\, x)^2} + \int_2^x u^{\frac{3}{2}} \frac{d}{du}\left(\frac{1}{u(\mathrm{Log}\, u)^2}\right) du\right\} \\ &= \mathrm{O}\left\{\frac{\sqrt{x}}{(\mathrm{Log}\, x)^2} + \int_2^x u^{\frac{1}{2}} \frac{du}{u(\mathrm{Log}\, u)^2}\right\} \\ &= \mathrm{O}\left\{\frac{\sqrt{x}}{(\mathrm{Log}\, x)^2} + \frac{x^{\frac{1}{4}}}{(\mathrm{Log}\, x)^2}\int_2^x u^{-\frac{3}{4}}\, du\right\} = \mathrm{O}\left(\frac{\sqrt{x}}{\mathrm{Log}^2\, x}\right). \end{aligned}$$

Hence we certainly have

$$\frac{Q(x)\,\mathrm{Log}\, x}{\sqrt{x}\,\mathrm{Log\,Log\,Log}\, x} - \frac{R(x)}{\sqrt{x}\,\mathrm{Log\,Log\,Log}\, x} = \mathrm{O}\left(\frac{1}{\mathrm{Log}\, x\,\mathrm{Log\,Log\,Log}\, x}\right).$$

Thus, the two expressions on the left hand side of the above equation have the same upper and the same lower limits as x tends to infinity. This proves the implication.

2.10 It remains to show that

$$\psi(x) - x = \Omega_{\pm}\left\{\sqrt{x}\,\mathrm{Log\,Log\,Log}\, x\right\}.$$

The proof of this relation is rather technical. Our starting point is theorem 5.8, namely

$$\int_a^b g(x) \left\{ \psi(x) - x - Z(0) + \frac{1}{2} \operatorname{Log}(1 - x^{-2}) \right\} dx = - \sum_\rho \int_a^b g(x) \frac{x^\rho}{\rho} dx .$$

where $1 \leqslant a < b < \infty$ and $g(x)$ is any function with a continuous derivative.

We take $g(x) = x^{-\frac{3}{2}}$ and make the change of variable $x = e^u$. Next we choose $a = \exp(\omega - \eta)$ and $b = \exp(\omega + \eta)$, where ω and η will be chosen explicitly later. The above equation then becomes

$$\int_{\omega-\eta}^{\omega+\eta} e^{-u/2} \left\{ \psi(e^u) - e^u + Z(0) + \frac{1}{2} \operatorname{Log}(1 - e^{-2u}) \right\} du =$$

$$- \sum_\rho \int_{\omega-\eta}^{\omega+\eta} \frac{e^{u(\rho - \frac{1}{2})}}{\rho} du . \qquad (6.7)$$

If we now define the function F by

$$F(u) = e^{-u/2} \left\{ \psi(e^u) - e^u + Z(0) + \frac{1}{2} \operatorname{Log}(1 - e^{-2u}) \right\} .$$

then it is clear that

$$F(u) = \Omega_\pm(\operatorname{Log} \operatorname{Log} u) \quad \Leftrightarrow \quad \psi(x) - x = \Omega_\pm(\sqrt{x} \operatorname{Log} \operatorname{Log} \operatorname{Log} x) .$$

We shall proof the first relationship.

2.11 We begin the proof that $F(u) = \Omega_\pm(\operatorname{Log} \operatorname{Log} u)$ by considering the right hand side of (6.7). Upon writing $\rho = \frac{1}{2} + i\gamma$, a trivial integration yields

$$\frac{1}{2\eta} \int_{\omega-\eta}^{\omega+\eta} F(u)\, du = - \sum_\rho \frac{\sin(\gamma\eta)}{\gamma\eta} \frac{e^{i\gamma\omega}}{\rho} .$$

Now we approximate the infinite sum by a finite sum of real functions.

Let T be a real number, to be chosen later, satisfying $T > \max(e^2, \gamma_1)$, where γ_1 is the imaginary part of the first zero of $\zeta(s)$ on the line $\mathfrak{R}(s) = \frac{1}{2}$. We write the summation over ρ as

$$\sum_\rho \frac{\sin(\gamma\eta)}{\gamma\eta} \frac{e^{i\gamma\omega}}{\rho} = \sum_{|\gamma| \leqslant T} \frac{\sin(\gamma\eta)}{\gamma\eta} \frac{e^{i\gamma\omega}}{\rho} + \sum_{|\gamma| > T} \frac{\sin(\gamma\eta)}{\gamma\eta} \frac{e^{i\gamma\omega}}{\rho} .$$

It is quite trivial that

$$\sum_{|\gamma|>T} \frac{\sin(\gamma\eta)}{\gamma\eta} \frac{e^{i\gamma\omega}}{\rho} = \mathrm{O}\left\{ \sum_{|\gamma|>T} (\eta\gamma^2)^{-1} \right\}.$$

and

$$\sum_{|\gamma|\leqslant T} \frac{\sin(\gamma\eta)}{\gamma\eta} \frac{e^{i\gamma\omega}}{\rho} = \sum_{|\gamma|\leqslant T} \frac{\sin(\gamma\eta)}{\gamma\eta} \frac{e^{i\gamma\omega}}{i\gamma} + \mathrm{O}\Big(\sum_{|\gamma|\leqslant T} \gamma^{-2} \Big)$$

$$= \sum_{|\gamma|\leqslant T} \frac{\sin(\gamma\eta)}{\gamma\eta} \frac{\cos(\gamma\omega)}{i\gamma} + \sum_{|\gamma|\leqslant T} \frac{\sin(\gamma\eta)}{\gamma\eta} \frac{\sin(\gamma\omega)}{\gamma} + \mathrm{O}\Big(\sum_{|\gamma|\leqslant T} \gamma^{-2} \Big)$$

$$= 0 + \sum_{|\gamma|\leqslant T} \frac{\sin(\gamma\eta)}{\gamma\eta} \frac{\sin(\gamma\omega)}{\gamma} + \mathrm{O}\Big(\sum_{|\gamma|\leqslant T} \gamma^{-2} \Big).$$

By appealing to theorem 5.6, which estimates the sums $\sum \gamma^{-2}$ and $\sum_{|\gamma|>T} \gamma^{-2}$ we now have

$$\frac{1}{2\eta} \int_{\omega-\eta}^{\omega+\eta} F(u)\,du = -2 \sum_{0<\gamma\leqslant T} \frac{\sin(\gamma\eta)}{\gamma\eta} \frac{\sin(\gamma\omega)}{\gamma} + \mathrm{O}(1) + \mathrm{O}\left(\frac{\operatorname{Log} T}{\eta T} \right),$$

$$= -2\,\mathrm{S}(\omega) + \mathrm{O}(1) + \mathrm{O}\left(\frac{\operatorname{Log} T}{\eta T} \right) \tag{6.8}$$

2.12 In our study of the sum $\mathrm{S}(\omega)$ we shall need the following well known elementary theorem of Dirichlet :

Let $\theta_1, \ldots, \theta_N$ be N real numbers and $q > 1$ a fixed real number. Then, for every $\tau > 0$ each interval of the form $[\tau, \tau q^N]$ contains an integer $\mathrm{U} = \mathrm{U}(\tau, q)$ such that

$$\| \mathrm{U}\theta_i \| < q^{-1} \quad \textit{for} \quad 1 \leqslant i \leqslant N.$$

There are precisely $\mathcal{N}(T)$ complex zeros of $\zeta(s)$ such that $0 < \gamma \leqslant T$. We apply Dirichlet's theorem to the $\mathcal{N}(T)$ real numbers $\gamma_i/2\pi$ and take $\tau = q^{\mathcal{N}(T)}$, where q will be chosen explicitly later. Thus, there is an integer U such that

$$q^{\mathcal{N}(T)} \leqslant \mathrm{U} \leqslant q^{\mathcal{N}(T)+1},$$

$$\left\| \mathrm{U} \frac{\gamma_i}{2\pi} \right\| < \frac{1}{q} \quad \text{for} \quad 1 \leqslant i \leqslant \mathcal{N}(T).$$

It now follows that for *all* real numbers v we have

$$| \pm \sin(\mathrm{U}\gamma_n \pm v\gamma_n) - \sin(v\gamma_n) | \leqslant \frac{2\pi}{q} \quad \text{for} \quad 1 \leqslant n \leqslant \mathcal{N}(T).$$

An immediate consequence of the above observation is

$$| \pm S(U \pm v) - S(v) | \leqslant \frac{2\pi}{q} \sum_{0<\gamma\leqslant T} \gamma^{-1}.$$

By theorem 5.6 we know that $\sum_{0<\gamma\leqslant T} \gamma^{-1} = O(\mathrm{Log}^2 T)$, hence

$$| \pm S(U \pm v) - S(v) | = O\left(\frac{\mathrm{Log}^2 T}{q}\right). \tag{6.9}$$

2.13 We are now going to make specific choices of ω in (6.8) and then use the above estimate for $S(\omega)$ to obtain an upper bound for the integral in (6.8).

Suppose that η satisfies $0 < \eta < \frac{1}{2}$. Successively choose ω in (6.8) to be $\omega = U + 2\eta$ and $\omega = 2\eta$. Simple changes of variables, a subtraction and use of (6.9) then yields

$$\frac{1}{2\eta} \int_{-\eta}^{+\eta} \{ \pm F(U \pm 2\eta + y) - F(2\eta + y) \} \, dy = O\left(\frac{\mathrm{Log}^2 T}{q}\right) + $$

$$+ O(1) + O\left(\frac{\mathrm{Log}\, T}{T\eta}\right).$$

Now we choose η and q as explicit functions of T, namely

$$\eta = \frac{\mathrm{Log}\, T}{T} \quad \text{and} \quad q = [\mathrm{Log}^2 T].$$

Since $T \geqslant \max(e^2, \gamma_1)$ we have $0 < \eta < \frac{1}{2}$ and $q \geqslant 4$, hence

$$\frac{1}{2\eta} \int_{-\eta}^{+\eta} + F(U \pm 2\eta + y) = \frac{1}{2\eta} \int_{-\eta}^{+\eta} F(2\eta + y) \, dy + O(1). \tag{6.10}$$

From the definition of F we see that as $\eta \to 0$ the following relations hold :

$$\begin{aligned} F(2\eta + y) &= \frac{1}{2} \mathrm{Log}\,(1 - e^{-4\eta - 2y}) + O(1) \\ &= \frac{1}{2} \mathrm{Log}\,(4\eta + 2y) + O(1) \\ &= \frac{1}{2} \mathrm{Log}\, \eta + O(1). \end{aligned}$$

Consequently as $T \to \infty$ we have

$$\frac{1}{2\eta}\int_{-\eta}^{+\eta} \pm F(\mathrm{U} \pm 2\eta + y)\, dy \leqslant \frac{1}{2}\operatorname{Log}\eta + A_2$$

for a suitable absolute constant A_1. Combining the above inequality with (6.10) we deduce that

$$\frac{1}{2\eta}\int_{-\eta}^{+\eta} \pm F(\mathrm{U} \pm 2\eta + y)\, dy \leqslant -\frac{1}{2}\operatorname{Log}\left(\frac{1}{\eta}\right) + A_2 \tag{6.11}$$

for a suitable absolute constant A_2.

2.14 Upon taking the + sign in the above inequality it follows that for some $u_0 \in [\mathrm{U} + \eta, \mathrm{U} + 3\eta]$ we have

$$F(u_0) \leqslant -\frac{1}{2}\operatorname{Log}\left(\frac{1}{\eta}\right) + A_2 . \tag{6.12}$$

Recall that

$$\mathrm{U} + \eta \leqslant u_0 \leqslant \mathrm{U} + 3\eta, \qquad q^{\mathcal{N}(T)} \leqslant \mathrm{U} \leqslant q^{\mathcal{N}(T)+1}$$

and

$$\mathcal{N}(T) \sim \frac{T}{2\pi}\operatorname{Log}\left(\frac{T}{2\pi}\right).$$

Thus, from the definitions of q and η the following asymptotic relations hold :

$$\operatorname{Log}\operatorname{Log}\mathrm{U} = \operatorname{Log}\mathcal{N}(T) + \mathrm{O}(\operatorname{Log}\operatorname{Log} q) \sim \operatorname{Log} T,$$

$$\operatorname{Log}\operatorname{Log} u_0 = \operatorname{Log}\operatorname{Log}\mathrm{U} + \mathrm{O}(1) \sim \operatorname{Log} T,$$

$$\operatorname{Log}\left(\frac{1}{\eta}\right) \sim \operatorname{Log} T.$$

Hence we have $\operatorname{Log}\operatorname{Log} u_0 \sim \operatorname{Log}\left(\frac{1}{\eta}\right)$, from which we can conclude, using (6.12), that

$$\liminf_{u\to\infty} \frac{F(u)}{\operatorname{Log}\operatorname{Log} u} \leqslant -\frac{1}{2}.$$

If we now take the $-$ sign in (6.11) and repeat the above argument we obtain

$$\limsup_{u\to\infty} \frac{F(u)}{\operatorname{Log}\operatorname{Log} u} \geqslant \frac{1}{2}.$$

Thus we have shown that $F(u) = \Omega_{\pm}(\text{Log Log } u)$ and as we remarked earlier, this is equivalent to $\psi(x) - x = \Omega_{\pm}(\sqrt{x} \text{ Log Log Log } x)$. ■

2.15 We have shown that each of the error terms $R(x)$, $Q(x)$ and $P(x)$ changes sign infinitely often as x tends to infinity. However, our methods do not answer such natural questions as :

" *When does the first sign change occur* ? "

or

" *How many sign changes occur in the interval* [X, X + Y] ? "

These questions, and some of the known results will be briefly discussed in the *Notes* at the end of the chapter.

As regards the magnitudes of $R(x)$, $Q(x)$ and $P(x)$ the results which we have obtained are not completely satisfactory. For example, if the Riemann Hypothesis is true the best known " Ω " result for $\psi(x)$ is given by theorem 6.3 and the gap between this and the result given by theorem 5.10, namely $\psi(x)-x=O(\sqrt{x}\,\text{Log}^2 x)$, is quite large. Reducing this gap has been a major unsolved problem for over sixty years.

We do know slightly more about $\psi_1(x) - \frac{1}{2}x^2$. If the Riemann Hypothesis is true, then from theorem 5.10,

$$\psi_1(x) - \frac{1}{2}x^2 = O(x^{\frac{3}{2}})$$

and it is an easy exercise to use theorem 6.1 to show

$$\psi_1(x) - \frac{1}{2}x^2 = \Omega_{\pm}(x^{\frac{3}{2}}).$$

Thus, if the Riemann Hypothesis is true, then we know the exact order of magnitude of $\psi_1(x) - \frac{1}{2}x^2$.

§ 3 THE CONJECTURES OF MERTENS AND POLYA

3.1 In this section we shall discuss a notorious conjecture of Mertens about the function $M(x)$. The conjecture states that for all $x > 1$ we have

$$|M(x)| = \left|\sum_{n\leqslant x} \mu(n)\right| < \sqrt{x}.$$

Mertens [1], [2] supplied copious numerical support for the conjecture and his computations were considerably extended by Neubauer [1] but no counterexample was found.

Polya enunciated a conjecture which seems to be very closely related to the conjecture of Mertens. If $\Omega(n)$ denotes the total number of primes which divide n, that is

$$\Omega(n) = \sum_{p^\alpha \| n} \alpha .$$

then we define $L(x)$ by

$$L(x) = \sum_{n \leqslant x} (-1)^{\Omega(n)} .$$

Polya's conjecture asserts that for all $x > 1$ we have $L(x) \leqslant 0$.

3.2 There were no very compelling theoretical reasons to suppose that either of the above conjectures were true. In fact Ingham has made the counterconjecture that

$$\left.\begin{array}{ll} \liminf_{x\to\infty} \dfrac{M(x)}{\sqrt{x}} = -\infty & \limsup_{x\to\infty} \dfrac{M(x)}{\sqrt{x}} = +\infty \\ \text{and} & \\ \liminf_{x\to\infty} \dfrac{L(x)}{\sqrt{x}} = -\infty & \limsup_{x\to\infty} \dfrac{L(x)}{\sqrt{x}} = +\infty \end{array}\right\} . \tag{6.13}$$

The theoretical justification for making the above conjecture lies in the following theorem of Ingham [2].

Theorem 6.4 *The four assertions* (6.13) *are all true if one of the following hypotheses is true.*

H1 *The Riemann Hypothesis is false.*

H2 *The Riemann Hypothesis is true and* $\zeta(s)$ *has a multiple complex zero.*

H3 (a) *The Riemann Hypothesis is true, all the complex zeros* $\frac{1}{2} + i\gamma_n$ *are simple and*

(b) *No finite non-trivial linear combination* $\sum_{n=1}^{N} r_n \gamma_n$ *is equal to* 0 *or to* $\pm \gamma_m$ *with* $1 \leqslant m \leqslant N$, *where* $r_n \in \{0, +1, -1\}$.

Proof. The theorem will be proved by showing that the assumption that any one of the relations (6.13) is false, to-gether with any of the three hypotheses of the theorem, leads to a contradiction.

The Dirichlet series corresponding to the functions $M(x)$ and $L(x)$ are given for $\Re(s) > 1$ by

$$\sum_{n=1}^{\infty} \frac{\mu(n)}{n^s} = \prod_p (1 - p^{-s}) = \frac{1}{\zeta(s)} \qquad \Re(s) > 1$$

and

$$\sum_{n=1}^{\infty} \frac{\lambda(n)}{n^s} = \prod_p (1 + p^{-s})^{-1} = \frac{\zeta(2\,s)}{\zeta(s)} \qquad \Re(s) > 1$$

where we have written $\lambda(n) = (-1)^{\Omega(n)}$. We shall handle the four assertions (6.13) simultaneously by introducing the following notation. Let $C(x)$ denote one of the four functions $\mp M(x)$, $\mp L(x)$ and let $f(s)$ denote the corresponding one of $\mp \zeta(s)^{-1}$, $\mp \zeta(2\,s)/\zeta(s)$.

Suppose that Ingham's conjecture is false, that is, at least one of the assertions (6.13) is false. Then there exists a positive constant A such that for all $x \geqslant 1$ we have

$$\frac{C(x)}{\sqrt{x}} \geqslant -A \tag{6.14}$$

where $C(x)$ is the appropriate one of $\mp M(x)$, $\mp L(x)$.

We now write

$$A(x) = C(x) + A\sqrt{x}$$

and note that (6.14) implies that $A(x) \geqslant 0$ for all $x \geqslant 1$. By the Abel summation formula, we have, for $\Re(s) > 1$,

$$\frac{f(s)}{s} + \frac{A}{s - \frac{1}{2}} = \int_1^{\infty} \frac{A(x)}{x^{s+1}}\, dx\,. \tag{6.15}$$

Since $\zeta(s)$ has no zeros on the segment $\Re(s) > \frac{1}{2}$ of the real axis and $\zeta(2\,s)$ is holomorphic in $\Re(s) > \frac{1}{2}$ we see that the left hand side of (6.15) can be analytically continued along the segment $\sigma > \frac{1}{2}$ of the real axis. Thus, by theorem 6.1 we know (6.15) is valid for $\Re(s) > \frac{1}{2}$ and

$$\left| \frac{f(\sigma + it)}{\sigma + it} + \frac{A}{\sigma - \frac{1}{2} + it} \right| \leqslant \frac{f(\sigma)}{\sigma} + \frac{A}{\sigma - \frac{1}{2}} \tag{6.16}$$

for $\sigma > \frac{1}{2}$ and all real t.

H1 The validity of (6.15) for $\Re(s) > \frac{1}{2}$ implies $f(s)$ is holomorphic for $\Re(s) > \frac{1}{2}$. Since $\zeta(2s)$ has no zeros in the half plane $\Re(s) > \frac{1}{2}$ we deduce that $\zeta(s)$ cannot have any zeros in this half plane. Thus, if Ingham's conjecture is false, then the Riemann Hypothesis is true. Hence the falsity of Ingham's conjecture and the hypothesis H1 of the theorem are in contradiction.

H2 Assuming now, as we may, that the Riemann Hypothesis is true let $\frac{1}{2} + i\gamma$ be any complex zero of $\zeta(s)$. Upon taking $t = \gamma$ in (6.16) and letting $\sigma \to \frac{1}{2}^{+}$ we deduce that

$$\frac{1}{\zeta(\sigma + i\gamma)} = O\left(\frac{1}{\sigma - \frac{1}{2}}\right)$$

which implies that the zero $\frac{1}{2} + i\gamma$ is simple. This shows that the falsity of Ingham's conjecture and hypothesis H2 of the theorem are in contradiction.

H3 The proof that the falsity of Ingham's conjecture and hypothesis H3 of the theorem are in contradiction is rather subtle. The argument is in two parts. Let us denote by R_n the residue of $f(s)/s$ at $s = \frac{1}{2} + i\gamma_n$ and consider the formal series $\sum | R_n |$. The first part of the argument consists in showing that the falsity of Ingham's conjecture and H3(*b*) implies that the series $\sum | R_n |$ is convergent. The second part of the argument is to show that H3(*a*) implies that the series $\sum | R_n |$ is divergent.

Before completing the first step of our argument we shall make some simple observations which will be of crucial importance to our proof. Let $\mathfrak{Q}$ denote the class of trigonometric polynomials which are non-negative on $\mathbb{R}$, that is

$$Q \in \mathfrak{Q} \Rightarrow \begin{cases} Q(u) \geqslant 0 \quad \text{for all} \quad u \in \mathbb{R} \\ \text{and} \\ Q(u) = \displaystyle\sum_{\substack{n \\ \text{finite}}} q_n\, e^{-iu\lambda_n}, \end{cases}$$

where $q_n \in \mathbb{C}$ and $\lambda_n \in \mathbb{R}$.

The observation we wish to make is that if $Q \in \mathfrak{Q}$ and $T > 0$ is any real number, then the function Q_T defined by

$$Q_T(u) = \sum_{|\lambda_n| < T} \left(1 - \frac{\lambda_n}{T}\right) q_n\, e^{-iu\lambda_n}$$

is also a member of $\mathfrak{Q}$. The proof is simple. It is evident that Q_T is a trigonometric polynomial. To show that $Q_T(u) \geqslant 0$ for all real u we recall the well known fact :

$$\int_{-\infty}^{+\infty} \left(\frac{\sin \frac{1}{2} y}{\frac{1}{2} y} \right)^2 e^{-ixy}\, dy = \begin{cases} 1 - |x| & \text{if } |x| \leqslant 1 \\ 0 & \text{if } |x| > 1 \end{cases}$$

Consequently we have

$$Q_T(u) = \sum_n q_n\, e^{-iu\lambda_n} \int_{-\infty}^{+\infty} \left(\frac{\sin \frac{1}{2} y}{\frac{1}{2} y} \right)^2 e^{-i\lambda_n y/T}\, dy$$

$$= \int_{-\infty}^{+\infty} \left(\frac{\sin \frac{1}{2} y}{\frac{1}{2} y} \right)^2 Q\left(u + \frac{y}{T}\right) dy \geqslant 0\,.$$

Now we are ready to prove that the falsity of Ingham's conjecture and H3(b) implies that the series $\sum |R_n|$ is convergent. It will be technically simpler if we change variables in (6.15) and write $x = e^u$. The equation then becomes

$$\frac{f(s)}{s} + \frac{A}{s - \frac{1}{2}} = \int_0^{\infty} A(e^u)\, e^{-su}\, du\,;$$

and is valid for all s satisfying $\mathfrak{R}(s) > \frac{1}{2}$. For convenience we shall define the function F by

$$F(s) = \frac{f(s)}{s} + \frac{A}{s - \frac{1}{2}}\,,$$

Suppose that $Q \in \mathfrak{Q}$, say

$$Q(u) = \sum_n q_n\, e^{-iu\lambda_n}$$

then upon multiplying $Q(u)$ by $A(e^u)\, e^{-\sigma u}$ and integrating we see that for $\sigma > \frac{1}{2}$ we have

$$\int_0^{\infty} Q(u)\, A(e^u)\, e^{-\sigma u}\, du = \sum_n q_n\, F(\sigma + i\lambda_n)$$

If we now multiply the above equation by $\sigma - \frac{1}{2}$ and then let $\sigma \to \frac{1}{2}^+$ we deduce that

$$0 \leqslant \sum_n q_n R(\lambda_n)$$

where $R(\lambda)$ denotes the residue of $F(s)$ at $s = \frac{1}{2} + i\lambda$. $\left(\text{Remember that } \frac{1}{2} + i\gamma \text{ is}\right.$ either a regular point or a simple pole.) From our observations about the class $\mathfrak{Q}$ we can also conclude that for every $T > 0$

$$0 \leqslant \sum_{|\lambda_n| < T} \left(1 - \frac{\lambda_n}{T}\right) q_n R(\lambda_n) . \tag{6.17}$$

Now we are going to choose a specific member of the class $\mathfrak{Q}$. Let N be a fixed positive integer, $0 < \gamma_1 < \cdots < \gamma_N$ the imaginary parts of the first N consecutive zeros of $\zeta(s)$, T a real number such that $\gamma_N < T \leqslant \gamma_{N+1}$ and $\theta_1, \ldots, \theta_N$ real numbers to be chosen explicitly later.

Consider the function Q defined by

$$Q(u) = \prod_{n=1}^{N} 2 \cos^2 \frac{1}{2} (u\gamma_n - \theta_n)$$

Clearly we have $Q(u) \geqslant 0$ for all real u and upon noting

$$2 \cos^2 \frac{1}{2} \varphi = 1 + \frac{1}{2}(e^{-i\varphi} + e^{i\varphi})$$

we see that $Q(u) \in \mathfrak{Q}$. In particular we can write $Q(u)$ as

$$Q(u) = 1 + \sum_{n=1}^{N} \frac{1}{2} \{ e^{i(u\gamma_n - \theta_n)} + e^{i(\theta_n - u\gamma_n)} \} + \mathrm{P}(u)$$

A typical term of $\mathrm{P}(u)$ is of the shape

$$2^{-r} e^{iL(\theta'\varepsilon) - iuL(\gamma'\varepsilon)} ,$$

where $L(\lambda, \varepsilon)$ is a linear form of the type

$$L(\lambda, \varepsilon) = \sum_{n=1}^{N} \varepsilon_n \lambda_n ,$$

with $\varepsilon_n \in \{ 0, 1, -1 \}$ and at least two non-zero terms.

Hypothesis H3(b) implies that for each form $L(\gamma, \varepsilon)$ satisfying $| L(\gamma, \varepsilon) | \leqslant T$ the point $\frac{1}{2} + iL(\gamma, \varepsilon)$ is *not* a pole of $F(s)$. Hence the residues at these points are all zero. Thus, with our choice of $Q(u)$ and hypothesis H3(b) we deduce from (6.17) that

$$0 \leqslant R(0) + \frac{1}{2} \sum_{n=1}^{N} \left(1 - \frac{\gamma_n}{T}\right) \{ e^{i\theta_n} R(\gamma_n) + e^{-i\theta_n} R(-\gamma_n) \} .$$

since $R(-\gamma_n) = \overline{R(\gamma_n)}$ and $R(0) = R_0 + A$, where R_0 is the residue of $f(s)/s$ at $s = \frac{1}{2}$, we have

$$0 \leqslant R_0 + A + \sum_{n=1}^{N} \left(1 - \frac{\gamma_n}{T}\right) \Re(e^{i\theta_n} R_n)$$

Now we choose the numbers $\{ \theta_n \}$ to satisfy

$$e^{i\theta_n} R_n = - | R_n | \quad \text{for} \quad n = 1, 2, \ldots, N .$$

Thus,

$$\sum_{n=1}^{N} \left(1 - \frac{\gamma_n}{T}\right) | R_n | \leqslant R_0 + A .$$

and so

$$\sum_{0 < \gamma_n < \frac{1}{2} T} \frac{1}{2} | R_n | \leqslant R_0 + A$$

which implies the convergence of the series $\sum_{n=1}^{\infty} | R(\gamma_n) |$.

To obtain our contradiction we will now show that H3(a) implies that the series $\sum | R(\gamma_n) |$ does *not* converge. The simplest way of proving this fact is by use of an explicit formula for the function $C_0(x)$, defined for $x > 1$ by

$$C_0(x) = \begin{cases} C(x) & \text{if } x \text{ is not an integer} \\ C(x) - \frac{1}{2} \mu(x) \left(\text{resp. } C(x) - \frac{1}{2} \lambda(x) \right) & \text{otherwise} . \end{cases}$$

The formula in question is

$$\frac{C_0(x)}{\sqrt{x}} = \frac{f(0)}{\sqrt{x}} + R_0 + 2 \sum_{n=1}^{\infty} \Re(R_n x^{i\gamma_n}) + J(x)$$

where

$$J(x) = \frac{1}{2\pi i} \int_{-1-i\infty}^{-1+i\infty} \frac{f(s)}{s} x^{s+\frac{1}{2}} \, ds .$$

The proof of this formula is very similar to the proof of theorem 5.9, so we only sketch the outline.

One first shows that

$$\frac{C_0(x)}{\sqrt{x}} = \lim_{T\to\infty} \frac{1}{2\,i\pi}\int_{2-iT}^{2+iT} \frac{f(s)}{s}\, x^{s-\frac{1}{2}}\, ds\,.$$

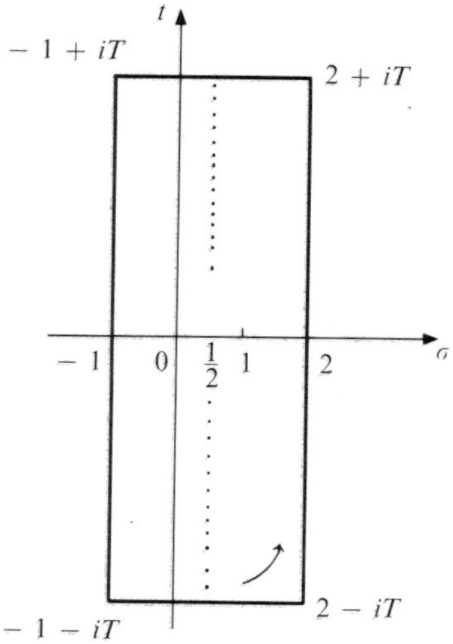

Fig. 12

In order to evaluate the integral one considers the function $f(x)\, x^{s-\frac{1}{2}}/s$ integrated around the contour $\mathcal{C}$ shown in figure 13. Hypothesis H3(a) implies that the only singularities of the integrand within $\mathcal{C}$ are simple poles at 0, the points $\left\{\frac{1}{2} + i\gamma_n\right\}$ with $|\gamma_n| < T$ and possibly at $\frac{1}{2}$.

Thus, by Cauchy's residue theorem

$$\frac{1}{2\,i\pi}\int_{\mathcal{C}} \frac{f(s)}{s}\, x^{s-\frac{1}{2}}\, ds = \frac{f(0)}{\sqrt{x}} + R_0 + \sum_{|\gamma_n|<T} R_n\, x^{i\gamma_n}\,.$$

As $T \to \infty$ one can show, using estimates for $\zeta(s)$ proved in chapter 5, that the integrals along the horizontal parts of $\mathcal{C}$ both tend to zero and so we obtain the stated explicit formula. Moreover one can show that the integral $J(x)$ is absolutely uniformly convergent and so represents a continuous function of x.

Suppose now that the series $\sum |R(\gamma_n)|$ is convergent. It implies that the series

$$\sum_{n=1}^{\infty} \mathfrak{R}(R_n\, x^{i\gamma_n})$$

is absolutely uniformly convergent and hence represents a continuous function of x. Consequently $C_0(x)$ is a continuous function. However, $C_0(x)$ has finite discontinuities whenever x is a squarefree integer. Thus we have a contradiction and so H3(a) implies that $\sum |R(\gamma_n)|$ diverges. This completes the proof of the theorem. ∎

3.3 It seems very plausible that one of the hypotheses of theorem 6.4 is true, with the consequence that both Mertens' and Polya's conjectures are false. Our proof of theorem 6.4 gives two possible ways of disproving the conjectures by numerical computation. The first way is to find an integer N such that

$$\sum_{n=1}^{N} \left(1 - \frac{\gamma_n}{\gamma_{N+1}}\right) |R_n| > R_0 + A\,,$$

where $A = 0$ for Polya's conjecture and $A = 1$ for Mertens'. Then one verifies that H3(b) is true for this choice of N. Unfortunately this does not seem to be a feasible computation to make; the value of N which one obtains is rather large. An alternative approach is to note that theorem 1 of Ingham [2] implies

$$\liminf_{x \to \infty} \frac{C(x)}{\sqrt{x}} \leqslant R_0 + A + 2 \sum_{0 < \gamma_n < T} \left(1 - \frac{\gamma_n}{T}\right) \mathfrak{R}(R_n\, x_0^{i\gamma_n}) \leqslant \limsup_{x \to \infty} \frac{C(x)}{\sqrt{x}}$$

for every x_0 and $T > 0$. By choosing x_0 and T in a sufficiently cunning manner one could hope to disprove both conjectures. In fact Haselgrove [1] and Lehman [1] have disproved Polya's conjecture in this way. They find, for example, that $L(906, 180, 359) = 1$.

No specific counterexample to Mertens conjecture is known, but Saffari [1] has reduced the problem to a finite amount of computation. A closely related conjecture of von Sterneck, namely

$$|\, M(x) \,| < \frac{1}{2}\sqrt{x} \quad \text{for all} \quad x > 200\,.$$

is known to be false. For example Neubauer [1] has shown that $M(7\,760\,000\,000) = 47\,465$ and this is greater than $\frac{1}{2}(7\,760\,000\,000)^{\frac{1}{2}} < 44\,046$.

§ 4 DISPROOF OF A CONJECTURE OF SHANKS

4.1 In this section we shall briefly discuss the distribution of prime numbers in the two arithmetic progressions $\{\, 4n + 1 \,\}$ and $\{\, 4n + 3 \,\}$. It is an elementary exercise to show that each progression contains an infinity of prime numbers. Let us define the two functions π_1 and π_3 by

$$\pi_r(x) = \text{card}\,\{\, p \leqslant x \quad \text{and} \quad p \equiv r (\text{mod } 4) \,\}\,, \qquad r = 1, 3\,.$$

Using the methods of chapter 2 it is not very difficult to show that for $r = 1, 3$

$$\pi_r(x) \sim \frac{x}{2 \operatorname{Log} x}$$

Thus, to a first approximation the primes are equally distributed between the two progressions. However, tables of prime numbers show a definite preponderance of primes of the form $4n + 3$ and one is led to suspect that, in some way, the progression $\{\, 4n + 3 \,\}$ contains " more " primes than the progression $\{\, 4n + 1 \,\}$. There are many ways of formulating this vague feeling as a precise mathematical assertion, which one can hope to prove. The most naïve way being to assert that for all $x > 2$

$$\Delta(x) = \pi_3(x) - \pi_1(x) \geqslant 0\,. \tag{6.18}$$

Tchebycheff [3] formulated two rather less simple assertions, namely :

(a) *There exists an infinite sequence of real numbers* $\{ x_r \}$ *with* $x_r \to \infty$ *such that*

$$\lim_{r\to\infty} \Delta(x_r) \frac{\operatorname{Log} x_r}{\sqrt{x_r}} = 1 .$$

(b) *Define the function* f *for* $\sigma > 0$ *by*

$$f(\sigma) = \sum_p (-1)^{\frac{p-1}{2}} e^{-\sigma p} .$$

then as $\sigma \to 0^+$ *the function* $f(\sigma)$ *tends to* $-\infty$.

It is known that (6.18) is false. For example Leech [1] observes that for $x = 26\,861$ we have $\pi_1(x) = 1\,473$ and $\pi_3(x) = 1\,472$ and consequently $\Delta(x) = -1$. In fact it is not very difficult to modify the proof of theorem 6.3 and show that

$$\Delta(x) = \Omega_{\pm}\left(\frac{\sqrt{x}}{\operatorname{Log} x} \operatorname{Log}\operatorname{Log}\operatorname{Log} x\right) \tag{6.19}$$

Thus, our naïve formulation of the empirical observation that there are " more " primes in the progression $\{ 4n + 3 \}$ than in the progression $\{ 4n + 1 \}$ is completely false.

The first of Tchebycheff's assertions is true, but really the statement is without interest, for it implies nothing about the behaviour of $\Delta(x)$ outside the special sequence $\{ x_r \}$. Indeed, after (6.19), it is an easy exercise to show that the " 1 " in Tchebycheff's assertion can be replaced by *any* real number and one still has a true statement.

Tchebycheff's second assertion is much more interesting. At the moment we do not know whether the statement is true or false. But, Hardy and Littlewood [3] show that if the function

$$L(s) = \sum_{n=0}^{\infty} \frac{(-1)^n}{(2n + 1)^s} \tag{6.20}$$

has no zeros in the half plane $\mathfrak{R}(s) > \frac{1}{2}$, then there exists $H > 0$ such that

$$\sum_p (-1)^{\frac{p-1}{2}} e^{-\sigma p} < -\frac{H}{\sqrt{\sigma} \operatorname{Log} \frac{1}{\sigma}} .$$

for all sufficiently small values of σ. That is, Tchebycheff's second assertion is a consequence of a generalisation of the Riemann Hypothesis. On the other hand, Landau [2] showed that if $L(s)$ does have a zero in the half plane $\mathfrak{R}(s) > \frac{1}{2}$, then

$$-\infty = \liminf_{\sigma\to 0^+} f(\sigma) \quad \text{and} \quad \limsup_{\sigma\to 0^+} f(\sigma) = +\infty .$$

4.2 Shanks [1], in attempting to formulate a precise version of the " many more " primes congruent to 3 modulo 4 phenomenon, considered the functions t and τ defined for all $x \geqslant 2$ by

$$t(x) = \Delta(x) \frac{\operatorname{Log} x}{\sqrt{x}} \quad \text{et} \quad \tau(x) = \Delta(x) \frac{\sqrt{x}}{\pi(x)}.$$

The function $t(x)$ arises naturally in a theoretical discussion of the problem, but the function $\tau(x)$, is for computational purposes, easier to calculate than $t(x)$. Clearly we have $\tau(x)/t(x) \to 1$ as x tends to infinity and one can hope to infer information about $t(x)$ from computations of $\tau(x)$.

Upon regarding the numerical information provided by the first two million primes with a sympathetic eye Shanks surmised that the function $\tau(x)$ has a normal distribution with mean value 1. However, this hypothesis was not subjected to any standard statistical analysis. On the basis of his calculations, together with some heuristic reasoning Shanks conjectured that

$$T(x) = \sum_{n \leqslant x} t(n) \sim x \qquad \text{and} \qquad S(x) = \sum_{n \leqslant x} \tau(n) \sim x$$

Unfortunately both conjectures are false. The proof of the falsity of the conjectures is relatively simple. It will be given in § 4.3. Afterwards we shall use theorem 6.1 to prove the following result, which tells us a little more about the behaviour of $T(x)$.

Theorem 6.5 *There exists an absolute constant* $\delta > 0$ *such that*

$$\liminf_{x \to \infty} \frac{T(x)}{x} \leqslant 1 - \delta < 1 + \delta \leqslant \limsup_{x \to \infty} \frac{T(x)}{x}.$$

A possible choice for δ *is* 0.027...

Obtaining unconditional refinements of the above theorem is a very difficult problem. It will be clear from our proof of theorem 6.5 that if the function $L(s)$ has a zero in the half plane $\mathfrak{R}(s) > \frac{1}{2}$, then

$$\liminf_{x \to \infty} \frac{T(x)}{x} = -\infty \quad \text{et} \quad \limsup_{x \to \infty} \frac{T(x)}{x} = +\infty.$$

However, if $L(s)$ is non-zero in the half plane $\mathfrak{R}(s) > \frac{1}{2}$, then one can show that there exist absolute constants γ and x_0 such that for all $x > x_0$ we have

$$\left| 1 - \frac{T(x)}{x} \right| < \gamma.$$

A possible choice for γ being $\frac{1}{5}$ (see exercise 6.9).

4.3 In order to discuss Shanks' conjectures we shall introduce some notation and state some extensions of theorems which we have already proved about $\zeta(s)$. We shall utilise the following properties of the function $L(s)$ defined for $\mathfrak{R}(s) > 0$ by (6.20) :

(i) $L(s)$ has an infinity of zeros in the half plane $\mathfrak{R}(s) \geqslant \frac{1}{2}$.

(ii) The zeros with smallest absolute value are approximately $\frac{1}{2} \pm i\, 6.020948...$

(iii) For real $s > 0$ we have $L(s) > 0$.

(iv) For s satisfying $\mathfrak{R}(s) > 1$ we have

$$L(s) = \prod_{p>2} \left\{ 1 - (-1)^{\frac{p-1}{2}} p^{-s} \right\}^{-1}.$$

The first two properties are relatively difficult to prove. One uses the methods of chapter 5 and adapts them to $L(s)$ rather than $\zeta(s)$. The proofs of (iii) and (iv) are trivial consequences of the definition of $L(s)$.

From (iv) it follows that $L(s) \neq 0$ if $\mathfrak{R}(s) > 1$. Hence, upon taking logarithms and using the Abel summation formula we obtain

$$\operatorname{Log} L(s) = \sum_{p>2} \sum_{\nu} \frac{(-1)^{\frac{1}{2}(p-1)\nu}}{\nu p^{\nu s}} \qquad \operatorname{Log} L(s) = s \int_1^{\infty} \frac{\Pi^*(x)\, dx}{x^{s+1}}$$

where the function Π^* is defined by

$$\Pi^*(x) = \sum_{\substack{p^\nu \leqslant x \\ p>2}} \frac{(-1)^{\frac{1}{2}(p-1)\nu}}{\nu}.$$

It is quite easy to see that

$$\begin{aligned} \Pi^*(x) &= \sum_{\substack{p \leqslant x \\ p \equiv 1(4)}} 1 - \sum_{\substack{p \leqslant x \\ p \equiv 3(4)}} 1 + \frac{1}{2} \sum_{p \leqslant \sqrt{x}} 1 + \sum_{\substack{p^\nu \leqslant x \\ \nu \geqslant 3}} \frac{(-1)^{\frac{1}{2}(p-1)^\nu}}{\nu} \\ &= -\Delta(x) + \frac{1}{2}\pi(\sqrt{x}) + O(x^{1/3}) \\ &= -\Delta(x) + \frac{\sqrt{x}}{\operatorname{Log} x}\{1 + o(1)\} = o\, \frac{x}{\operatorname{Log} x}, \end{aligned}$$

and consequently

$$\Pi^*(x) \frac{\operatorname{Log} x}{\sqrt{x}} = -t(x) + 1 + o(1).$$

Upon defining the functions A and A_1 for all $u \geqslant 1$ by

$$A(u) = \Pi^*(u)\frac{\operatorname{Log} u}{\sqrt{u}} \quad \text{and} \quad A_1(u) = \int_1^u A(v)\,dv,$$

we have

$$\begin{aligned} A_1(x) + \mathrm{o}(x) &= \sum_{n \leqslant x} A(n) = \sum_{n \leqslant x} \{ - t(n) + 1 + \mathrm{o}(1) \} \\ &= - T(x) + x + \mathrm{o}(x) . \end{aligned} \tag{6.21}$$

Thus the first of Shanks' conjectures is *equivalent* to the assertion that $A_1(x) = \mathrm{o}(x)$. It is this latter assertion which we shall disprove.

4.4 Disproof of Shanks' conjectures. Let us define the function f in the half plane $\Re(s) > 1$ by

$$f(s) = -\frac{d}{ds}\left(\frac{\operatorname{Log} L(s)}{s}\right) = -\frac{1}{s}\frac{L'(s)}{L(s)} + \frac{1}{s^2}\operatorname{Log} L(s) .$$

Then on the one hand we have

$$f(s) = -\frac{1}{s}\frac{L'(s)}{L(s)} + \frac{1}{s^2}\operatorname{Log} L(s)$$

and on the other hand

$$\begin{aligned} f(s) &= -\frac{d}{ds}\int_1^\infty \frac{\Pi^*(x)\,dx}{x^{s+1}} = \int_1^\infty \frac{\Pi^*(x)\operatorname{Log} x}{x^{s+1}}\,dx \\ &= \int_1^\infty \frac{A(x)}{x^{s+\frac{1}{2}}}\,dx = \left(s + \frac{1}{2}\right)\int_1^\infty \frac{A_1(x)}{x^{s+\frac{3}{2}}}\,dx \end{aligned}$$

We have now shown that for $\Re(s) > 1$

$$\int_1^\infty \frac{A_1(x)}{x^{s+\frac{3}{2}}}\,dx = -\frac{1}{s\left(s+\frac{1}{2}\right)}\frac{L'(s)}{L(s)} + \frac{1}{s^2\left(s+\frac{1}{2}\right)}\operatorname{Log} L(s) = \frac{f(s)}{s+\frac{1}{2}} . \tag{6.22}$$

The hypothesis $A_1(x) = \mathrm{o}(x)$ implies that the above integral converges for $\Re(s) > \frac{1}{2}$, thus providing an analytic continuation of $f(s)$ into this half plane. We see that $f(s)$ is holomorphic in the half plane $\Re(s) > \frac{1}{2}$ and consequently $L(s) \neq 0$

for $\mathcal{R}(s) > \frac{1}{2}$. Moreover (6.22) and the hypothesis $A_1(x) = o(x)$ implies that as $\sigma \to \frac{1}{2}^+$ we have, uniformly in t, the following estimate :

$$\frac{f(\sigma + it)}{\sigma + \frac{1}{2} + it} = o\left(\frac{1}{\sigma - \frac{1}{2}}\right).$$

The above relation implies that $L(s)$ has no zeros *on* the line $\mathcal{R}(s) = \frac{1}{2}$. Thus, we have shown that the first of Shanks' conjectures implies that $L(s)$ is non zero on the closed half plane $\mathcal{R}(s) \geqslant \frac{1}{2}$. This is a contradiction, hence the first of Shanks' conjectures is false.

The falsity of the second of Shanks' conjectures is quickly demonstrated. One defines the functions B and B_1 for $u \geqslant 2$ by

$$B(u) = \Pi^*(u)\frac{\sqrt{u}}{\pi(u)} \quad \text{and} \quad B_1(u) = \int_2^u B(v)\, dv$$

and it is easily seen that the assertion " $S(x) \sim x$ " is equivalent to the assertion " $B_1(x) = o(x)$ ". Now observe that

$$A_1(x) - A_1\left(\frac{x}{2}\right) = \int_{x/2}^{x} \frac{\pi(u)}{u} \operatorname{Log} u\, B(u)\, du$$

and by appealing to the Prime Number Theorem we have

$$\frac{\pi(u)}{u} \operatorname{Log} u = 1 + o(1)$$

consequently

$$A_1(x) - A_1\left(\frac{x}{2}\right) = \int_{x/2}^{x} (1 + o(1))\, B(u)\, du$$

$$= B_1(x) - B_1\left(\frac{x}{2}\right) + o\left\{ | B_1(x) | + \left| B_1\left(\frac{x}{2}\right) \right| \right\}.$$

It is now clear that the hypothesis $B_1(x) = o(x)$ implies

$$A_1(x) - A_1\left(\frac{x}{2}\right) = o(x)$$

This latter relationship trivially implies that $A_1(x) = o(x)$. Thus, we have shown that the truth of the second conjecture implies the truth of the first conjecture. Since the first conjecture is false the second conjecture must be false also.

4.5 Proof of theorem 6.5. It is now a routine matter to use theorem 6.1 and prove theorem 6.5. From (6.21) we know that

$$\limsup_{x\to\infty}\left(\frac{T(x)}{x}-1\right)=\limsup_{x\to\infty}\left(-\frac{A_1(x)}{x}\right)$$

We shall only consider lim sup $(-A_1(x)/x)$; the treatment of lim inf $(-A_1(x)/x)$ is completely analogous.

If lim sup $(-A_1(x)/x) = +\infty$ there is nothing more to prove, so we shall suppose that there exists constants K and x_0 such that for all $x > x_0$ we have

$$-A_1(x) \leqslant Kx .$$

Consequently there exists $C \geqslant 0$ such that for all $x \geqslant 1$

$$A_1(x) + Kx + C \geqslant 0 .$$

Now consider the function g, defined for $\mathcal{R}(s) > 1$ by

$$g(s) = \int_1^\infty \frac{A_1(x) + Kx + C}{x^{s+\frac{3}{2}}}\, dx .$$

It follows from (6.26) that

$$g(s) = -\frac{1}{s\left(s+\frac{1}{2}\right)}\frac{L'(s)}{L(s)} + \frac{1}{s^2\left(s+\frac{1}{2}\right)}\operatorname{Log} L(s) + \frac{K}{s-\frac{1}{2}} + \frac{C}{s+\frac{1}{2}} .$$

Now $L(s)$ is non-zero along the segment $\sigma > \frac{1}{2}$ of the real axis, hence we can continue $g(s)$ along this segment. All the conditions of theorem 6.1 are satisfied and so we can conclude that :

(i) $g(s)$ is holomorphic in the half plane $\mathcal{R}(s) > \frac{1}{2}$,

(ii) $|\, g(\sigma + it)\,| \leqslant g(\sigma)$ for $\sigma > \frac{1}{2}$ and all t.

From (i) we deduce that $L(s) \neq 0$ if $\Re(s) > \frac{1}{2}$. If $\rho_1 = \frac{1}{2} + i\gamma_1$ is a zero of $L(s)$ with multiplicity m we deduce from (ii), in the usual manner, that

$$\frac{m}{\left|\rho_1\left(\rho_1 + \frac{1}{2}\right)\right|} \leqslant K$$

Thus we have shown that

$$\limsup_{x\to\infty}\left(\frac{T(x)}{x} - 1\right) \geqslant \frac{m}{\left|\rho_1\left(\rho_1 + \frac{1}{2}\right)\right|} = \delta\,.$$

and in the same way, one can show that

$$\liminf_{x\to\infty}\left(\frac{T(x)}{x} - 1\right) \leqslant \frac{-m}{\left|\rho_1\left(\rho_1 + \frac{1}{2}\right)\right|} = -\delta\,.$$

To obtain a numerical value for δ one can take $\rho_1 = \frac{1}{2} + i\,6.020948\ldots$, which is a simple zero of $L(s)$. ∎

Notes to chapter 6

THE DIFFERENCE $\pi(x) - \mathrm{Li}(x)$

We proved, as theorem 6.3 that

$$\pi(x) - \mathrm{Li}(x) = \Omega_{\pm}\left\{\frac{\sqrt{x}}{\mathrm{Log}\, x}\,\mathrm{Log\,Log\,Log}\, x\right\}$$

which has the consequence that $\pi(x) > \mathrm{Li}(x)$ for an infinity of integral values of x. However no specific value of x is known for which the difference is positive. If we denote the first such value of x by x_0, then Rosser and Schoenfeld [3] proved that $x_0 > 10^8$ and Skewes [1], [2] has shown that

$$x_0 < \exp\exp\exp\exp(7.705)\ .$$

This enormous upper bound for x_0 was reduced by Cohen and Mayhew [1] to

$$x_0 < 10^{10529.7}$$

but this upper bound is still far beyond the reach of simple minded checking the value of $\pi(x) - \mathrm{Li}(x)$ in order to find x_0 explicitly. The problem of finding x_0 was brought (slightly) closer to being solved by Lehman [2], who proved that between 1.53×10^{1165} and $1.65 \times 10^{1165.4}$ there are more than 10^{500} consecutive integers for which $\pi(x) > \mathrm{Li}(x)$.

The analogous problem occurs in the study of primes in the ring of Gaussian integers $\mathbb{Z}[i]$. In this situation the problem is solved. If we denote by $\pi_{\sqrt{-1}}(x)$ the number of prime Gaussian integers $a + ib$ which satisfy $a \geqslant 1$, $b \geqslant 0$ and $a^2 + b^2 \leqslant x$, then since each prime $p \equiv 3 \pmod 4$ is a Gaussian prime, each prime $p \equiv 1 \pmod 4$ gives rise to two Gaussian primes and one obtains all Gaussian primes, except 2, in this way we have

$$\pi_{\sqrt{-1}}(x) = 2\,\pi_1(x) + \pi_3(\sqrt{x}) + 1\ .$$

Thus, it follows that

$$\pi_{\sqrt{-1}}(x) \sim \mathrm{Li}(x)\ .$$

and it can be shown that " on average " $\pi_{\sqrt{-1}}(x) < \mathrm{Li}(x)$. However the analogue of theorem 6.3 is true and

$$\pi_{\sqrt{-1}}(x) - \mathrm{Li}(x) = \Omega_{\pm}\left\{\frac{\sqrt{x}}{\mathrm{Log}\, x}\,\mathrm{Log\,Log\,Log}\, x\right\}.$$

Explicit values of x are known such that $\pi_{\sqrt{-1}}(x) > \mathrm{Li}(x)$. For example, Leech [1] notes that when $x_0 = 617\,537$ we have

$$\pi_{\sqrt{-1}}(x_0) = 50\,509 = \mathrm{Li}(x_0) + 19.5...$$

Regarding the *number* of sign changes of $\pi(x) - \mathrm{Li}(x)$ in the interval $[1, T]$, which we denote by $v(T)$, Knapowski [1] proved

$$v(T) \geqslant e^{-35}\,\mathrm{Log\,Log\,Log\,Log}\, T$$

provided

$$T \geqslant \exp\exp\exp\exp\exp(35).$$

An important point in connection with the difference $\pi(x) - \mathrm{Li}(x)$ which we have not discussed is the question of effectiveness. One cannot use our proof of theorem 6.3 to deduce any information about the sign changes of $\pi(x) - \mathrm{Li}(x)$. The proof can be modified to give effective results such as those mentioned above. In fact, there is a general technique from logic which allows one to replace certain types of ineffective proofs by effective proofs. Unfortunately an adequate discussion of this technique is beyond out scope. We refer the reader to Kreisel [1] where the method is used to construct an effective proof of theorem 6.3.

SIGN CHANGES OF ARITHMETIC FUNCTIONS

Our principal tool for studying the oscillatory behaviour of arithmetic functions was theorem 6.1. This result is difficient in two respects. Firstly, it does not allow as to deduce an explicit lower bound for the number of sign changes in a given interval and secondly, we can deduce nothing about the amplitude of oscillations in a given interval. By a more careful analysis it is possible to refine theorem 6.1 and obtain some answers to the above problems. For example Katai [1], [2] proves :

Let A be a real valued function satisfying $A(x) = \mathrm{O}(x^{\theta})$ and such that

$$f(s) = \int_1^{\infty} \frac{dA(x)}{x^s}$$

is absolutely convergent in the half plane $\Re(s) > \sigma_1$. If $f(s)$ has an analytic continuation into the half plane $\Re(s) > \sigma_2$ and a pole of order k at $s = \theta_1 + i\gamma$, where

$\theta_1 > \sigma_2$ and $\gamma \neq 0$, then provided certain mild regularity conditions are satisfied we have

$$\max_{T \leqslant x \leqslant T^\alpha} \frac{A(x)}{x^{\theta_1}(\operatorname{Log} x)^{k-1}} > \delta$$

and

$$\min_{T \leqslant x \leqslant T^\alpha} \frac{A(x)}{x^{\theta_1}(\operatorname{Log} x)^{k-1}} < -\delta$$

for all $T > c_1$, where δ, κ and c_1 are effective constants.

Steinig [1] extends theorem 6.1 in a different way. He studies the problem of sign changes in the error term associated with the coefficient sum of a Dirichlet series which satisfies a functional equation. In particular he obtains lower bounds for the number of sign changes in a given interval.

In § 4 we made some remarks on the difference $\pi_3(x) - \pi_1(x)$. The more general problem of studying

$$\pi(x\,;k, l_1) - \pi(x\,;k, l_2)$$

where $\pi(x, k, l)$ denotes the number of primes in the progression $\{kn + l\}$ which are less than x, belongs to " comparative prime number theory ". For an introduction to this vast topic we refer the reader to Knapowski and Turán [5], ..., [10] and, for a brief survey of oscillation problems, to Groswald [1].

Exercises to chapter 6

6.1 Prove that $\zeta(s)$ has no zeros on the line $\Re(s) = 1$ by considering the function $\zeta^2(s)$ $\zeta(s - ia)\,\zeta(s + ia)$ in conjunction with theorem 6.1

6.2 (i) Prove that

$$\psi_1(x) - \frac{1}{2}x^2 = \Omega_\pm(x^{3/2}).$$

(ii) What are the best estimates that you can obtain for

$$\liminf_{x\to\infty} \frac{\psi_1(x) - \frac{1}{2}x^2}{x^{3/2}} \quad \text{and} \quad \limsup_{x\to\infty} \frac{\psi_1(x) - \frac{1}{2}x^2}{x^{3/2}} \ ?$$

if the Riemann Hypothesis is assumed ?

6.3 Define the functions g and G as follows :

$$g(x) = \sum_{n \leqslant x} \frac{\mu(n)}{n} \quad \text{et} \quad G(x) = \sum_{m \leqslant x} g(m).$$

Let K be any fixed real number. Prove that as x tends to infinity the function $G(x) - K$ changes sign infinitely often. (Lehmer and Selberg [2] give numerical information in the case $K = 2$.)

6.4 Prove that $g(x) = \sum_{n \leqslant x} \frac{\mu(n)}{n}$ satisfies $g(x) = \Omega_\pm(x^{-\frac{1}{2}})$ as x tends to infinity.

6.5 Let k be a positive integer; define the function $\rho_k(n)$ by

$$\rho_k(n) = \begin{cases} 1 & \text{if } n \text{ is } k\text{th power free} \\ 0 & \text{otherwise}. \end{cases}$$

Prove that $\sum_{n \leqslant x} \rho_k(n) = \frac{[x]}{\zeta(k)} + E_k(x)$, where the function E_k satisfies :

(i) $E_k(x) = O(x^{\frac{1}{k}})$ and (ii) $E_k(x) = \Omega_{\pm}(x^{\frac{1}{2}k})$.

6.6 (i) Show that the Riemann Hypothesis implies that for every $\varepsilon > 0$, $M(x) = O(x^{\frac{1}{2}+\varepsilon})$. (In fact the Riemann Hypothesis implies that

$$M(x) = O(x^{\frac{1}{2}+\rho(x)})$$

where $\rho(x) = A(\text{Log Log } x)^{-1}$, see Titchmarsh [3], page 316.)

(ii) Prove unconditionally that $M(x) = \Omega_{\pm}(\sqrt{x})$.

6.7 (i) Prove the following assertion, which is a special case of a theorem of Riesz [1]. Let f be defined in the half plane $\mathfrak{R}(s) > 1$ by

$$f(s) = \sum_{n=1}^{\infty} \frac{a_n}{n^s}$$

where $\{ a_n \}$ is a sequence of real numbers such that $\sum_{n \leqslant x} a_n = O(x)$. If $f(s)$ is holomorphic at $s = 1$, then

$$\sum_{n \leqslant x} \frac{a_n}{n} = O(1)\,.$$

(ii) Prove that the following two conjectures are either both true or both false :

$$C_1 : M(x) = O(\sqrt{x})\,; \quad C_2 : \sum_{n \leqslant x} \frac{\mu(n)}{n^{1/2}} = O(1)\,.$$

6.8 Give a detailed proof that

$$\pi_3(x) - \pi_1(x) = \Omega_{\pm}\left(\frac{\sqrt{x}}{\text{Log } x} \text{ Log Log Log } x\right).$$

6.9 With the notation of 4.4, show that for $c > 1$

$$\frac{A_1(x)}{x} = \frac{1}{2\,i\pi}\int_{c-i\infty}^{c+i\infty} \frac{f\left(s + \frac{1}{2}\right)}{s + 1}\, x^s\, ds\,.$$

Assume that $L(s)$ has no zeros in the half plane $\mathcal{R}(s) > \frac{1}{2}$ and obtain an " explicit formula " for $A_1(x)$. Now prove that

$$\limsup_{x\to\infty} \left| 1 - \frac{T(x)}{x} \right| \leqslant \sum_{\rho} \frac{1}{\left|\rho\left(\rho + \frac{1}{2}\right)\right|}$$

where the summation is over the complex zeros of $L(s)$.

CHAPTER 7

Dirichlet L-functions (I)

§ 1 INTRODUCTION

In this chapter we begin to study a class of functions which are a generalisation of the Riemann zeta function. These functions, introduced by Dirichlet [1], are called L-functions and form an essential tool for the study of prime numbers in arithmetic progressions. As an application of our preliminary study of L-functions we shall prove the famous theorem of Dirichlet which asserts that if the integers a and b are coprime, then the arithmetic progression $\{ an + b \}$ contains an infinity of prime numbers.

Before defining L-functions and investigating their elementary properties in § 3 we must study, in some detail, characters of finite abelian groups. Our discussion of this topic will be directed towards explicit and constructive representation of characters.

§ 2 CHARACTERS OF FINITE ABELIAN GROUPS

2.1 In general the notion of group characters is connected with the study of topological groups. But for our needs it will suffice to study the very special case when the group is finite and abelian. We refer the reader to Feit [1] for an introduction to the important study of the characters of finite non-abelian groups. This topic is required for the study of a generalisation of L-functions due to Artin (see Cassels and Frohlich [4], Brauer [1], [2]).

From now on G will denote a finite abelian group written in multiplicative notation. A character χ of G is a homomorphism of G into the multiplicative group $\mathbb{C}^*$ of complex numbers. The set of characters of the group G, form a group Hom $(G, \mathbb{C}^*)$, which we denote by $\hat{G}$ and call the dual group of G. The identity element χ_0 of $\hat{G}$ is called the *principal character* of G. It has the property that $\chi_0(g) = 1$ for all $g \in G$. A trivial, but important, observation as as follows. Let $n = | G |$ and $g \in G$, then for any $\chi \in \hat{G}$ we have $\chi(g)$ is an nth root of unity, since

$$\{ \chi(g) \}^n = \chi(g^n) = \chi(1) = 1$$

Our first proposition shows that a character of a subgroup H of G can be extended to a character of G. This result will be used in the proof of a decomposition for finite abelian groups. It is this latter result which will enable us to give a very explicit representation of the characters of G when they are required.

Theorem 7.1. *Let H be a subgroup of G and denote the index of H in G by $(G : H)$. If χ is a character of H, then χ can be extended to a character of G in precisely $(G : H)$ ways.*

Proof. We may suppose that G can be obtained from H by the adjunction of a single element $g_1 \in G$, because we can get from H to G by a finite number of such adjunctions. If $(G : H) = m$, then $g_1^m = h_1 \in H$ and every element of G can be written in the form hg_1^r, where $h \in H$ and $0 \leqslant r < m$.

Suppose for the moment that χ_1 is a character of G which coincides with χ on H. We necessarily have

$$\chi_1(g) = \chi_1(hg_1^r) = \chi_1(h)\,\chi_1^r(g_1) = \chi(h)\,\chi_1^r(g_1)\,.$$

now clearly

$$\chi_1^m(g_1) = \chi_1(g_1^m) = \chi_1(h_1) = \chi(h_1) = \omega$$

where ω is a fixed root of unity. There are now m possible values for $\chi_1(g_1)$, namely the mth roots of ω. Let us call them $\varepsilon_1, \ldots, \varepsilon_m$. Thus we have shown that if χ_1 is an extension of χ, then χ_1 is of the form

$$\chi_1(hg_1^r) = \chi(h)\,\varepsilon_i^r\,.$$

Conversely, we can define m functions χ_i by

$$\chi_i(hg_1^r) = \chi(h)\,\varepsilon_i^r\,.$$

and we see that each χ_i is a character of G which coincides with χ on J. ■

We can now prove the decomposition theorem for finite abelian groups.

Theorem 7.2 *A finite abelian group is isomorphic to a direct product of cyclic groups.*

Proof. Let h = l.c.m. of the orders of the elements of G. If $h = p_1^{\alpha_1} \ldots p_s^{\alpha_s}$, then for $1 \leqslant i \leqslant s$ the group G contains an element g_i of order $p_i^{\alpha_i}$. Consequently the order of $g = g_1 \ldots g_s$ is exactly h. We are going to use theorem 7.1 with $H = \{ g \}$, the cyclic group generated by g, and χ the character on H defined by

$$\chi(g^r) = \exp\left(\frac{2\,i\pi}{h}\,r\right) = \omega^r$$

where $\omega = e^{2i\pi/h}$. Let χ_1 be any extension of χ to G. Because $x^h = 1$ for each $x \in G$ it follows that $\chi_1(x)$ is a power of ω. Hence $\chi_1(G)$ is the group of hth roots of unity. If K is the kernal of χ_1, then G/K has order h and $H \cap K = \{1\}$. Because H has order h we must have $G \simeq H \times K$. Thus, we see that a cyclic subgroup of maximal order is a direct factor of G. The theorem now follows by repeated application of this observation. ■

It is now possible to deduce the facts about $\hat{G}$ which will be required for our applications to prime number theory.

Theorem 7.3 (a) *A finite abelian group G has precisely $|G|$ characters.*

(b) *The group $\hat{G}$ is (non canonically) isomorphic to G.*

(c) *The following " orthogonality relations " hold :*

$$\sum_{\chi \in \hat{G}} \chi(g) = \begin{cases} |G| & \text{if } g = 1 \\ 0 & \text{otherwise} \end{cases}$$

$$\sum_{g \in G} \chi(g) = \begin{cases} |G| & \text{if } \chi = \chi_0 \\ 0 & \text{otherwise.} \end{cases}$$

Proof. (*a*) From theorem 7.2 we know that G is a direct product of cyclic groups, hence there exist $g_1, \ldots, g_s \in G$ such that each $g \in G$ can be written in the form

$$g = g_1^{r_1} g_2^{r_2} \ldots g_s^{r_s}, \tag{7.1}$$

where $0 \leqslant r_i < h_i$ for $1 \leqslant i \leqslant s$ and the integers h_i depend only on G and satisfy $h_1 \ldots h_s = |G|$.

Let χ be any character of G. Then for $1 \leqslant i \leqslant s$ we have

$$\chi(g_i)^{h_i} = \chi(g_i^{h_i}) = \chi(1) = 1 .$$

Thus we see that $\chi(g_i) = \omega_i$, an h_ith root of unity. Consequently it follows that

$$\chi(g) = \omega_1^{r_1} \omega_2^{r_2} \ldots \omega_s^{r_s}, \tag{7.2}$$

Hence there are at most $h_1 \ldots h_s = |G|$ characters of the group G.

Conversely for each of the $|G|$ possible choices of the omega's in (7.2) the corresponding function does define a character of G.

(*b*) This is an immediate consequence of (7.1) and (7.2) which can be used to construct an isomorphism between G and $\hat{G}$.

(*c*) The proofs of the two " orthogonality " relations are very similar. We shall only prove the first relation.

If $g = 1$, then $\chi(g) = 1$ for all χ and so $\sum_{\chi} \chi(g) = |G|$. Now if $g \neq 1$ there exists a character χ' such that $\chi'(g) \neq 1$. The character $\chi\chi'$ runs through all the characters of G if χ does, hence

$$\begin{aligned} S(g) &= \sum_{\chi \in \hat{G}} \chi\chi_1(g) = \sum_{\chi \in \hat{G}} \chi_1(g)\,\chi(g) \\ &= \chi_1(g) . S(g) . \end{aligned}$$

Because $\chi'(g) \neq 1$ it follows that $S(g) = 0$. ■

2.2 Characters modulo *k*. The groups which occur in elementary prime number theory arise when one looks at the integers modulo k and considers the multiplicative group of reduced residues, $G(k)$. As we shall see in § 3 and in chapter 8 the characters of $G(k)$, form an indespensible tool in the study of primes which lie in the arithmetic progressions with difference k. Consequently we shall study these characters in considerable detail.

Let χ be a character of $G(k)$. It will be very convenient for our later analytic work to abuse our notation and consider χ as defined on $\mathbb{Z}$ rather than $G(k)$. This is done as follows :

$$\chi(n) = \begin{cases} \chi(A) & \text{if } n \in A \in G(k) \\ 0 & \text{if } (n, k) > 1 . \end{cases}$$

The resulting arithmetic function is called a *character modulo k*. The useful properties of these functions are summarised in the following theorem. The proof of which is a simple consequence of theorem 7.3 and the definition of characters modulo k.

Theorem 7.4 (a) *If* $(k, n) = 1$, *then* $\chi(n)$ *is a* $\varphi(k)$*th root of unity* (φ *is Euler's function*).

(b) *A character modulo k is a completely multiplicative function.*

(c) χ *has period k, that is* $\chi(n + k) = \chi(n)$ *for all n.*

(d) *The following orthogonality relations hold* :

$$\sum_{n=1}^{k} \chi(n) = \begin{cases} \varphi(k) & \text{if } \chi = \chi_0 \\ 0 & \text{otherwise} \end{cases}$$

$$\sum_{\chi} \chi(n) = \begin{cases} \varphi(k) & \text{if } n \equiv 1 (\text{mod } k) \\ 0 & \text{otherwise.} \end{cases}$$

(e) *If* $\bar{\chi}$ *is the conjugate character to* χ (*i.e.* $\bar{\chi}(n) = \overline{\chi(n)}$) *and* χ' *is a given character, then*

$$\sum_{n=1}^{k} \chi(n)\,\chi'(n) = \begin{cases} \varphi(k) \text{ if } \bar{\chi} = \chi' \\ 0 \quad \text{otherwise} \end{cases}$$

$$\sum_{\chi} \bar{\chi}(m)\,\chi(n) = \begin{cases} \varphi(k) \text{ if } (m, k) = 1 \text{ et } m \equiv n (\text{mod } k) \\ 0 \quad \text{and} \quad m \equiv n \ (\text{mod } k) \text{ otherwise} . \end{cases}$$

2.3 The explicit representation of characters modulo *k*. From the decomposition theorem for $\hat{G}(k)$ it is not difficult to see that in order to investigate the group of characters modulo k it will suffice to consider the case when k is a prime power. The cases p^α, $p > 2$ and 2^α will be considered separately.

(a) $k = p^\alpha$, $p > 2$. Let g be a primitive root (mod p^α). If n is not a multiple of p, then n can be uniquely represented in the form

$$n \equiv g^\nu (\text{mod } p^\alpha) \quad \text{where} \quad 1 \leqslant \nu \leqslant \varphi(p^\alpha)\,.$$

We now define for each a satisfying $1 \leqslant a \leqslant \varphi(p^\alpha)$ the function χ^α by

$$\chi^a(n) = \begin{cases} \exp\left(\dfrac{2\,i\pi a\nu}{\varphi(p^\alpha)}\right) & \text{if} \quad (n, k) = 1 \\ 0 & \text{if} \quad (n, k) > 1\,. \end{cases} \tag{7.3}$$

This obviously gives the $\varphi(p^\alpha)$ characters modulo p^α.

(b) $k = 2^\alpha$. There are three separate cases to consider : $\alpha = 1$, $\alpha = 2$ and $\alpha \geqslant 3$. If $\alpha = 1$, then there is only one character modulo 2, namely the principal character. When $\alpha = 2$ there are two characters modulo 4. First we have the principal character, then the character χ_4 which is defined by

$$\chi_4(n) = \begin{cases} -1 & \text{if} \quad n \equiv -1 (\text{mod } 4) \\ +1 & \text{if} \quad n \equiv 1 (\text{mod } 4) \\ 0 & \text{if} \quad n \equiv 0 (\text{mod } 2)\,. \end{cases} \tag{7.4}$$

The case when $\alpha \geqslant 3$ is slightly more complicated since the reduced residues (mod 2^α) do not form a cyclic group. In fact the group is isomorphic to $C_2 \times C_2 C_{2^{\alpha-1}}$. More precisely if $n \equiv 1$ (mod 4), then n can be uniquely represented in the form

$$n \equiv 5^\nu (\text{mod } 2^\alpha) \quad \text{where} \quad 0 \leqslant \nu < 2^{\alpha-2}$$

and if $n \equiv -1$ (mod 4), then n can be uniquely represented in the form

$$n \equiv -5^\nu (\text{mod } 2^\alpha) \quad \text{where} \quad 0 \leqslant \nu < 2^{\alpha-2}\,.$$

Thus each odd integer n has a unique representation in the form

$$n \equiv (-1)^{\nu'}\, 5^\nu (\text{mod } 2^\alpha)\,,$$

where $\nu' \in \{ +1, 0 \}$ and $\nu \in \{ 0, 1, \ldots, 2^{\alpha-2} - 1 \}$. For each integer a satisfying $0 \leqslant a < \varphi(2^\alpha)$ we define the function χ^a by

$$\chi^a(n) = \begin{cases} \exp\left(\pi i a \nu' + \dfrac{\pi i a \nu}{2^{\alpha-3}}\right) & \text{if} \quad n \equiv 1 (\text{mod } 2) \\ 0 & \text{otherwise}\,. \end{cases} \tag{7.5}$$

This obviously gives the $\varphi(2^\alpha)$ characters modulo 2^α.

2.4 Proper and improper characters. The remainder of § 2 will not be required in this chapter. As we shall see in chapter 8 many properties of L-functions depend on the concept of proper and improper characters. It is convenient to define and study the properties of these characters now.

We say that a character χ modulo k and a character χ' modulo k' are *equivalent* if $(n, kk') = 1$ implies that $\chi(n) = \chi'(n)$. It is easy to verify that " equivalence " defines an equivalence relation on the set of characters, which we write as $\chi \sim \chi'$.

For convenience we introduce the notation χ_k to denote a character modulo k. If χ_k is a non-principal character modulo k and $\chi_k \sim \chi_{k'}$, where $k' < k$, then we say that χ_k is an *improper character.* A *proper character* is a non-principal character which is not improper. If χ is a non-principal character, then the *conducter* f of χ is defined by

$$f = \min \{ k \geqslant 2 \mid \chi_k \sim \chi \} .$$

The following theorem collects together several simple facts which will be needed in chapters 8 and 9. The proof is left as an easy exercise for the reader.

Theorem 7.5 (a) *If* $\chi_r \sim \chi_s$ *and* $(r, s) = t$, *then there exists a character* χ_t *such that* $\chi_r \sim \chi_t$.

(b) *Any non-principal character* χ_m *is equivalent to a unique proper character* χ_n *and* $n \mid m$.

(c) *Given any character* χ_m *and an integer* $n \equiv 0 \pmod m$, *there exists a character* χ_n *such that* $\chi_n \sim \chi_m$.

Under what conditions is a character modulo k a proper character ? The answer is simple. Suppose that $k = p_1^{\alpha_1} \dots p_s^{\alpha_s}$, then the character χ is proper if and only if for $1 \leqslant i \leqslant s$ the restriction of χ to the reduced set of residues modulo $p_i^{\alpha_i}$ is a proper character. Thus we need only investigate characters to prime power moduli.

If $p > 2$, then the character defined by (7.3) is proper if and only if $a \not\equiv 0 \pmod p$, for if $a = rp$ then $\chi_{p^\alpha}^a(n) = \chi_{p^{\alpha-1}}^r(n)$.

The case $p = 2$ is trivial; there are three cases to consider : $\alpha = 1$, $\alpha = 2$ and $\alpha \geqslant 3$. If $\alpha = 1$ there is only the principal character, which we have left unclassified. When $\alpha = 2$ the only non-principal character is χ_4 which is proper. Finally, if $\alpha \geqslant 3$, then the characters defined by (7.5) are proper if and only if $a \not\equiv 0 \pmod 2$.

2.5 Real proper characters. In chapter 8 proper characters which are real valued functions of n will have a special significance. It is convenient to describe them explicitly now. As usual it suffices to investigate the case of prime power moduli.

When $k = p^\alpha$ with $p > 2$ we recall that the general character modulo k is given by

$$\chi^a(n) = \begin{cases} \exp \left\{ 2\, i\pi \dfrac{av}{\varphi(p^\alpha)} \right\} & \text{if } \; p \nmid n \\ 0 & \text{if } \; p \mid n \end{cases}$$

If χ^a is a proper character, then $(a, p) = 1$. In addition if χ^a is real valued for all n, then we must have

$$av \equiv 0 \bmod \left(\frac{1}{2}(p-1)\, p^{\alpha-1}\right)$$

for all v. This implies that $a = \frac{1}{2}(p-1)\, p^{\alpha-1}$ and since $(a, p) = 1$ we necessarily have $\alpha = 1$. Thus, a character modulo p^α is real and proper if and only if $\alpha = 1$ and $a = \frac{1}{2}(p-1)$. In this case the character is

$$\chi(n) = \begin{cases} (-1)^v = \left(\dfrac{n}{p}\right) & \text{if } p \nmid n \\ 0 & \text{if } p \mid n \end{cases}$$

where $(n \mid p)$ is the Legendre symbol.

The case $k = 2^\alpha$ requires a discussion of the four cases : $\alpha = 1, 2, 3$ and $\alpha \geqslant 4$.

If $\alpha = 1$, then there is only the principal character, which we have left unclassified. When $\alpha = 2$ the only non-principal character is χ_4, defined by (7.4), which is real and proper. The case $\alpha = 3$ is a little more complicated. There are precisely three non-principal characters, which we write as

$$\chi(n) = (-1)^{\varepsilon(n)}, \quad \chi(n) = (-1)^{\omega(n)}, \quad \chi(n) = (-1)^{\varepsilon(n)+\omega(n)}$$

where ε and ω are defined as follows :

$$\varepsilon(n) \equiv \frac{n-1}{2} \pmod 2$$

and

$$\omega(n) \equiv \frac{n^2-1}{8} \pmod 2 .$$

The first character is equivalent to χ_4, hence it is improper. The remaining two characters are easily seen to be proper. Finally, when $\alpha \geqslant 4$ the " general " character modulo 2^α is

$$\chi(n) = \exp\left(\pi i a v' + \frac{\pi i a v}{2^{\alpha-3}}\right).$$

If this character is real valued for all n, then we must have $2^{\alpha-3} \mid a$, in which case the character is improper. Thus, for $\alpha \geqslant 4$ there are no real proper characters modulo 2^α.

The preceding results are collected together in the following theorem, which is stated in terms of the Jacobi symbol, defined for odd integers b and any integer a prime to b, by

$$\left(\frac{a}{b}\right) = \prod_{p^\alpha \parallel b} \left(\frac{a}{p}\right)^\alpha .$$

Theorem 7.6 *Real proper characters modulo k exist if and only if k has one of the following forms* :

(a) $k = p_1 \dots p_s$, *where* $p_i > 2$ *for* $1 \leqslant i \leqslant s$ *and the unique character is*

$$\chi(n) = \begin{cases} \left(\dfrac{n}{k}\right) & \text{if} \quad (n, k) = 1 \\ 0 & \text{if} \quad (n, k) > 1 \,. \end{cases}$$

(b) $k = 4$ *or* $k = 4\,p_1 \dots p_s$, *where* $p_i > 2$ *for* $1 \leqslant i \leqslant s$ *and the unique characters are respectively*

$$\chi(n) = \begin{cases} \left(\dfrac{-1}{n}\right)\left(\dfrac{n}{p_1\,p_2 \cdots p_s}\right), \quad \left(\text{resp. } \left(\dfrac{-1}{n}\right)\right) & \text{if} \quad (n, k) = 1 \\ 0 & \text{if} \quad (n, k) > 1 \,. \end{cases}$$

(c) $k = 8$ *or* $k = 8\,p_1 \dots p_s$, *where* $p_i > 2$ *for* $1 \leqslant i \leqslant s$. *There are two real proper characters, namely*

$$\chi(n) = \begin{cases} \left(\dfrac{2}{n}\right)\left(\dfrac{n}{p_1\,p_2 \cdots p_s}\right), \quad \left(\text{resp. } \left(\dfrac{2}{n}\right)\right) & \text{if} \quad (n, k) = 1 \\ 0 & \text{if} \quad (n, k) > 1 \end{cases}$$

and

$$\chi'(n) = \begin{cases} \left(\dfrac{-2}{n}\right)\left(\dfrac{n}{p_1\,p_2 \cdots p_s}\right), \quad \left(\text{resp. } \left(\dfrac{-2}{n}\right)\right) & \text{if} \quad (n, k) = 1 \\ 0 & \text{if} \quad (n, k) > 1 \,. \end{cases}$$

§ 3 PRIMES IN ARITHMETIC PROGRESSIONS

3.1 Dirichlet's *L*-functions. We are now in a position to define *L*-functions and to investigate some of their more elementary properties. Let χ be a character modulo k. The function $L(s, \chi)$ is defined for s satisfying $\mathfrak{R}(s) > 1$ by

$$L(s, \chi) = \sum_{n=1}^{\infty} \frac{\chi(n)}{n^s} \,.$$

In this chapter all series, both simple and double, are absolutely convergent in the half plane $\mathfrak{R}(s) > 1$. The simple series will be uniformly convergent in any fixed half plane $\mathfrak{R}(s) > 1 + \delta > 1$ and so represent holomorphic functions in the half plane $\mathfrak{R}(s) > 1$. As a consequence formal manipulations such as rearrangements, term by term differentiation and term by term integration over finite intervals etc., can be justified. Also the logarithms of functions of s will always be holomorphic in the half plane $\mathfrak{R}(s) > 1$ and tend to zero as s tends to infinity along the real axis.

Our first result is the analogue of theorem 2.1 and provides us with an analytic continuation of $L(s, \chi)$ into the half plane $\Re(s) > 0$.

Theorem 7.7 *The function $L(s, \chi)$ exists as a holomorphic function in the half plane $\Re(s) > 0$, except when χ is a principal character. In this case there is a simple pole at $s = 1$ with residue $\varphi(k)/k$.*

Proof. It will be convenient to define the functions C and E by

$$C(x) = \sum_{n \leqslant x} \chi(n)$$

$$E(\chi) = \begin{cases} 1 & \text{if} \quad \chi = \chi_0 \\ 0 & \text{otherwise}. \end{cases}$$

From theorem 7.4(d) it follows that

$$C(x) = E(\chi)\, \varphi(k) \frac{x}{k} + R(x)$$

where $|R(x)| \leqslant \varphi(k)$.

By using the Abel summation formula we deduce for $x > 1$

$$\sum_{n \leqslant x} \frac{\chi(n)}{n^s} = \frac{C(x)}{x^s} + s \int_1^x \frac{C(u)}{u^{s+1}}\, du$$

$$= E(\chi) \frac{\varphi(k)}{k} \left\{ \frac{s}{s-1} - \frac{x^{1-s}}{s-1} \right\} + \frac{R(x)}{x^s} + s \int_1^x \frac{R(u)}{u^{s+1}}\, du .$$

For a given s satisfying $\Re(s) > E(\chi)$ the right hand side of the above expression tends to a finite limit as x tends to infinity. Hence the left hand side tends to the finite limit

$$E(\chi) \frac{\varphi(k)}{k} \frac{s}{s-1} + s \int_1^\infty \frac{R(u)}{u^{s+1}}\, du \tag{7.6}$$

Since the above integral is absolutely, uniformly convergent for $\Re(s) \geqslant \delta > 0$ we see that (7.6) defines an analytic continuation of $L(s, \chi)$ into the half plane $\Re(s) > 0$ with the properties stated in the theorem. ∎

3.2 Just as the distribution of the zeros of $\zeta(s)$ gives information about the primes, the location of zero free regions for the functions $L(s, \chi)$ can be used to deduce information about the distribution of primes in arithmetic progressions. In this chapter we shall only prove the simplest result, which is however strong enough to prove Dirichlet's theorem on primes in arithmetic progressions. Much more precise results will be proved in chapter 8. A discussion of Dirichlet's original proof of the following theorem can be found in Davenport [4].

Theorem 7.8 (a) *Let χ be a character modulo k, then $L(s,\chi)$ does not vanish in the half plane $\Re(s) > 1$.*

(b) *If χ is not a principal character, then $L(1, \chi) \neq 0$.*

Proof. (*a*) Since χ is a completely multiplicative function and the Dirichlet series representation of $L(s, \chi)$ is absolutely convergent in the half plane $\Re(s) > 1$ we can appeal to theorem 1.2 and deduce that

$$L(s, \chi) = \prod_p \left(1 - \frac{\chi(p)}{p^s}\right)^{-1}.$$

in the half plane $\Re(s) > 1$. From the above representation we see that $L(s, \chi) \neq 0$ if $\Re(s) > 1$. This proves the first part of the theorem.

(*b*) Our proof will be in two parts. First we shall consider the case when χ is a complex character, then the more difficult case when χ is a real character.

(i) From the Euler product representation of $L(s, \chi)$ we deduce that if $\Re(s) > 1$, then

$$\begin{aligned} \operatorname{Log} L(s, \chi) &= -\sum_p \operatorname{Log}\left(1 - \frac{\chi(p)}{p^s}\right) \\ &= \sum_p \sum_{\nu=1}^{\infty} \frac{\chi(p^\nu)}{\nu p^{\nu s}}. \end{aligned} \tag{7.7}$$

If l and k are coprime positive integers and χ is a character modulo k, then from theorem 7.4(*e*) and (7.7) it follows that

$$\sum_{\chi \in G(k)} \bar{\chi}(l) \operatorname{Log} L(s, \chi) = \sum_p \sum_{\nu=1}^{\infty} \frac{1}{\nu p^{\nu s}} \sum_{\chi \in G(k)} \bar{\chi}(l)\, \chi(p^\nu), \tag{7.8}$$

Upon taking $l = 1$ in the above equation and taking exponentials it follows that for real $s > 1$ we have

$$\prod_\chi L(s, \chi) \geqslant 1. \tag{7.9}$$

Suppose that χ' is a complex character such that $L(1, \chi') = 0$. Then $\bar{\chi}' \neq \chi'$ and

$$L(1, \bar{\chi}') = \overline{L(1, \chi')} = 0.$$

For real $s > 1$ we have

$$\prod_{\chi} L(s, \chi) = L(s, \chi_0) \mid L(s, \chi') \mid^2 \prod{}' L(s, \chi),$$

where Π' denotes the product over the remaining characters modulo k. From theorem 7.7 we deduce that as $s \to 1^+$

$$\prod_{\chi} L(s, \chi) = \mathrm{O}\{(s-1)^{-1}(s-1)^2\} = \mathrm{o}(1). \tag{7.10}$$

This latter relation contradicts (7.9), hence $L(1, \chi') \neq 0$.

(ii) Let χ be a real character and define the function g in the half plane $\Re(s) > 1$ by

$$g(s) = \frac{\zeta(s) L(s, \chi)}{\zeta(2s)}.$$

The Euler product representations of $\zeta(s)$ and $L(s, \chi)$ gives

$$g(s) = \prod_p \frac{1 + p^{-s}}{1 - \chi(p) p^{-s}} = \prod_p \left(1 + \frac{1}{p^s}\right) \sum_{r=0}^{\infty} \frac{\chi^r(p)}{p^{rs}}$$

$$= \prod_p \left\{1 + \sum_{r=1}^{\infty} \frac{\chi^r(p) + \chi^{r-1}(p)}{p^{rs}}\right\} = \prod_p \left\{1 + \sum_{r=1}^{\infty} \frac{b(p^r)}{p^{rs}}\right\}$$

where $b(p\mathbf{r}) \geqslant 0$. Hence the Dirichlet series representation of $g(s)$ is of the form

$$g(s) = \sum_{n=1}^{\infty} \frac{b(n)}{n^s} \tag{7.11}$$

where $b(n) \geqslant 0$ and $b(1) = 1$.

If $L(1, \chi) = 0$, then from the definition of g we see that g is holomorphic along the portion $s > \frac{1}{2}$ of the real axis. Since $b(n) \geqslant 0$ for all n it now follows from theorem 6.1 that the series (7.11) is convergent in the half plane $\Re(s) > \frac{1}{2}$. This leads to an immediate contradiction, for we certainly have $g(s) \geqslant b(1) = 1$ for real $s > \frac{1}{2}$ and from the definition of g it follows that $g(s)$ tends to zero as $s \to \frac{1}{2}^+$. Hence, if χ is a real character, then $L(1,\chi) \neq 0$. ■

3.3 Many results concerning L-functions take a much simpler form when the character χ is proper. Consequently it is interesting to know the relationship between the corresponding. L-functions when the characters χ_a and χ_b are equivalent. This relationship is easily determined by first considering the case when $\Re(s) > 1$ and using the Euler product representation of $L(s, \chi_a)$ and $L(s, \chi_b)$. Because $\chi_a \sim \chi_b$ it follows that $\chi_a(p) = \chi_b(p)$ if $p \nmid ab$; hence we have

$$\prod_{p}\left(1 - \frac{\chi_a(p)}{p^s}\right)^{-1} \prod_{p|ab}\left(1 - \frac{\chi_a(p)}{p^s}\right) = \prod_{p}\left(1 - \frac{\chi_b(p)}{p^s}\right)^{-1} \prod_{p|ab}\left(1 - \frac{\chi_b(p)}{p^s}\right)$$

Consequently for $\Re(s) > 1$ we conclude that

$$L(s, \chi_a) \prod_{p|ab}\left(1 - \frac{\chi_a(p)}{p^s}\right) = L(s, \chi_b) \prod_{p|ab}\left(1 - \frac{\chi_b(p)}{p^s}\right).$$

This equality holds, by analytic continuation, into the half plane $\Re(s) > 0$. In particular we observe that $L(s, \chi_a)$ and $L(s, \chi_b)$ have precisely the same zeros in $\Re(s) > 0$.

3.4 Now we have enough information about L-functions to prove the famous theorem of Dirichlet which asserts that if the positive integers k and l are coprime, then the arithmetic progression $\{ kn + l \}$ contains an infinity of primes. This theorem is an immediate consequence of the following result.

Theorem 7.9 *Let k and l be positive integers. If $(k, l) = 1$, then the following series diverges*

$$\sum_{p \equiv l(k)} p^{-1}.$$

Proof. For real $s > 1$ we have, by (7.8),

$$\sum_{\nu=1}^{\infty} \sum_{p^\nu \equiv l(k)} \nu^{-1} p^{-\nu s} = \frac{1}{\varphi(k)} \sum_{\chi} \bar{\chi}(l) \operatorname{Log} L(s, \chi). \tag{7.12}$$

Consider first the right hand side of the above equation. Each logarithm is a continuous function of s for $s > 1$ and when $s \to 1^+$ the function

$$(s - 1)^{E(\chi)} L(s, \chi)$$

tends to a finite *non zero* limit. Consequently, as $s \to 1^+$

$$E(\chi) \operatorname{Log}(s - 1) + \operatorname{Log} L(s, \chi)$$

Thus, as $s \to 1^+$ the right hand side of (7.12) is

$$\frac{1}{\varphi(k)} \sum_{\chi} \bar{\chi}(l) \operatorname{Log} L(s, \chi) = -\frac{1}{\varphi(k)} \operatorname{Log}(s - 1) + O(1).$$

Now let us consider the left hand side of equation (7.12). We have

$$\sum_{\nu=1}^{\infty} \sum_{p^\nu \equiv l(k)} \nu^{-1} p^{-\nu s} = \sum_{p \equiv l(k)} p^{-s} + \sum_{\nu=2}^{\infty} \sum_{p^\nu \equiv l(k)} \nu^{-1} p^{-\nu s} .$$

Since we trivially have

$$\sum_{\nu=2}^{\infty} \sum_{p^{\nu} \equiv l(k)} \nu^{-1} p^{-\nu s} \leqslant \sum_{\nu, p} \frac{1}{2 p^{\nu s}} = \sum_{p} \frac{1}{2} \frac{1}{p^{2s}} \frac{1}{1 - p^{-s}}$$
$$\leqslant \sum_{n=2}^{\infty} \frac{1}{2\, n(n-1)} < \infty .$$

if follows that as $s \to 1^+$

$$\mathrm{P}(s) = \sum_{p \equiv l(k)} p^{-s} = \frac{1}{\varphi(k)} \operatorname{Log} \frac{1}{s-1} + \mathrm{O}(1) . \tag{7.13}$$

If P(1) has a finite value, then we should have $\mathrm{P}(s) \leqslant \mathrm{P}(1)$ as $s \to 1^+$, which contradicts (7.13). ■

Remark. Equation (7.13) suggests a possible way of measuring the " density " of primes in the progression $\{ kn + l \}$, for it is easy to see that for $s > 1$

$$\sum_{p} p^{-s} = \operatorname{Log} \left(\frac{1}{s-1} \right) + \mathrm{O}(1)$$

with the consequence that

$$\lim_{s \to 1^+} \frac{\sum_{p \equiv l(k)} p^{-s}}{\sum_{p} p^{-s}} = \frac{1}{\varphi(k)} .$$

In general, if A is any set of primes one can define the *analytic density* of A as

$$\lim_{s \to 1^+} \frac{\sum_{p \in A} p^{-s}}{\sum_{p} p^{-s}} ,$$

and the natural density as

$$\lim_{x \to \infty} \frac{1}{\pi(x)} \operatorname{card} \{ p \leqslant x \mid p \in A \} .$$

One can prove, using the techniques of chapter 2, that the natural density of primes in the progression $\{ kn + l \}$ is equal to $1/\varphi(k)$. Thus, the two types of density are equal in this case. Indeed, it is not difficult to show that if a set of primes has a natural density, then it has an analytic density. However, the converse is not true in general (see exercises 7.15, 7.16).

Notes to chapter 7

LOWER BOUNDS FOR $L(1, \chi)$

In our proof of theorem 7.9 it was of crucial importance to know that $L(1, \chi) \neq 0$ and as theorem 7.8 we gave an indirect proof of this fact. However, it is often very useful to have explicit lower bounds for $L(1, \chi)$. We shall need such a lower bound in the proof of theorem 9.5. The first, and in a sense the most precise theorem is due to Dirichlet. It is known as " Dirichlet's class number formula " and can be stated as follows.

Theorem 7.10 *Let χ be a real proper character modulo k.*

(a) *If $\chi(-1) = -1$, then we have*

$$L(1, \chi) = \frac{2\pi h(-k)}{\omega\sqrt{k}} = -\frac{\pi}{k^{3/2}} \sum_{n=1}^{k} n\chi(n)$$

where $h(-k)$ denotes the class number and ω the number of roots of unity in the field $Q(\sqrt{-k})$.

(b) *If $\chi(-1) = 1$, then we have*

$$L(1, \chi) = \frac{h(k) \operatorname{Log} \varepsilon}{\sqrt{k}} = -\frac{1}{k^{1/2}} \sum_{n=1}^{k} \chi(n) \operatorname{Log} \sin\left(\frac{n\pi}{k}\right)$$

where $h(k)$ denotes the class number and $\varepsilon > 1$ the fundamental unit of the field $Q(\sqrt{k})$.

A sketch of the proof is given by Davenport [4] and a complete proof in Landau [3]. Dirichlet used the fact that $h(\pm k) \geqslant 1$ to deduce that $L(1, \chi) \neq 0$ and thence to prove his theorem on primes in arithmetic progressions.

In chapter 8 we shall prove a theorem of Siegel which affirms :

For any $\varepsilon > 0$ there exists $c(\varepsilon)$ such that $L(1, \chi) > c(\varepsilon)\, k^{-\varepsilon} > 0$.

A consequence of this theorem is the inequality :

$$h(-k) > \frac{c(\varepsilon)}{2\pi} k^{\frac{1}{2}-\varepsilon}. \tag{7.14}$$

Hence, on choosing ε less than $\frac{1}{2}$, we see that $h(-k)$ tends to infinity with k. Unfortunately Siegel's theorem has a serious defect. The constant $c(\varepsilon)$ is *non-effective* in the sense that given a specific value of ε we are unable to assign a numerical value to $c(\varepsilon)$. Thus, we cannot use (7.14) to find all complex quadratic number fields with a given class number. The best that can be done, using the techniques of chapter 8, is to prove a theorem of the following type, which is due to Iatayawa [1].

Theorem 7.11 *Let $h_0 \geqslant 1$ be a fixed integer. If k is the fundamental discriminant of a complex quadratic field and satisfies*

$$k \geqslant 2\,100\, h_0^2 \operatorname{Log}^2 (13\, h_0)\,,$$

then apart from at most one exceptional value of k we have $h(-k) \geqslant h_0$.

Returning now to the behaviour of $L(1, \chi)$ we first mention an interesting observation due to Davenport [1]. If χ is a real proper character, then we know that

$$L(1, \chi) = \sum_{n=1}^{\infty} \frac{\chi(n)}{n} > 0\,.$$

Since χ is periodic with period k it is natural to divide the above sum into blocks of length k, say

$$L(1, \chi) = \sum_{v=0}^{\infty} S_v\,,$$

where

$$S_v = \sum_{n=vk+1}^{(v+1)k} \frac{\chi(n)}{n}$$

and to consider the sums S_v. Davenport proves that if $\chi(-1) = +1$, then $S_v > 0$ for all v whilst if $\chi(-1) = -1$, then $S_0 > 0$, $S_v > 0$ if $v > v(k)$ and that for any integer $r \geqslant 1$ there exist values of k for which

$$S_1 < 0, \ldots, S_r < 0\,.$$

If we consider $L(1, \chi)$ as a function of k, then one would like to know about the asymptotic behaviour of $L(1, \chi)$ as k tends to infinity. Not a great deal is known. Elliot [1] has proved that there exist absolute constants c_1 and c_2 such that $L(1, \chi)$ " usually " lies between $c_1(\operatorname{Log}\operatorname{Log} k)^{-1}$ and $c_2 \operatorname{Log}\operatorname{Log} k$ and Joshi [1] shows that if k is prime, then

$$\limsup_{k\to\infty} \frac{L(1, \chi)}{\operatorname{Log}\operatorname{Log} k} \geqslant e^{\gamma}$$

and

$$\liminf_{k\to\infty} L(1, \chi) \operatorname{Log}\operatorname{Log} k \leqslant \frac{\pi}{6} e^{-\gamma}$$

where γ is Euler's constant. One would expect the above limits to be finite and non-zero respectively.

We shall now give a general theorem, which as we shall see, is easy to prove and contains as a special case an explicit lower bound for $L(1, \chi)$. Before stating the theorem we need to introduce some notation. Let g be a real valued function defined on the positive integers. We define the functions S and f by

$$\mathrm{S}(x) = \sum_{n \leqslant x} g(n) \quad \text{et} \quad f(n) = \sum_{m \mid n} g(m) .$$

Theorem 7.12 *Let g, f and* S *be the functions defined above. Suppose that for all integers $n \geqslant 1$ we have*

$$f(n) \geqslant 0, f(n^2) \geqslant 1 \text{ and } \mathrm{S}(n) \geqslant -K$$

where $K \geqslant 1$ is a fixed real number. If the series $\sum_{n=1}^{\infty} g(n)\, n^{-1}$ is convergent, then its sum is at least $(16\, K)^{-1}$.

Proof. Since $\sum g(n)\, n^{-1}$ is convergent we have $g(n) = o(n)$, thus $f(n) = O(n^2)$ and the series

$$\sum_{m=1}^{\infty} f(m)\, e^{-xm} = F(x)$$

is convergent for each $x > 0$. Let $\delta \in (0, 1)$ be a real parameter and $N \geqslant 1$ an integer, both of which will be chosen explicitly later.

It is clear from the hypotheses on f that

$$F(\delta) \geqslant \sum_{v=1}^{\infty} e^{-\delta v^2} > \sum_{v=1}^{N} e^{-\delta v^2} > N\, e^{-\delta N^2} \tag{7.15}$$

and from the definition of F and f we have

$$\begin{aligned} F(\delta) &= \sum_{r,n=1}^{\infty} g(n)\, e^{-\delta n r} = \sum_{n=1}^{\infty} g(n)\,(e^{\delta n} - 1)^{-1} \\ &= \frac{L}{\delta} - G(\delta) \end{aligned} \tag{7.16}$$

where L denotes the sum of the series $\sum_{n=1}^{\infty} g(n)\, n^{-1}$ and the function G is defined by

$$G(\eta) = \sum_{n=1}^{\infty} g(n) \left\{ \frac{1}{\eta n} - (e^{\eta n} - 1)^{-1} \right\} .$$

The Abel summation formula now gives us

$$G(\delta) = -\int_{\delta}^{\infty} S\left(\frac{x}{\delta}\right) \left\{ -\frac{1}{x^2} + e^x(e^x - 1)^{-2} \right\} dx .$$

From the hypothesis : $S(x/\delta) \geqslant -K$ for all $x \geqslant \delta$, since

$$\{ -x^{-2} + e^x(e^x - 1)^{-2} \} \leqslant 0 ,$$

it follows that

$$G(\delta) \geqslant -K \int_{\delta}^{\infty} \{ x^{-2} - e^x(e^x - 1)^{-2} \} \, dx$$

$$= -K \left\{ \frac{1}{\delta} - (e^{\delta} - 1)^{-1} \right\} \geqslant -\frac{K}{2} . \tag{7.17}$$

From (7.15), (7.16) and (7.17) we conclude

$$L = \delta \{ F(\delta) + G(\delta) \}$$

$$\geqslant \delta \left\{ N e^{-\delta N^2} - \frac{K}{2} \right\} .$$

If we choose $N = [2K]$ and $\delta = \frac{1}{2} N^{-2}$, then

$$L \geqslant \frac{1}{8K^2} \left\{ \frac{[2K]}{\sqrt{e}} - \frac{K}{2} \right\} \geqslant \frac{1}{16K} . \qquad \blacksquare$$

It is clear that no attempt has been made to obtain the best lower bound for L which the method is capable of giving. We merely sought a tidy final result.

If we take $g = \chi$ a real proper character modulo k, then it is an easy exercise to show that all the hypotheses of the theorem are satisfied and that one can take $K = \frac{1}{2} k$ by a trivial estimate. Thus, theorem 7.12 implies that

$$L(1, \chi) \geqslant \frac{1}{8k} .$$

A less trivial estimate for K is given by the Polya-Vinogradov theorem which asserts that for any non-principal character modulo k we have

$$\left| \sum_{n \leqslant x} \chi(n) \right| < \sqrt{k} \operatorname{Log} k$$

(for a proof of this theorem see the *Notes* to chapter 9), and so we can conclude that

$$L(1, \chi) \geqslant \frac{1}{16\sqrt{k}\,\mathrm{Log}\,k}.$$

We may remark that Paley [1] proved

$$\left|\sum_{n\leqslant x} \chi(n)\right| = \Omega(\sqrt{k}\,\mathrm{Log\,Log}\,k),$$

and Montgomery and Vaughan have shown that the generalised Riemann hypothesis implies that $\sum_{n\leqslant x} x(n) \ll \sqrt{k}\log\log k$.

The generalisation and abstraction of Dirichlet's theorem that $L(1, \chi) \neq 0$ has attracted the attention of several authors. We now describe some of the known results. The first is due to Bateman [2]. It is of the same general character as theorem 7.12 but harder to prove.

Theorem 7.13 *If f and g satisfy the hypotheses of theorem 7.12 and if for all* $n \geqslant 1$

$$S(n) \geqslant -A\sqrt{n} - B$$

where $0 \leqslant A < \left|\zeta\left(\frac{1}{2}\right)\right|^{-1}$ *and* $B \geqslant 0$, *then*

$$\liminf_{x\to\infty} \sum_{n\leqslant x} \frac{g(n)}{n} \geqslant \frac{16}{75}\left\{1 + A\zeta\left(\frac{1}{2}\right)^2\right\}\left(1 + \frac{1}{2}B\right)^{-1}.$$

A different type of generalisation of Dirichlet's theorem has been given by Baker, Birch and Wirsing who prove the following result, first conjectured by Erdös.

Theorem 7.14 *Let k be a positive integer. Suppose that the function f, defined on the positive integers is periodic with period k and such that*

$$f(n) \in \{-1, +1\} \text{ pour } 1 \leqslant n < k \quad et \quad f(n) = 0 \quad if \quad 1 < (n, k) < k.$$

Then the series $\sum_{n=1}^{\infty} f(n)n^{-1}$ *does not converge to zero.*

Perhaps the most straightforward generalisation of Dirichlet's theorem was given by Čudakov and Rodosskiĭ [3]. They defined a generalised Dirichlet character to be a completely multiplicative function g with the properties :

(i) $|g(n)| = 0$ or 1,

(ii) $\max_x \left|\sum_{n\leqslant x} g(n)\right| \leqslant M$.

It is clear that characters modulo k are generalised Dirichlet characters. Čudakov and Rodosskiĭ then proved that for a generalised Dirichlet character g one has

$$\sum_{n=1}^{\infty} \frac{g(n)}{n} > \frac{1}{14\,M}.$$

The real interest in such characters comes when one tries to give an example of a generalised character which is *not* a character modulo k. Like any completely multiplicative function a generalised character g is completely determined by its values at the primes. Thus, we call the set of primes p for which $g(p) \neq 0$ a *basis* for g. It is clear that there exist generalised characters with a basis consisting of one prime, for example

$$g_p(n) = \begin{cases} e^{i\alpha r} & \text{if } n = p^r \\ 0 & \text{otherwise}, \end{cases}$$

where α is any fixed real number. Characters modulo k provide examples of generalised characters with an infinite basis. However, Čudakov and Linnik [2] proved the remarkable theorem that there do not exist generalised characters with a finite basis of more than one element. It seems to be unknown whether or not there exist generalised characters which have an infinite basis and which are *not* Dirichlet characters or of the form $\gamma(n)\, n^{it}$.

PRIMES IN SEQUENCES

If $\{ a_n \}$ is any infinite sequence of positive integers, then one can ask whether or not the sequence contains an infinity of prime numbers. In general one can say very little, for example it is unknown whether or not the sequences :

$$\{ n^2 + 1 \}, \qquad \{ 2^n + 1 \}, \qquad \{ 2^n - 1 \}$$

all contain an infinity of prime numbers. However, certain special sequences are more tractable. We have already considered the case when $\{ a_n \}$ is an arithmetic progression, now we shall briefly discuss two other sequences where positive results are known.

Let $c > 1$ be a non-integral positive number. Consider the two sequences :

(i) $\{ [n^c] \mid n = 1, 2, 3, \dots \}$ and (ii) $\{ [cn] \mid n = 1, 2, 3, \dots \}$

We denote by $\pi_1(x, c)$, (resp. $\pi_2(x, c)$) the number of primes less than x and in the first (resp. second) sequence. It is conjectured that

$$\pi_1(x, c) \sim \frac{1}{c} \frac{x}{\operatorname{Log} x}.$$

but it has only been proved for values of c which satisfy $1 < c < \frac{10}{9}$ (cf. Kolesnik [1] and Pyateski-Shapiro [1]). However, Deshouillers [1] has proved that for $c > 1$ we have

$$\pi_1(x, c) = O\left(\frac{x}{\operatorname{Log} x}\right),$$

that for almost all $c > 1$ we have

$$\lim_{x \to \infty} \pi_1(x, c) = +\infty,$$

and finally, the set of $c > 1$ for which

$$\pi_1(x, c) = o\left(\frac{x}{\operatorname{Log} x}\right).$$

has measure zero.

Our knowledge about the distribution of primes in the sequence $[cn]$ is a little more complete. When c is rational it is a simple consequence of Dirichlet's theorem on primes in arithmetic progressions that the sequence $[cn]$ contains an infinity of primes. In the case when c is irrational the existence of an infinity of primes follows from the fact that the sequence $\{ cp \}$ of real numbers is uniformly distributed modulo 1. This latter result will be proved in the *Notes* to chapter 9, as also will the asymptotic formula

$$\pi_2(x, c) \sim \frac{1}{c} \frac{x}{\operatorname{Log} x},$$

valid when c is irrational.

Exercises to chapter 7

7.1 Let $\mathbb{N}$ denote the set of natural numbers. Define a topology $\mathfrak{J}$ on $\mathbb{N}$ by taking as a basis for the open sets the collection of arithmetic progressions $\{ an + b \}$ with $(a, b) = 1$. Show that :

(i) The topology is Hausdorff and connected, but not regular.

(ii) Dirichlet's theorem on primes in arithmetic progressions is equivalent to the fact that primes are everywhere dense in $\mathfrak{J}$.

7.2 Let χ be an improper character modulo k. If χ is equivalent to the proper character χ^* prove that

$$\chi(n) = \sum_{\substack{rs=n \\ r \mid k}} \mu(r)\, \chi^*(r)\, \chi^*(s) .$$

7.3 Let p be an odd prime, r and s positive integers with $s \leqslant r$. If χ is a proper character modulo p^r, then there exists $f_s(x) \in Q(x)$ with the following properties :

(i) $\chi(1 + up^s) = \exp \{ 2\, \pi i f_s(u) \}$,

(ii) the denominator of the coefficient of x^k is $p^{r-sk+\kappa}$, where $p^\kappa \parallel k$,

(iii) if $s > 1$, then the degree of $f_s(x)$ is at most $r(s - 1)$.

7.4 If χ is a non-principal character modulo k, prove that

$$L(s, \chi) = k^{-s} \sum_{n=1}^{k} \chi(n)\, \zeta\left(s, \frac{n}{k} \right)$$

where $\zeta(s, \alpha)$ is the Hurwitz zeta function, defined by

$$\zeta(s, \alpha) = \sum_{n=0}^{\infty} (n + \alpha)^{-s} .$$

7.5 Let f be a real valued piecewise continuous and monotone function. Prove the following result, known as the Poisson summation formula :

$$\sum_{A \leqslant n \leqslant B}' f(n) = \sum_{\nu=-\infty}^{+\infty} \int_A^B f(x)\, e^{2i\pi\nu x}\, dx$$

where $\sum'$ means that the terms corresponding to $n = A$ and $n = B$ are to be replaced by $\frac{1}{2} f(A)$ and $\frac{1}{2} f(B)$ respectively.

[Hint : Let $f_1(x)$ coincide with $f(x)$ for $0 \leqslant x < 1$ and be defined by periodicity with period 1 for all x. Compute the Fourier expansion of $f_1(x)$ and deduce the special case $A = 0$, $B = 1$ of the formula.]

7.6 Let p be a prime and χ a character modulo p. Denote by $\omega(\chi)$ the sum

$$\omega(\chi) = \sum_{n=1}^{p} \chi(n)\, e^{2i\pi n/p} .$$

(*a*) By considering $\omega(\chi)\,\overline{\omega(\chi)}$ prove that $|\,\omega(\chi)\,| = \sqrt{p}$.

(*b*) Write $\omega(\chi) = \varepsilon(p)\sqrt{p}$. Prove that $\varepsilon(p)$ is a root of unity if and only if χ is a real character (see Mordell [2]).

(*c*) If χ is a real character prove that

$$\omega(\chi) = \sum_{n=0}^{p-1} e^{2i\pi n^2/p}$$

and that

$$\omega(\chi) = \begin{cases} \sqrt{p} & \text{if } \; p \equiv 1 (\text{mod } 4) \\ i\sqrt{p} & \text{if } \; p \equiv 3 (\text{mod } 4) . \end{cases}$$

[Hint : Use exercise 7.5 with $f(x)$ taken to be $\cos\left(\dfrac{2\pi x^2}{p}\right)$ and $\sin\left(\dfrac{2\pi x^2}{p}\right)$ to deduce that

$$\omega(\chi) = \frac{1 + i^{-p}}{1 + i^{-1}} \cdot \sqrt{p} \, . \Bigg]$$

7.7 Let χ be a non-principal character. Show that the following limit exists :

$$\lim_{s \to 1^+} \sum_p \frac{\chi(p)}{p^s} .$$

7.8 For each non-principal character χ prove that the following series is convergent

$$\sum_p \frac{\chi(p)}{p}$$

[Hint : Prove that

$$\sum_{n \leq x} \frac{\chi(n)}{n} \operatorname{Log} n = L(1, \chi) \sum_{m \leq x} \frac{\chi(m)}{m} \Lambda(m) + \mathrm{O}\left(\frac{1}{x} \psi(x)\right).$$

then deduce

$$\sum_{m \leq x} \frac{\chi(m) \Lambda(m)}{m} = \mathrm{O}(1)$$

which, by use of the Abel summation formula implies that $\sum_p \frac{\chi(p)}{p}$ is convergent.]

7.9 Prove that if χ is a non-principal character, then

$$L(1, \chi) = \sum_{n=1}^{\infty} \frac{\chi(n)}{n} = \prod_p \left(1 - \frac{\chi(p)}{p}\right)^{-1}.$$

7.10 Prove that as x tends to infinity

(i) $$\sum_{\substack{p \leq x \\ p \equiv l(k)}} \frac{1}{p} = \frac{1}{\varphi(k)} \operatorname{Log} \operatorname{Log} x + c + \mathrm{O}\left(\frac{1}{\operatorname{Log} x}\right)$$

where $A(k, l)$ is a constant depending only on k and l,

(ii) $$\sum_{\substack{n \leq x \\ n \equiv l(k)}} \frac{\Lambda(n)}{n} = \frac{1}{\varphi(k)} \operatorname{Log} x + \mathrm{O}(1)$$

(iii) $$\sum_{\substack{p \leq x \\ p \equiv l(k)}} \frac{\operatorname{Log} p}{p} = \frac{1}{\varphi(k)} \operatorname{Log} x + \mathrm{O}(1).$$

7.11 Prove that $L(s, \chi)$ has no zeros on the line $\Re(s) = 1$.

7.12 If k and l are positive coprime integers, then prove that as x tends to infinity

$$\pi(x\,; k, l) \sim \frac{1}{\varphi(k)} \frac{x}{\operatorname{Log} x} \quad \text{and} \quad \psi(x\,; k, l) \sim \frac{1}{\varphi(k)} x.$$

7.13 (*a*) Let χ be a character modulo k. Write

$$f(x) = \sum_{m=1}^{k-1} \chi(m)\, x^{m-1} .$$

Verify that for $|x| < 1$,

$$\sum_{n=1}^{\infty} \chi(n)\, x^n = \frac{1}{1-x^k} \sum_{m=1}^{k-1} \chi(m)\, x^m = \frac{xf(x)}{1-x^k}$$

and use the known formula

$$\Gamma(s)\, n^{-s} = \int_0^1 x^{n-1}\, |\operatorname{Log} x|^{s-1}\, dx ,$$

to deduce that

$$L(s,\chi) = -\frac{1}{\Gamma(s)} \int_0^1 \frac{f(x)}{x^k - 1}\, |\operatorname{Log} x|^{s-1}\, dx ,$$

which provides an analytic continuation for $L(s, \chi)$ over the whole complex plane.

(*b*) When k is a prime and χ is a real proper character take $s = 1$ in the above integral representation of $L(s, \chi)$. By expressing the integrand as a sum of partial fractions deduce that

$$L(1,\chi) = -\frac{1}{G} \sum_{m=1}^{k-1} \chi(m) \left[\operatorname{Log}\left(2 \sin \frac{\pi m}{k}\right) + i\left(\frac{\pi m}{k} - \frac{\pi}{2}\right)\right]$$

where

$$G = \sum_{m=1}^{k-1} \chi(m)\, e^{2i\pi m/k} .$$

(*c*) Show that if $k \equiv 3 \pmod 4$, then

$$L(1,\chi) = -\frac{\pi}{k^{3/2}} \sum_{m=1}^{k-1} m\chi(m) \quad \text{if} \quad k \equiv 3 \pmod 4$$

and that if $k \equiv 1 \pmod 4$, then

$$L(1,\chi) = -\frac{1}{\sqrt{k}} \sum_{m=1}^{k-1} \chi(m) \operatorname{Log}\left(2 \sin \frac{\pi m}{k}\right).$$

7.14 Let χ be a generalised character in the sense of Gelfond and Tchudakov.

(i) Prove that the series

$$L(\chi) = \sum_{n=1}^{\infty} \frac{\chi(n)}{n}$$

is convergent.

(ii) Prove that $L(\chi) \neq 0$.

[Hint : When χ is real consider the function U, defined for $0 < x < 1$ by

$$U(x) = \sum_{n=1}^{\infty} \frac{\chi(n)\, x^n}{1 - x^n} = \sum_{n=1}^{\infty} f(n)\, x^n$$

Show that $f(n) \geqslant 0$ and $f(n^2) \geqslant 1$ for all positive integers n, then prove that

$$U(x) \geqslant \frac{\sqrt{\pi}}{2\,|\,\mathrm{Log}\, x\,|}.$$

Next consider the function R defined by

$$R(x) = U(x) - \frac{L(\chi)}{1 - x}$$

Write $R(x)$ as

$$R(x) = \sum_{n=1}^{\infty} \chi(n) \left\{ \frac{x^n}{1 - x^n} - \frac{x^n}{n(1 - x)} \right\} - \sum_{n=0}^{\infty} S_{n+1}\, x^n$$

where $S_n = \sum_{r=n}^{\infty} \frac{\chi(r)}{r}$. Use partial summation to show that

$$R(x) = O\left(\mathrm{Log}\, \frac{1}{1 - x}\right).$$

7.15 If θ is a positive real number we denote the fractional part of θ by $\{\theta\}$.

(*a*) Prove that the sequence $\{\log p\}$ is dense modulo 1.

(*b*) Show that the sequence $\{\log p\}$ is not uniformly distributed modulo 1. More precisely prove that for $0 \leqslant \xi \leqslant 1$

$$\liminf_{x \to \infty} \frac{\mathrm{card}\,(p : \log p \leqslant x \text{ and } \{\log p\} \leqslant \xi)}{\mathrm{card}\,(p : \log p \leqslant x)} = \frac{e^{\xi} - 1}{e - 1}$$

and

$$\limsup_{x \to \infty} \frac{\mathrm{card}\,(p : \log p \leqslant x \text{ and } \{\log p\} \leqslant \xi)}{\mathrm{card}\,(p : \log p \leqslant x)} = \frac{1 - e^{-\xi}}{1 - e^{-1}}.$$

(*c*) Prove that the set of prime numbers with first digit 1 does not have a natural density.
[Hint (*b*) : Write $N(x) = \mathrm{card}\,(p : \log p \leqslant x)$ and $\varphi(x, \xi) = \mathrm{card}\,(p : \log p \leqslant x$ and $\{\log p\} \leqslant \xi)$. Show that

$$\varphi(x, \xi) = N(1 + \xi) - N(1) + \cdots + N(\min\,(x, [x] + \xi)) - N([x]).$$

Use the Prime Number Theorem to prove that

$$\sum_{k=1}^{m} N(k + \xi) \sim \frac{e^{m+1+\xi}}{m(e - 1)}$$

and obtain an asymptotic formula for $\varphi(x, \xi)/N(x)$. For any $\alpha \in [0, 1]$ take a sequence of primes such that $\{ \log p_n \} \to \alpha$ and choose $x = n + \{ \log p_n \}$ in the expression for $\varphi(x, \xi)/N(x)$.]

7.16 (*a*) Prove that if $0 \leqslant \xi \leqslant 1$ and $s \to 1^+$, then

(i) $$\sum_{\{\log p\} \leqslant \xi} \frac{\log p}{p^s} = \frac{\xi}{s-1} + O(1),$$

and

(ii) $$\sum_{\{\log p\} \leqslant \xi} p^{-s} = -\xi \log(s-1) + O(1).$$

(*b*) Prove that the set of prime numbers with first digit 1 has an analytic density equal to $\log_{10} 2 = 0{,}3010300\ldots$

[Hint : Write

$$\sum_{\{\mathrm{Log}\, p\} \leqslant \xi} \frac{\mathrm{Log}\, p}{p^s} = \sum_{n=1}^{\infty} \left(\frac{1}{n^s} - \frac{1}{(n+1)^s} \right) \sum_{\substack{p \leqslant n \\ \{\mathrm{Log}\, p\} \leqslant \xi}} \mathrm{Log}\, p,$$

then use the Prime Number Theorem to show that

$$\sum_{\substack{p \leqslant n \\ \{\mathrm{Log}\, p\} \leqslant \xi}} \mathrm{Log}\, p = \frac{e}{e-1} (e^{\xi} - 1)\, e^{[\mathrm{Log}\, n]} + O\left(\frac{n}{\mathrm{Log}^2 (n+2)} \right) - \begin{cases} 0 \text{ if } \{\mathrm{Log}\, n\} > \xi \\ e^{\xi + [\mathrm{Log}\, n]} \text{ if } \{\mathrm{Log}\, n\} \leqslant \xi . \end{cases}$$

Next, prove that

$$\sum_{n=1}^{\infty} \left(\frac{1}{n^s} - \frac{1}{(n+1)^s} \right) e^{[\mathrm{Log}\, n]} = \frac{e-1}{e} \frac{1}{s-1} + O(1)$$

and

$$\sum_{\{\mathrm{Log}\, n\} \leqslant \xi} \left(\frac{1}{n^s} - \frac{1}{(n+1)^s} \right) (e^{\xi + [\mathrm{Log}\, n]} - n) =$$
$$\frac{e^{\xi} - 1}{s-1} + O(1) - \sum_{m=0}^{\infty} \sum_{e^m \leqslant n < e^{m+\xi}} n \left(\frac{1}{n^s} - \frac{1}{(n+1)^s} \right);$$

Finally show that this last double sum is $\dfrac{\xi}{s-1} + O(1).$]

CHAPTER 8

L-functions (II)

§ 1 INTRODUCTION

1.1 We continue our study of L-functions in this chapter. The results which we shall obtain will be analogous to the theorems about $\zeta(s)$ proved in chapter 4. The deeper properties of L-functions, such as the functional equation, the distribution of zeros in the " critical strip " etc., will not be considered in this book. For an introduction to these topics the reader is referred to Davenport [4], Prachar [1] and Montgomery [1].

The principal objective of this chapter is a study of the Prime Number Theorem for arithmetic progressions with non-trivial error terms. Our proof of such theorems will be very similar to the proof of theorem 4.7, thus necessitating a study of " zero free regions " and estimates for $L(s, \chi)$ and $L'(s, \chi)$ strictly to the left of the line $\Re(s) = 1$.

1.2 A serious defect in several of our theorems is one that bedevils a large part of prime number theory, namely the problem of ineffective constants; that is to say, absolute constants which are proved to exist, but for which we are unable to assign numerical values. This means that some theorems which have every appearance of being quantitative statements are seen, on closer examination, to be only qualitative assertions. For example, we shall prove as theorem 8.8 that for any fixed $H > 1$ and any positive integers k, l with $(k, l) = 1$ we have

$$\psi(x; k, l) = \sum_{\substack{n \leqslant x \\ n \equiv l(k)}} \Lambda(n) = \frac{x}{\varphi(k)} + O\left(\frac{x}{\operatorname{Log}^H x}\right),$$

where the implied " O " constant does *not* depend on k or l. However, given a specific value of H, one cannot compute a numerical value for the " O " constant and our rather precise looking asymptotic formula cannot be used to derive any quantitative information about $\psi(x; k, l)$.

The reason for the ineffectiveness is that the " constants " depend on certain hypothetical real zeros of L-functions formed with real characters. It is conjectured that these zeros do not exist. If this is true then the theorems can be made completely effective.

1.3 The contents of the chapter are as follows. In § 2 we find estimates for $L(s, \chi)$ and $L'(s, \chi)$. These are used in § 3 to establish the existence of certain zero free regions of $L(s, \chi)$ and in § 4 to prove a theorem of Siegel. Siegel's theorem says that $L(s, \chi)$ does not vanish on a certain segment of the real axis and also gives a lower bound for $L(1, \chi)$. Finally, in § 5, we prove the Prime Number Theorem for arithmetic progressions.

§ 2 ESTIMATES FOR $L(s, \chi)$ AND $L'(s, \chi)$

2.1 Our first theorem is the analogue of theorem 2.3. It gives upper bounds for $|L(s, \chi)|$ and $|L'(s, \chi)|$ in a region strictly to the left of the line $\mathcal{R}(s) = 1$. The theorem will be used often in this chapter and the next. By using the methods of chapter 10 the estimates could be improved upon, but for most purposes the easily derived estimates of theorems 8.1 and 8.2 suffice.

We recall the following notation, introduced in § 3.2 of chapter 7 :

$$E(\chi) = \begin{cases} 1 & \text{if } \chi \text{ is a principal character} \\ 0 & \text{otherwise.} \end{cases}$$

Theorem 8.1 *Let χ be a character modulo k and θ a real number satisfying $0 < \theta < 1$. For $\mathcal{R}(s) \geqslant \theta$ we have*

$$\left| L(s, \chi) - \frac{\varphi(k)}{k} \frac{E(\chi)}{s - 1} \right| < 3 \frac{(k\,|\,s\,|)^{1-\theta}}{\theta(1 - \theta)}$$

and

$$\left| L'(s, \chi) + \frac{\varphi(k)}{k} \frac{E(\chi)}{(s - 1)^2} \right| < \frac{6(k\,|\,s\,|)^{1-\theta}}{\theta(1 - \theta)} \left[\frac{1}{\theta} + \operatorname{Log}\left(\frac{k\,|\,s\,|}{\theta} \right) \right].$$

Proof. For convenience we shall write

$$\mathrm{X} = E(\chi) \frac{\varphi(k)}{k} \quad \text{and} \quad R(x) = \sum_{n \leqslant x} \chi(n) - \mathrm{X}x .$$

thus $0 \leqslant x < 1$.

If $\mathcal{R}(s) > 0$ and $x \geqslant 1$ is a parameter to be chosen explicitly later, then from (7.6)

$$L(s, \chi) - \sum_{n \leqslant x} \frac{\chi(n)}{n^s} = \mathrm{X} \frac{x^{1-s}}{s - 1} + s \int_x^{\infty} \frac{R(u)}{u^{s+1}}\, du - \frac{R(x)}{x^s},$$

We write the above as :

$$L(s, \chi) - \frac{X}{s-1} = \sum_{n \leqslant x} \frac{\chi(n)}{n^s} - X \int_1^x \frac{du}{u^s} + s \int_x^\infty \frac{R(u) - R(x)}{u^{s+1}} \, du \,. \tag{8.1}$$

Upon taking absolute values and recalling that for all x and u greater than 1 we have

$$| R(u) - R(x) | \leqslant k$$

it follows that

$$\left| L(s, \chi) - \frac{X}{s-1} \right| \leqslant \sum_{n \leqslant x} \frac{1}{n^\sigma} + X \int_1^x \frac{du}{u^\sigma} + | s | \int_x^\infty \frac{k\, du}{u^{\sigma+1}} \,. \tag{8.2}$$

Thus, for $\sigma \geqslant \theta$ we certainly have

$$\left| L(s, \chi) - \frac{X}{s-1} \right| \leqslant (1 + X) \int_0^x \frac{du}{u^\theta} + k \, | s | \int_x^\infty \frac{du}{u^{\theta+1}}$$

$$\leqslant 2 \frac{x^{1-\theta}}{1-\theta} + k \, | s | \frac{x^{-\theta}}{\theta} \,.$$

If we now take $x = k \, | s |/\theta$ and recall that $0 < \theta < 1$ we obtain

$$\left| L(s, \chi) - \frac{X}{s-1} \right| \leqslant \left(\frac{2}{1-\theta} + 1 \right) \left(\frac{k \, | s |}{\theta} \right)^{1-\theta} .$$

which implies the first part of the theorem, since $\theta^\theta < 1$.

Now differentiate (8.1) with respect to s to obtain :

$$L'(s, \chi) + \frac{X}{(s-1)^2} = - \sum_{n \leqslant x} \frac{\chi(n) \operatorname{Log} n}{n^s} + X \int_1^x \frac{\operatorname{Log} u}{u^s} \, du$$

$$- \int_x^\infty \frac{1 - s \operatorname{Log} u}{u^{1+s}} (R(x) - R(u)) \, du \,. \tag{8.3}$$

Hence, upon taking absolute values and noting that $| s | \geqslant \mathcal{R}(s) = \sigma \geqslant \theta$ we find

$$\left| L'(s, \chi) + \frac{X}{(s-1)^2} \right| \leqslant \sum_{n \leqslant x} \frac{\operatorname{Log} n}{n^\theta} + X \int_1^x \frac{\operatorname{Log} u}{u^\theta} \, du \, +$$

$$\int_x^\infty \frac{| s | (\theta^{-1} + \operatorname{Log} u)}{u^{1+\theta}} k . du$$

$$\leqslant 2 \int_1^x \frac{\operatorname{Log} u}{u^\theta} \, du + k \, | s | \int_x^\infty (\theta^{-1} + \operatorname{Log} u) \frac{du}{u^{1+\theta}}$$

$$\leqslant \frac{2 \, x^{1-\theta}}{1-\theta} \operatorname{Log} x + k \, | s | \frac{x^{-\theta}}{\theta^2} (2 + \theta \operatorname{Log} x) \,.$$

If we choose $x = k\,|\,s\,|/\theta$, then

$$\left| L'(s, \chi) + \frac{X}{(s-1)^2} \right| \leqslant \left(\frac{2}{1-\theta} + 1\right)\left(\frac{k\,|\,s\,|}{\theta}\right)^{1-\theta} \operatorname{Log}\frac{k\,|\,s\,|}{\theta} + \frac{2}{\theta}$$

$$\leqslant \frac{3}{1-\theta}\left(\frac{k\,|\,s\,|}{\theta}\right)^{1-\theta}\left(\operatorname{Log}\frac{k\,|\,s\,|}{\theta} + \frac{2}{\theta}\right)$$

$$\leqslant \frac{6}{\theta(1-\theta)}(k\,|\,s\,|)^{1-\theta}\left(\operatorname{Log}\frac{k\,|\,s\,|}{\theta} + \frac{1}{\theta}\right).$$

which implies the second part of the theorem since $\theta^\theta < 1$. ■

The next theorem gives estimates for $L(s, \chi)$ and $L'(s, \chi)$ in the half plane $\mathfrak{R}(s) \geqslant 1$. Of course both $L(s, \chi)$ and $L'(s, \chi)$ are represented by absolutely convergent Dirichlet series in the half plane $\mathfrak{R}(s) > 1$ and both functions are O(1) in any fixed half plane $\mathfrak{R}(s) \geqslant 1 + \delta$. The theorem is useful because it gives information about the behaviour of these functions on the line $\mathfrak{R}(s) = 1$.

Theorem 8.2 *If $\chi \neq \chi_0$, then for $\mathfrak{R}(s) \geqslant 1$ we have*

(a) $|\,L(s, \chi)\,| < 2 + \operatorname{Log}(k\,|\,s\,|)$,

(b) $|\,L'(s, \chi)\,| < (2 + \operatorname{Log}(k\,|\,s\,|))^2$.

(c) *If χ is any character modulo k and $\sigma > 1$, then*

$|\,L(s, \chi)\,|^{-1} < \zeta(\sigma) < \dfrac{\sigma}{\sigma - 1}$.

Proof. (*a*) If $\chi \neq \chi_0$, then $k \geqslant 3$ and $\kappa = 0$. From (8.2) we have

$$|\,L(s, \chi)\,| \leqslant \sum_{n \leqslant x} n^{-1} + |\,s\,|\,k \int_x^\infty u^{-2}\,du$$

$$\leqslant 1 + \operatorname{Log} x + \frac{k\,|\,s\,|}{x}$$

$$\leqslant 2 + \operatorname{Log}(k\,|\,s\,|),$$

if we choose $x = k\,|\,s\,| \geqslant 3 > e$.

(*b*) From (8.3) it follows that if $\mathfrak{R}(s) \geqslant 1$, then

$$|\,L'(s, \chi)\,| \leqslant \sum_{n \leqslant x} \frac{\operatorname{Log} n}{n} + k\,|\,s\,| \int_x^\infty \frac{1 + \operatorname{Log} u}{u^2}\,du$$

$$< \operatorname{Log} x(\operatorname{Log} x + 1) + k\,|\,s\,|\,\frac{2 + \operatorname{Log} x}{x}.$$

$$|\,L'(s, \chi)\,| < \{\operatorname{Log}(k\,|\,s\,|)\}^2 + 2\operatorname{Log}(k\,|\,s\,|) + 2$$

$$< \{\operatorname{Log}(k\,|\,s\,|) + 2\}^2$$

if we choose $x = k\,|\,s\,| \geqslant 3 > e$.

(*c*) From the Euler product representation of $L(s, \chi)$ we have

$$|L(s, \chi)|^{-1} = \prod_p \left| 1 - \frac{\chi(p)}{p^s} \right| \leqslant \prod_p \left(1 + \frac{1}{p^\sigma} \right)$$

$$< \zeta(\sigma) < 1 + \int_1^\infty u^{-\sigma}\, d\sigma = \frac{\sigma}{\sigma - 1}.$$

■

§ 3 ZERO FREE REGIONS

3.1 For $\zeta(s)$ we were able to establish, as theorem 4.5, the existence of a zero free region strictly to left of the line $\mathfrak{R}(s) = 1$. One would like to prove an analogous theorem for *L*-functions. For a specific numerical value of k one can use the proof of theorem 8.3 to construct, for all *L*-functions formed with characters modulo k, a zero free region of the kind

$$\mathfrak{R}(s) \geqslant 1 - \frac{c(k)}{\text{Log}\, \tau(k, t)}$$

where $\tau(k, t)$ is a simple explicit function of k and t and $c(k)$ is a very complicated function of k. Indeed, we know nothing about the behaviour of $c(k)$ other than $c(k) > 0$ for all k.

For our applications we need to know zero free regions which are *explicit* functions of k and t. We would like to prove that there exists an absolute constant c_1 such that $c(k) \geqslant c_1 > 0$ for all k. Unfortunately we cannot quite prove this. The best that we can do is to prove that there exists an absolute constant c_2 such that within the region

$$\mathfrak{R}(s) \geqslant 1 - \frac{c_2}{\text{Log}\, \tau(K, t)}$$

there is at most one zero of all *L*-functions formed with character modulo k for all $k \leqslant K$. It is conjectured that this possible zero does not exist. Indeed, it is conjectured that no $L(s, \chi)$ has a zero in the half plane $\mathfrak{R}(s) > \frac{1}{2}$. This latter conjecture is known as the *generalised Riemann Hypothesis*.

3.2 Before we can state the main theorem of this section we need to introduce some notation.

(i) Let k be a positive integer and t a real number, then

$$\tau(k, t) = \max \{ |t|, k + 2 \}.$$

(ii) Let c satisfy $0 < c < 1$ and denote by $R_c(k)$ the set

$$R_c(k) = \left\{ s = \sigma + it \mid 1 - \frac{c}{\operatorname{Log} \tau(k, t)} \leqslant \sigma \right\}.$$

Theorem 8.3 *There exists an effectively computable absolute constant* α *with the following properties.*

(i) *If K is any positive integer, then the set*

$\{ L(s, \chi_k) : 1 \leqslant k \leqslant K, \chi_k$ *proper* $\}$

contains at most one L-function $L(s, \chi_{q(K)})$ *which has a zero* $\beta(\alpha, K)$ *in* $R_{2\alpha}(K)$;

(ii) *if* $\beta(\alpha, K)$ *exists it is a real, simple zero of* $L(s, \chi_{q(K)})$ *and* $\chi_{q(K)}$ *is a real character*;

(iii) *the L-function* $L(s, \chi_{q(K)})$ *has no other zeros in* $R_{2\alpha}(K)$.

Proof. Let k be any positive integer and χ a character modulo k. The proof of the theorem is in three main sections.

(*a*) For χ a complex character we establish the existence of an absolute constant a such that $L(s, \chi) \neq 0$ in $R_{2a}(k)$.

(*b*) For χ a real character we establish the existence of an absolute constant b such that $L(s, \chi)$ has at most one zero in $R_{2b}(k)$ and if such a zero exists it is real and simple.

(*c*) Parts (*a*) and (*b*) are combined to construct a region $R_{2\alpha}(K)$ with the properties stated in the theorem.

All constants which are going to occur in the proof are absolute and can be explicitly determined, however we shall not assign numerical values to them.

Our main tool will be theorem 4.2 and this will require us to choose various parameters. To a certain extent the choice is fairly arbitrary; we will make a choice which will simplify later computations rather than an optimal choice.

Because $L(s, \chi) \neq 0$ for $\Re(s) > 1$ and $\overline{L(\bar{s}, \bar{\chi})} = L(s, \chi)$ we need only consider possible zeros $\rho = \beta + i\gamma$, where $\beta \leqslant 1$ and $\gamma \geqslant 0$.

3.3 Case (*a*), χ a complex character. Let λ, μ be real numbers, to be chosen explicitly later, such that

$$\lambda > 0, \qquad \mu > 0, \qquad \text{and} \quad \lambda + \mu < \frac{1}{10}.$$

Suppose that $\rho = \beta + i\gamma$ is a zero of $L(s, \chi)$ which satisfies

$$1 - \lambda \leqslant \beta \leqslant 1 \quad \text{and} \quad \gamma \geqslant 0.$$

We shall use theorem 4.2 to show that the existence of such a zero implies the existence of a lower bound for λ which is a function of k and γ. Thus, if we choose $\lambda = \lambda_1(k, \gamma)$ *less* than this lower bound it follows that $L(s, \chi)$ can have no zeros $\rho = \beta + i\gamma$ which satisfy $1 - \lambda_1(k, \gamma) \leqslant \beta \leqslant 1$. By letting γ vary we will obtain a zero free region of the type $R_{2a}(k)$.

For any $\gamma \geqslant 0$ we define the function f by

$$f(s) = L^4(s + i\gamma, \chi)\, L(s + 2\, i\gamma, \chi^2)$$

We are going to apply theorem 4.2 with $s_0 = \frac{11}{10}$, $R = \frac{3}{10}$ and $q = 2$ to $f(s)$.

We must first verify that the hypotheses of that theorem are satisfied by f. Since χ is a complex character, χ^2 is non-principal, so by theorem 7.7 both $L(s + i\gamma, \chi)$ and $L(s + i\,2\,\gamma, \chi^2)$ are holomorphic in the half plane $\Re(s) > 0$. Thus, $f(s)$ is holomorphic in the region $| s - s_0 | \leqslant 2\,R = \frac{3}{5}$. From the Euler product representation of L-functions when $\Re(s) > 1$ we know that $L(s_0 + i\gamma, \chi) \neq 0$ and $L(s_0 + i\,2\,\gamma, \chi^2) \neq 0$, thus $f(s_0) \neq 0$. Finally, using theorem 8.1 with $\theta = \frac{1}{2}$ and theorem 8.2(*c*) we see that in the region $| s - s_0 | \leqslant \frac{3}{5}$ we have

$$\left| \frac{f(s)}{f(s_0)} \right| < \{\, 11.12\, k^{1/2}(2 + 2\,\gamma)^{1/2} \,\}^5 = e^{\mathrm{U}}$$

where $\mathrm{U} = \frac{5}{2} \operatorname{Log} k(1 + \gamma) + 5 \operatorname{Log}(132\sqrt{2})$.

If $\rho = \beta + i\gamma$ is a zero of $L(s, \chi)$, then f has a zero of order at least 4 at $s = \beta$. Moreover, $f(s)$ has no zeros in the half plane $\Re(s) > 1$. Thus, by theorem 4.2(*c*) with $r = d = \frac{1}{10}$ it follows that for $| s - s_0 | \leqslant \frac{1}{10}$ we have

$$\begin{aligned} \Re\left(\frac{4}{s - \beta}\right) &\leqslant \frac{2\,R\mathrm{U}}{(R - d)^2} + \Re\left(\frac{f'(s)}{f(s)}\right) \\ &= 15\,\mathrm{U} + \Re\left(\frac{f'(s)}{f(s)}\right). \end{aligned}$$

We now obtain an upper bound for $\Re(f'(s)/f(s))$. By logarithmic differentiation of f and use of (7.7) it follows that for $\Re(s) > 1$

$$\begin{aligned} \frac{f'(s)}{f(s)} &= 4\,\frac{L'(s + i\gamma, \chi)}{L(s + i\gamma, \chi)} + \frac{L'(s + 2\,i\gamma, \chi^2)}{L(s + 2\,i\gamma, \chi^2)} \\ &= -\sum_{n=1}^{\infty} \left(\frac{4\,\chi(n)}{n^{s+i\gamma}} + \frac{\chi^2(n)}{n^{s+2i\gamma}} \right) \Lambda(n)\,. \end{aligned} \tag{8.4}$$

Thus we certainly have

$$\frac{f'(s)}{f(s)} - 3\,\mathrm{Z}(s) = -\sum_{n=1}^{\infty} \frac{\Lambda(n)}{n^s}\,(3 + 4\,\chi(n)\,n^{-i\gamma} + \chi^2(n)\,n^{-2i\gamma})\,.$$

For any complex number z satisfying $|\,z\,| \leqslant 1$ it is easy to see that

$$\Re(3 + 4\,z + z^2) \geqslant 0$$

By using this inequality in conjunction with (8.4) we deduce that for real $s > 1$

$$\Re\left(\frac{f'(s)}{f(s)}\right) \leqslant 3\,\Re(\mathrm{Z}(s)) = 3\,\mathrm{Z}(s)\,.$$

If we now write $s = 1 + \mu$, then

$$0 < s - \beta < \lambda + \mu\,.$$

and for some effectively computable constant A_1

$$\mathrm{Z}(1 + \mu) < \frac{1}{\mu} + A_1\,.$$

(In fact a possible value for A_1 is zero; see the *Notes* to chapter 5.) Thus we conclude

$$\frac{4}{\lambda + \mu} < 15\,\mathrm{U} + 3\left(\frac{1}{\mu} + A_1\right)\,.$$

Upon taking $\mu = 6\,\lambda$ and supposing λ is chosen so that $0 < 7\,\lambda < \frac{1}{10}$ we deduce

$$\frac{1}{14\,\lambda} < 15\,\mathrm{U} + 3\,A_1\,.$$

Recalling that $\tau(k, t) = \max\,\{\,|\,t\,|, k + 2\,\}$ it follows from the definition of U that there exists an effective absolute constant A_2 such that for all k and γ

$$14.15\,\mathrm{U} + 3.14\,A_1 < A_2\,\mathrm{Log}\,\tau(k, \gamma)\,.$$

If we now choose $\lambda_1(k, \gamma) = \{\,A_3\,\mathrm{Log}\,\tau(k, \gamma)\,\}^{-1}$, where $A_3 = \max(70, A_2)$, then $\lambda_1(k, \gamma) < \frac{1}{70}$ and $L(s, \chi)$ does *not* have a zero $\beta + i\gamma$ with $\beta \geqslant 1 - \lambda_1(k, \gamma)$. This will hold for each $\gamma \geqslant 0$ and so $L(s, \chi)$ has no zeros in the region $R_{2a}(k)$, where $2\,a$ is any real number satisfying $0 < 2\,a \leqslant A_3^{-1}$. This completes the discussion of the case when χ is a complex character.

3.4 Case (*b*), χ a real character. The discussion of case (*b*) is divided into two parts. First we shall consider the possible zeros which are " near " the real axis, then we shall consider the location of zeros which are " far " from the real axis.

3.5 Zeros " near " the real axis. Let λ, μ be real numbers, to be chosen explicitly later, satisfying

$$\lambda > 0, \qquad \mu > 0 \qquad\qquad 2\lambda + \mu \leqslant \frac{1}{10}.$$

We shall suppose that either :

(i) $L(s, \chi)$ has a *complex* zero $\rho = \beta_1 + i\gamma_1$ in the rectangle

$$1 - \lambda \leqslant \beta \leqslant 1, \quad |\gamma| \leqslant \lambda$$

(and therefore another zero $\bar{\rho} = \beta_1 - i\gamma_1$ in this rectangle), or

(ii) $L(s, \chi)$ has two *real* zeros β_1 and β_2 satisfying

$$1 - \lambda \leqslant \beta_1 \leqslant \beta_2 \leqslant 1.$$

If χ is the principal character, then from our discussion in § 3.3 of chapter 7 it follows that $L(s, \chi)$ has precisely the same zeros as $\zeta(s)$ in the half plane $\Re(s) > 0$. Because $\zeta(s)$ is holomorphic except for a simple pole at $s = 1$ it follows that there exists an absolute constant A_4 such that $\zeta(s)$ and hence $L(s, \chi)$ has no zeros in the rectangle

$$1 - A_4^{-1} \leqslant \sigma \leqslant 1, \quad |t| \leqslant A_4^{-1}.$$

From exercise 2.5 it follows that a possible choice for A_4 is $\sqrt{2}$.

For the rest of this paragraph we shall suppose that χ is not the principal character. We are going to apply theorem 4.2 to $L(s, \chi)$ and deduce a lower bound for λ. In theorem 4.2 we choose $s_0 = \frac{11}{10}$, $R = \frac{3}{10}$ and $q = 2$.

The hypotheses of theorem 4.2 are easily verified. By theorem 7.8(*a*), theorem 7.7, theorem 8.1 with $\theta = \frac{1}{2}$ and theorem 8.2(*c*) it follows that $L(s_0, \chi) \neq 0$, and in the region $|s - s_0| \leqslant \frac{3}{5}$ the function $L(s, \chi)$ is holomorphic and satisfies

$$\left| \frac{L(s, \chi)}{L(s_0, \chi)} \right| \leqslant 11.12 \left(k \frac{17}{10} \right)^{1/2} = e^{\mathrm{U}}$$

where

$$\mathrm{U} = \frac{1}{2} \operatorname{Log} k + \operatorname{Log} \left(132 \sqrt{\frac{17}{10}} \right).$$

By hypothesis $L(s, \chi)$ has at least two zeros ρ_1 and ρ_2 within the disc $|s - s_0| < R = \frac{3}{10}$ and we know that it has no zeros in the region $\Re(s) > 1$. Thus, we can apply theorem 4.2(*c*) with $r = d = \frac{1}{10}$ to deduce that for $| s - s_0 | < \frac{1}{10}$

$$\Re\left(\frac{1}{s - \rho_1} + \frac{1}{s - \rho_2}\right) \leqslant 15\,\mathrm{U} + \Re\left\{\frac{L'(s, \chi)}{L(s, \chi)}\right\}.$$

If we write $s = 1 + \mu$, then for $i = 1, 2$

$$| s - \rho_j | \leqslant \{ (\lambda + \mu)^2 + \lambda^2 \}^{1/2}$$

and upon noting that $\rho_1 + \rho_2$ is real and at most equal to 2 we have

$$\Re\left(\frac{1}{s - \rho_1} + \frac{1}{s - \rho_2}\right) = \Re\left\{\frac{2\,s - \rho_1 - \rho_2}{(s - \rho_1)(s - \rho_2)}\right\} \geqslant \frac{2\,\mu}{(\lambda + \mu)^2 + \lambda^2}.$$

Combining the above results it follows that

$$\frac{2\,\mu}{(\lambda + \mu)^2 + \lambda^2} \leqslant 15\,\mathrm{U} + \Re\left(\frac{L'(1 + \mu, \chi)}{L(1 + \mu, \chi)}\right). \tag{8.5}$$

For $\Re(s) > 1$ we have the representation

$$\frac{L'(s, \chi)}{L(s, \chi)} = - \sum_{n=1}^{\infty} \frac{\chi(n)\, \Lambda(n)}{n^s}$$

and so

$$\frac{L'(s, \chi)}{L(s, \chi)} - \mathrm{Z}(s) = - \sum_{n=1}^{\infty} \frac{1 + \chi(n)}{n^s} \Lambda(n).$$

Since $\chi(n)$ takes only the values $-1, 0, +1$ it follows that for real $s > 1$

$$\frac{L'(s, \chi)}{L(s, \chi)} \leqslant \mathrm{Z}(s).$$

Upon taking $s = 1 + \mu$ in the above inequality it follows from (8.5) that

$$\frac{2\,\mu}{(\lambda + \mu)^2 + \lambda^2} < 15\,\mathrm{U} + \mathrm{Z}(1 + \mu) \leqslant 15\,\mathrm{U} + \frac{1}{\mu} + A_1.$$

Choosing $\mu = 6\,\lambda$ and supposing that λ is such that $0 < 8\,\lambda < \frac{1}{10}$ we obtain

$$\frac{11}{150\,\lambda} < 15\,\mathrm{U} + A_1.$$

It is clear there exists an effective, absolute constant A_5 such that for all k

$$15\,\mathrm{U} + A_1 \leqslant A_5 \operatorname{Log}(k + 2).$$

Thus, if we take $A_6 = \max(A_4, 80, 150\, A_5/11)$ and choose

$$\lambda_1^{-1}(k) = A_6 \operatorname{Log}(k+2)$$

it follows that $L(s, \chi)$ has no complex and at most one real zero in the rectangle

$$\begin{cases} 1 - \dfrac{1}{A_5 \operatorname{Log}(k+2)} \leqslant \sigma \leqslant 1 \\ |t| \leqslant \dfrac{1}{A_5 \operatorname{Log}(k+2)} \end{cases}$$

Because $A_6 \operatorname{Log}(k+2) > A_4$ it follows that when χ is the principal character $L(s, \chi)$ has no zeros in the above rectangle.

This completes the discussion of the first part of case (b).

3.6 Zeros " far " from the real axis. We now assume that if $\rho = \beta + i\gamma$ is a zero of $L(s, \chi)$, then

$$\gamma \geqslant \{ A_5 \operatorname{Log}(k+2) \}^{-1} .$$

We would like to mimic the discussion of case (a) by considering the function $f(s) = L^4(s + i\gamma, \chi)\, L(s + i\,2\,\gamma, \chi^2)$, but now we have $\chi^2 = \chi_0$ and possibly even $\chi = \chi_0$. Thus, $f(s)$ has a pole at $s = 1 - i\,2\,\gamma$ and perhaps a pole of order 4 at $s = 1 - i\gamma$. Hence the condition that f is holomorphic within certain discs need not be satisfied. To circumvent these difficulties we shall define a modified function f_1 by

$$f_1(s) = (s + i\gamma - 1)^{E(\chi)}\,(s + i\,2\,\gamma - 1)\, f(s) .$$

We are going to apply theorem 4.2 with $s_0 = \dfrac{11}{10}$, $R = \dfrac{3}{10}$ and $q = 2$ to $f_1(s)$.

It is easily seen that $f_1(s_0) \neq 0$ and that on the disc $|s - s_0| \leqslant \dfrac{3}{5}$ the function $f_1(s)$ is holomorphic. To obtain an upper bound for $|f(s)/f(s_0)|$ on this disc we first observe that for $r = 1$ or 2 we have

$$\left| \frac{s - 1 + 2\,i\gamma}{s_0 - 1 + 2\,i\gamma} \right| = \left| 1 + \frac{s - s_0}{s_0 - 1 + 2\,i\gamma} \right| \leqslant 1 + \frac{2\,R}{s_0 - 1} = 7 ,$$

upon using theorems 8.1 and 8.2(c) we then conclude in the usual way that

$$\left| \frac{f_1(s)}{f_1(s_0)} \right| \leqslant 7 \left| \frac{f(s)}{f(s_0)} \right| \leqslant 7 \{ 11.12\, k^{1/2}(2 + 2\,\gamma)^{1/2} \}^5 = e^{\mathrm{U}}$$

where $\mathrm{U} = A_7 \operatorname{Log} \tau(k, \gamma)$ and A_7 is an effectively computable absolute constant.

Let λ, μ be positive real numbers such that $\lambda + \mu < \frac{1}{10}$ and suppose that $L(s, \chi)$ has a zero $\rho = \beta + i\gamma$ such that $1 - \lambda \leqslant \beta \leqslant 1$. By hypothesis we also have $\gamma^{-1} \leqslant A_6 \operatorname{Log}(k + 2)$. Thus, $f_1(s)$ has at least a zero of order 4 at $s = \beta$ and no zeros in the half plane $\mathfrak{R}(s) > 1$. We now apply theorem 4.2(*c*) with $r = d = \frac{1}{10}$ and conclude that for $|s - s_0| \leqslant \frac{1}{10}$

$$\frac{4}{\lambda + \mu} \quad 4\,\mathfrak{R}\left(\frac{1}{s - \beta}\right) \leqslant 15\,\mathrm{U} + \mathfrak{R}\left(\frac{f_1'(s)}{f_1(s)}\right).$$

We next obtain an upper bound for $\mathfrak{R}(f_1'(s)/f_1(s))$. It is clear that

$$\frac{f_1'(s)}{f_1(s)} = \frac{1}{s + i\,2\,\gamma - 1} + \frac{4\,E(\chi)}{s + i\gamma - 1} + \frac{f'(s)}{f(s)}.$$

If we write $s = 1 + \mu$ and consider $\mathfrak{R}(1/(s + ir\gamma - 1))$ as a function of μ, then the maximum occurs when $\mu = r\gamma$ and we certainly have

$$\mathfrak{R}\left(\frac{f_1'(s)}{f_1(s)}\right) \leqslant \frac{9}{4\,\gamma} + \mathfrak{R}\left(\frac{f'(s)}{f(s)}\right).$$

As in case (*a*) we have

$$\mathfrak{R}\left(\frac{f'(1 + \mu)}{f(1 + \mu)}\right) \leqslant 3\left(\frac{1}{\mu} + A_1\right),$$

and so

$$\frac{4}{\lambda + \mu} - \frac{3}{\mu} \leqslant \frac{9}{4\,\gamma} + 3\,A_1 + 15\,\mathrm{U} < \frac{9\,A_6}{4} \operatorname{Log}(k + 2) + 3\,A_1 + 15\,\mathrm{U}$$
$$< A_8 \operatorname{Log} \tau(k, \gamma),$$

for a suitable effective, absolute constant A_8. If we now choose $\mu = 6\,\lambda$ and λ is such that $0 < 7\,\lambda < \frac{1}{10}$, then

$$\frac{1}{\lambda} < 14\,A_8 \operatorname{Log} \tau(k, \gamma).$$

Upon taking $A_9 = \max(70, 14\,A_8)$ and choosing

$$\lambda_1^{-1}(k, \gamma) = A_9 \operatorname{Log} \tau(k, \gamma)$$

we conclude that $L(s, \chi)$ has no zeros $\beta + i\gamma$ which satisfy

$$1 - 1/A_9 \operatorname{Log} \tau(k, \gamma) \leqslant \beta \leqslant 1 \quad \text{and} \quad |\gamma| \geqslant 1/A_6 \operatorname{Log}(k + 2).$$

Putting both parts of case (b) together we conclude that if b is any real number satisfying $0 < 2b < \min(A_6^{-1}, A_9^{-1})$, then in the region $R_{2b}(k)$ the function $L(s, \chi)$ has no complex and at most one real zero. If this real zero does exist it is simple and χ is non-principal.

3.7 Case (c). We now construct a region $R_{2\alpha}(K)$ with the properties stated in the theorem. If c is any real number satisfying $0 < c \leqslant \min(a, b)$ and $k \leqslant K$, then $R_{2c}(K) \subseteq R_{2a}(k) \cap R_{2b}(k)$.

By cases (a) and (b) it follows that for any character χ_k the function $L(s, \chi_k)$ has no complex zeros in $R_{2c}(K)$ and if χ_k is complex $L(s, \chi_k)$ does not have any real zeros in $R_{2c}(K)$.

Suppose that $L(s, \chi_k)$ and $L(s, \chi_r)$, where $k, r \leqslant K$, both have real zeros in $R_{2c}(K)$ say β_k and β_r, where $\beta_k < \beta_r$. The characters χ_k, χ_r must both be real and non-principal. If $\chi_k \sim \chi_r$, then it follows that $L(s, \chi_k)$ has two distinct real zeros in $R_{2b}(k)$, which contradicts the results of § 3.5. Thus, the characters χ_k and χ_r are *not* equivalent.

Hence, to complete the proof of the theorem we must show that it is possible to choose a value of c such that if χ_k and χ_r are any two non-equivalent, real, non-principal characters, then both the functions $L(s, \chi_k)$ and $L(s, \chi_r)$ do not have real zeros in the region $R_{2c}(K)$.

Let us suppose that χ_k, χ_r are non-principal, real, non-equivalent characters such that

$$L(\beta_k, \chi_k) = L(\beta_r, \chi_r) = 0,$$

where $1 - \lambda \leqslant \beta_k \leqslant \beta_r \leqslant 1$. We observe that $\chi_k \chi_r$ is a non-principal, real character modulo kr and that if we define the function $f(s)$ by

$$f(s) = L(s, \chi_k)\, L(s, \chi_r)\, L(s, \chi_k \chi_r)$$

then we can use theorem 4.2 in the usual way to deduce that

$$\frac{1}{\lambda} < A_9 \operatorname{Log}(K + 2).$$

where A_{10} is an effective, absolute constant.

Thus, if we choose $\alpha = \min\left(a, b, \frac{1}{2} A_{10}^{-1}\right)$, then the region $R_{2\alpha}(K)$ has all the required properties. ■

§4 SIEGEL'S THEOREM

4.1 In the *Notes* to chapter 7 we discussed the importance of obtaining a lower bound for $L(1, \chi)$ in the case when χ is a real, non-principal character. We now return to this problem. The result which will be obtained in theorem 8.4 asserts that $L(1, \chi) > c_1(\varepsilon)\, k^{-\varepsilon}$ for any $\varepsilon > 0$. However $c_1(\varepsilon)$ will depend on the possible real zeros of $L(s, \chi)$ and we do not know whether such zeros exist. Consequently, for a given value of ε we are unable to compute a numerical value for $c_1(\varepsilon)$. This means that all later theorems which depend on theorem 8.4 are " ineffective ". We shall always be careful to isolate the ineffective parts of such theorems and to state explicitly which constants are effective and which are ineffective.

Theorem 8.4 *For each $\varepsilon > 0$ there exist constants $c(\varepsilon) > 0$ and $c_1(\varepsilon) > 0$ such that if χ is a real, non-principal character modulo k, then*

(a) $L(s, \chi) \neq 0$ *for real* $s > 1 - c(\varepsilon)\, k^{-\varepsilon}$ *and*
(b) $L(1, \chi) > c_1(\varepsilon)\, k^{-\varepsilon}$.
The constants $c(\varepsilon)$ and $c_1(\varepsilon)$ are not effective.

Proof. We first introduce some notation :

$$\Omega = \{\, \chi : \chi \text{ real and non-principal} \,\},$$

$$\delta = \min\left(\frac{\varepsilon}{6}, \frac{1}{10}\right).$$

If for some $\chi \in \Omega$, $L(s, \chi)$ has a real zero β satisfying $\beta \geqslant 1 - \delta$, then we define

$$k_0 = \min_k \{\, k \geqslant 3 \mid \exists \chi_k \text{ with } L(\beta_k, \chi_k) = 0 \text{ and } \beta_k \geqslant 1 - \delta \,\}$$

and

$$\beta_0 = \max_\beta \left\{ \beta \in \mathbb{R} \mid \prod_{\chi (\mathrm{mod}\ k_0)} L(\beta, \chi) = 0 \right\}.$$

It is clear that β_0 and k_0 are well defined. Let χ' be a character modulo k_0 such that $L(\beta_0, \chi') = 0$. This choice of χ' will remain fixed throught the discussion.

The proof is divided into the following cases :

(i) N_0 $L(s, \chi)$ has a real zero β satisfying $\beta \geqslant 1 - \delta$,

(ii) β_0 and k_0 exist and we prove the theorem for all $L(s, \chi_k)$ with $k \leqslant k_0$,

(iii) β_0 and k_0 exist and we prove the theorem for all $L(s, \chi_k)$ where $\chi_k \sim \chi'$ and $k > k_0$,

(iv) β_0 and k_0 exist and we prove the theorem for all $L(s, \chi_k)$ where $\chi_k \nsim \chi'$ and $k > k_0$.

4.2 In order to settle cases (i), (ii) and (iv) we require the following rather technical lemma, the proof of which will be given in § 4.3.

Lemma 8.1 *Let χ be a real, non-principal character modulo k. There exists an effectively computable constant A_0 such that for each η satisfying $0 < \eta \leqslant 1/10$ the following two statements are true* :

(a) $\frac{1}{2} < \frac{1}{\eta} L(1, \chi) (A_0 k)^{2\eta} + \zeta(1 - \eta) L(1 - \eta, \chi)$;

(b) *if χ_1 is a non-principal real character modulo k_1 and $\chi_1 \nsim \chi$, then*

$$\frac{1}{2} < \frac{1}{\eta} L(1, \chi) L(1, \chi_1) L(1, \chi\chi_1) (A_0 \sqrt{kk'})^{2\eta} +$$
$$+ \zeta(1 - \eta) L(1 - \eta, \chi) L(1 - \eta, \chi_1) L(1 - \eta, \chi\chi_1) .$$

We will now use the above lemma to prove the theorem.

Cases (i) *and* (ii). Statement (*a*) of the theorem holds trivially for any $c(\varepsilon)$ satisfying $0 < c(\varepsilon) \leqslant \delta$.

To obtain part (*b*) we will employ the lemma with $\eta = \delta$. If χ satisfies the hypothesis of case (i) or case (ii) we know that $L(s, \chi) \neq 0$ for $s \geqslant 1 - \delta$. Since $L(s, \chi) \to 1$ as $s \to \infty$ it follows that $L(s, \chi) > 0$ for all $s \geqslant 1 - \delta$. In addition, from equation (2.3) we know that $\zeta(1 - \delta) < 0$. Thus, using the first part of the lemma we conclude

$$\frac{1}{2} < \frac{1}{\delta} L(1, \chi) (A_0 k)^{2\delta}$$

Since $2\delta < \varepsilon$ we have

$$L(1, \chi) > \frac{\delta}{2} (A_0 k)^{-2\delta} \geqslant \frac{\delta}{2 A_0^{2\delta}} \cdot k^{-\varepsilon}$$

Thus part (*b*) of the theorem follows with any $c_1(\varepsilon)$ satisfying

$$0 < c_1(\varepsilon) \leqslant \frac{\delta}{2 A_0^{2\delta}} .$$

Case (iii). If $k > k_0$ and $\chi_k \sim \chi'$ it follows from the definition of β_0 that

$$L(s, \chi_k) \neq 0 \quad \text{for} \quad s > \beta_0 = 1 - (1 - \beta_0) .$$

From our discussion of equivalent characters in § 3.3 of chapter 7 we have

$$\begin{aligned} 0 < \frac{L(1, \chi')}{L(1, \chi_k)} &= \prod_{p \mid kk_0} \frac{1 - \chi_k(p)\, p^{-1}}{1 - \chi'(p)\, p^{-1}} \\ &\leqslant \prod_{p \mid kk_0} \frac{1 + p^{-1}}{1 - p^{-1}} \leqslant \prod_{p \mid kk_0} \frac{3}{p^{\varepsilon}} \cdot p^{\varepsilon} \\ &\leqslant A(\varepsilon)\, (kk_0)^{\varepsilon} \end{aligned}$$

where $A(\varepsilon) = \prod_{p^{\varepsilon} < 3} 3\, p^{-\varepsilon}$. Thus statements (*a*) and (*b*) of the theorem hold for any $c(\varepsilon)$ and $c_1(\varepsilon)$ which satisfy

$$0 < c(\varepsilon) \leqslant 1 - \beta_0$$

and

$$0 < c_1(\varepsilon) \leqslant \frac{L(1, \chi')}{A(\varepsilon)\, k_0^{\varepsilon}}$$

Case (iv). We now suppose that $k > k_0$ and $\chi_k \nsim \chi'$. Part (*b*) of lemma 8.1 will be used with $\chi_1 = \chi', \chi = \chi_k$ and $\eta = 1 - \beta_0$. Since $L(1 - \eta, \chi') = L(\beta_0, \chi') = 0$ we have

$$\frac{1}{2} < L(1, \chi_k)\, L(1, \chi')\, L(1, \chi_k \chi')\, \frac{(A_0 \sqrt{k_0\, k})^{2(1 - \beta_0)}}{1 - \beta_0} .$$

From theorem 8.2 we have the estimates

$$L(1, \chi') < 2 + \operatorname{Log} k_0 < \frac{k_0^{\delta}}{\delta}$$

and similarly

$$L(1, \chi_k \chi') < \frac{(k_0\, k)^{\delta}}{\delta} ,$$

since $\delta \geqslant 1 - \beta_0$ we can conclude that

$$L(1, \chi_k) > \frac{1}{2\, A_0^{2(1 - \beta_0)}} \cdot \frac{(1 - \beta_0)\, \delta^2}{k_0^{2\delta + 1 - \beta_0}} \cdot \frac{1}{k^{\delta + 1 - \beta_0}}$$

$$L(1, \chi_k) > \frac{c_1(\varepsilon)}{k^{\varepsilon}} .$$

Thus part (*b*) of the theorem holds for any $c_1(\varepsilon)$ which satisfies

$$0 < c_1(\varepsilon) \leqslant \frac{(1 - \beta_0)\,\delta^2}{2\,A_0^{2\delta}\,k_0^{3\delta}}\,.$$

It remains now to compute an upper bound for $c(\varepsilon)$. Using theorem 8.1 with $\theta = 1 - \frac{1}{2}\delta$ we have for all s satisfying $1 - \frac{1}{2}\delta \leqslant s \leqslant 1$ the following inequalities :

$$L'(s, \chi_k) < 14\,\frac{k^{\delta/2}}{\delta}\,(2 + \operatorname{Log} 2\,k)$$

$$< 14\,\frac{k^{\delta/2}}{\delta}\left(\frac{2\,k}{\frac{1}{2}\,\delta}\right)^{\delta/2} < 20\,\frac{k^{\delta}}{\delta}\,.$$

By the mean-value theorem we have for some $s_1 \in (s, 1)$:

$$L(1, \chi_k) - L(s, \chi_k) = (1 - s)\,L'(s_1, \chi_k) \tag{8.6}$$

$$\leqslant (1 - s)\,\frac{20\,k^{\delta}}{\delta}\,,$$

and so

$$L(s, \chi_k) \geqslant L(1, \chi_k) - 20(1 - s)\,\frac{k^{\delta}}{\delta}\,.$$

Using our lower bound for $L(1, \chi_k)$ it follows that

$$L(s, \chi_k) > \frac{c_1(\varepsilon)}{k^{\varepsilon}} - 20(1 - s)\,\frac{k^{\delta}}{\delta}$$

$$\geqslant c_1(\varepsilon)\,k^{-2\delta} - (1 - s)\,\frac{24}{\delta^2}\,k^{\delta}\,.$$

This last expression is strictly positive if

$$c_1(\varepsilon)\,\frac{\delta^2}{24}\,k^{-3\delta} > 1 - s$$

and so

$$L(s, \chi_k) > 0 \quad \text{for} \quad s > 1 - c(\varepsilon)\,k^{-\varepsilon}\,,$$

where $c(\varepsilon)$ is any real number satisfying

$$0 < c(\varepsilon) \leqslant \frac{\delta^4}{48}(1 - \beta_0)(A_0^2 k_0^3)^{-\delta}.$$

Combining cases (i), (ii), (iii) and (iv) we have the result for

$$0 < c(\varepsilon) \leqslant \min\left\{\delta, 1 - \beta_0, \frac{\delta^4}{48}(1 - \beta_0)(A_0^2 k_0^3)^{-\delta}\right\}$$

and

$$0 < c_1(\varepsilon) \leqslant \min\left\{\frac{1}{2}\delta A^{-2\delta}, \frac{L(1, \chi')}{A(\varepsilon) k_0^\varepsilon}, \frac{1}{2}\delta^2(1 - \beta_0)(A_0^2 k_0^3)^{-\delta}\right\}. \qquad \blacksquare$$

4.3 Proof of lemma 8.1. We can handle parts (*a*) and (*b*) simultaneously if we define functions f, K and g as follows :

$$f(s) = \begin{cases} L(s, \chi) & \text{in case } (a) \\ L(s, \chi)\, L(s, \chi_1)\, L(s, \chi\chi_1) & \text{in case } (b) \end{cases}$$

$$K = \begin{cases} k & \text{in case } (a) \\ \sqrt{kk'} & \text{in case } (b) \end{cases}$$

$$g(s) = \zeta(s)\, f(s)$$

and prove that

$$\frac{1}{2} < \frac{f(1)}{\eta}(A_0 K)^{2\eta} + g(1 - \eta).$$

The lemma will be proved by choosing a suitable contour and applying Cauchy's residue theorem to the function

$$h(s) = g(1 - \eta + s)\frac{x^s}{s(s + 1)}$$

where x is a parameter to be chosen later. This function has simple poles at $s = 0$ and $s = \eta$ with residues

$$g(1 - \eta) \quad \text{and} \quad f(1)\frac{x^\eta}{\eta(1 + \eta)}.$$

Let $\mathcal{C}$ be the contour shown in figure 14, where α is an absolute constant, to be chosen later, satisfying $0 < \eta \leqslant \alpha < \frac{1}{8}$. By the residue theorem we have

$$\frac{1}{2\pi i}\int_{\mathcal{C}} g(1-\eta+s)\frac{x^s}{s(s+1)}\,ds$$
$$= g(1-\eta) + \frac{f(1)\,x^\eta}{\eta(\eta+1)},$$

provided of course that the integral exists. In fact the integral does exist and we will prove it is equal to

$$\frac{1}{2\pi i}\int_{1-i\infty}^{1+i\infty} g(1-\eta+s)\frac{x^s}{s(s+1)}\,ds - \frac{1}{2\pi i}\int_{-\alpha-i\infty}^{-\alpha+i\infty} g(1-\eta+s)\frac{x^s}{s(s+1)}\,ds. \quad (8.6)$$

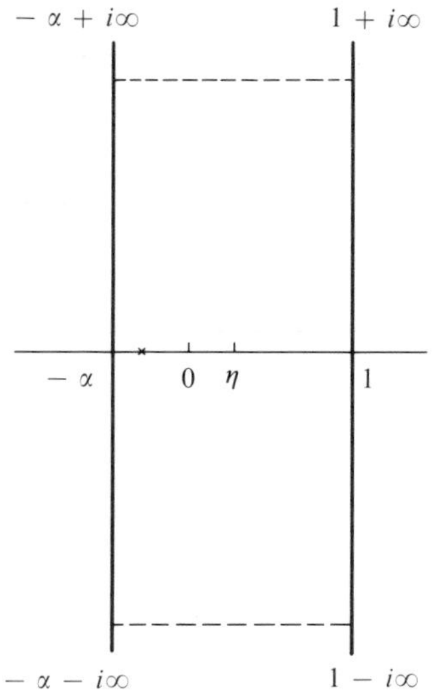

Fig. 13

Observing that $1-\eta+\sigma \geqslant 1-2\alpha$ on and between the lines $\mathcal{R}(s) = -\alpha$ and $\mathcal{R}(s) = 1$ then employing theorems 8.1 and 2.3 with $\theta = 1-2\alpha$, we deduce that the integrand is uniformly

$$O\{(|t|^{2\alpha})^4\,|t|^{-2}\} = o(1)$$

as $t \to \infty$ or as $t \to -\infty$. Thus the contribution of the horizontal segments of the contour to the integral is zero and both integrals (8.6) are convergent. So we have

$$\frac{f(1)\,x^\eta}{\eta(\eta+1)} + g(1-\eta) = \frac{1}{2i\pi}\int_{1-i\infty}^{1+i\infty} h(s)\,ds - \frac{1}{2i\pi}\int_{-\alpha-i\infty}^{-\alpha+i\infty} h(s)\,ds$$
$$= I(1) - I(-\alpha).$$

On the line $\mathcal{R}(s) = -\alpha$ we have $1-\eta+\sigma < 1-\alpha$ and on using theorems 8.1 and 2.3 with $\theta = 1-2\alpha$ we see that since $2-8\alpha > 1$

$$|I(-\alpha)| < \frac{1}{2\pi}\int_{-\alpha-i\infty}^{-\alpha+i\infty} A_1(\alpha)\,K^{2\alpha}\frac{|s|^{4\alpha}}{|s(s+1)|}\,x^{-\alpha}\,|ds|$$
$$< A_2(\alpha)\frac{K^{2\alpha}}{x^\alpha}.$$

where $A_1(\alpha)$, $A_2(\alpha)$ are effectively computable functions of α.

We next obtain a lower bound for $I(1)$. By considering the Euler product representation of $g(s)$ we see, as in the proof of theorem 7.8(b), that for $R(s) > 1$

$$g(s) = \sum_{n=1}^{\infty} b_n . n^{-s}$$

where $b_1 = 1$ and $b_n \geqslant 0$ for all $n \geqslant 1$. Hence we have using theorem 2.6

$$I(1) = \frac{1}{2\,i\pi} \int_{1-i\infty}^{1+i\infty} \left(\sum_{n=1}^{\infty} b_n . n^{\eta - 1 - s} \right) \frac{x^s\, ds}{s(s+1)} .$$

$$\begin{aligned} I(1) &= \frac{x^{-1}}{2\,i\pi} \sum_{n=1}^{\infty} b_n\, n^{\eta} \int_{1-i\infty}^{1+i\infty} \frac{(x/n)^{s+1}\, ds}{s(s+1)} \\ &= \frac{1}{x} \sum_{n \leqslant x} b_n\, n^{\eta} \left(\frac{x}{n} - 1 \right) \\ &= \sum_{n \leqslant x} \frac{b_n}{n^{1-\eta}} \left(1 - \frac{n}{x} \right) \\ &\geqslant 1 - \frac{1}{x} . \end{aligned}$$

if we take $x \geqslant 4$.

We now have

$$\begin{aligned} \frac{f(1)\, x^{\eta}}{\eta(1+\eta)} + g(1-\eta) &= I(1) - I(-\alpha) \\ &\geqslant \frac{3}{4} - A_2(\alpha)\, K^{2\alpha}\, x^{-\alpha} . \end{aligned}$$

On taking $\alpha = \dfrac{1}{10}$ and $x = \left(4\, A_2\!\left(\dfrac{1}{10}\right)\right)^{10} K^2$ we conclude that for η satisfying $0 < \eta \leqslant \dfrac{1}{10}$ we have

$$\frac{1}{2} < \frac{f(1)}{\eta(1+\eta)} (A_0\, K)^{2\eta} + g(1-\eta)$$

where $A_0 = \left(4\, A_2\!\left(\dfrac{1}{10}\right)\right)^5$. This completes the proof of the lemma. ■

§ 5 PRIMES IN ARITHMETIC PROGRESSIONS

5.1 We showed in chapter 7 that every arithmetic progression $\{ kn + l \}$ with $(k, l) = 1$ contains an infinity of primes. In this section we shall study the density of primes in arithmetic progressions and prove the analogue of the Prime Number Theorem with error term. The following notation is a natural extension of the familiar functions π, θ, ψ, ψ_1.

Let k and l be coprime, positive integers, the for all $x \geqslant 0$ we write

$$\pi(x; k, l) = \sum_{\substack{p \leqslant x \\ p \equiv l(k)}} 1 ,$$

$$\theta(x; k, l) = \sum_{\substack{p \leqslant x \\ p \equiv l(k)}} \operatorname{Log} p ,$$

$$\psi(x; k, l) = \sum_{\substack{n \leqslant x \\ n \equiv l(k)}} \Lambda(n) ,$$

$$\psi_1(x; k, l) = \int_0^x \psi(u; k, l)\, du = \sum_{\substack{n \leqslant x \\ n \equiv l(k)}} (x - n)\, \Lambda(n) .$$

It is a fairly straightforward exercise, using the methods of chapter 2, to prove that as x tends to infinity

$$\pi(x; k, l) \sim \frac{1}{\varphi(k)} \frac{x}{\operatorname{Log} x} .$$

However, our object is to obtain more precise expressions for $\pi(x; k, l)$ etc. Before we can state our main theorem we need to introduce some more notation.

Let α be an absolute constant with the property stated in theorem 8.3. If k and l are coprime, positive integers, then we shall define $b(k)$ and $\lambda(k, l)$ as follows :

(i) If there exists a character χ_k such that $L(s, \chi_k)$ has a real zero $\beta(k)$ in the region $R_{2\alpha}(k)$, then

$$b(k) = \beta(k) , \qquad \lambda(k, l) = \chi_k(l) ;$$

(ii) If no such character exists, then

$$b(k) = 1 , \qquad \lambda(k, l) = 0 .$$

Theorem 8.5 *For $x \geqslant 2$ and k, l coprime, positive integers we have the following asymptotic relations* :

(a) $$\psi_1(x\,;k,l) = \frac{x^2}{2\,\varphi(k)} - \frac{\lambda x^{b+1}}{b(b+1)\,\varphi(k)} + O(x^2\,e^{-2a\sqrt{\operatorname{Log} x}})$$

(b) $$\psi(x\,;k,l) = \frac{x}{\varphi(k)} - \frac{\lambda x^b}{b\varphi(k)} + O(x\,e^{-a\sqrt{\operatorname{Log} x}})$$

(c) $$\theta(x\,;k,l) = \frac{x}{\varphi(k)} - \frac{\lambda x^b}{b\varphi(k)} + O(x\,e^{-a\sqrt{\operatorname{Log} x}})$$

(d) $$\pi(x\,;k,l) = \frac{\operatorname{Li}(x)}{\varphi(k)} - \frac{\lambda\operatorname{Li}(x^b)}{\varphi(k)} + O(x\,e^{-a\sqrt{\operatorname{Log} x}})\,.$$

where a and the implied "O" constants are absolute and effective.

The proof of the above theorem is rather similar to the proof of theorem 4.6. First we shall deduce the expression for $\psi_1(x; k, l)$, then by the usual differencing argument obtain the formula for $\psi(x; k, l)$. Finally, the formulae for $\pi(x; k, l)$ and $\theta(x; k, l)$ will be deduced by a partial summation argument. We now give a brief outline of the essential ideas behind the proof.

Our starting point will be equation (7.8) which asserts that for $\mathcal{R}(s) > 1$

$$\sum_\chi \bar{\chi}(l)\operatorname{Log} L(s,\chi) = \varphi(k)\sum_{\nu=1}^{\infty}\sum_{p^\nu\equiv l(k)} \nu^{-1}p^{-\nu s}\,.$$

Differentiating with respect to s we obtain

$$-\frac{1}{\varphi(k)}\sum_\chi \bar{\chi}(l)\frac{L'(s,\chi)}{L(s,\chi)} = \sum_{n\equiv l(k)}\frac{\Lambda(n)}{n^s}\,. \tag{8.7}$$

If we denote the left hand side of the above equation by $G(s)$ and define $I(x)$, for $x \geqslant 1$ by

$$I(x) = \frac{1}{2\,i\pi}\int_{2-i\infty}^{2+i\infty}\frac{G(s)}{s(s+1)}\,x^{s+1}\,ds\,.$$

then a routine integration will show that

$$I(x) = \psi_1(x\,;k,l)\,.$$

It will then remain to evaluate the integral $I(x)$. This will be done by distorting the line of integration and moving it into the region $R_{2\alpha}(k)$ and then using information about the behaviour of $L'(s, \chi)$ and $L(s, \chi)$ in $R_{2\alpha}(k)$ to estimate the integral. As the reader can see the proof is very similar in structure to the proof of theorem 4.6. However, there are one or two annoying technicalities which arise from the possible real zero in $R_{2\alpha}(k)$. The precise results which we shall require about $L'(s, \chi)$ and $L(s, \chi)$ are collected together in the next two theorems.

5.2 Before we state the theorems we shall introduce some further notation which will hide the technicalities and permit us to handle various special cases in a uniform way.

First of all, let α be the absolute constant of theorem 8.3 and χ any character modulo k. We define the function f as follows :

(i) If $L(s, \chi)$ has no zeros in $R_{2\alpha}(k)$, then

$$f(s, \chi) = (s - 1)^{E(\chi)} L(s, \chi),$$

(ii) If $L(s, \chi)$ has a zero $\beta \in R_{2\alpha}(k)$, then

$$f(s, \chi) = \frac{L(s, \chi)}{s - \beta}.$$

We can now state the required properties of $L'(s, \chi)$ and $L(s, \chi)$ in terms of $f(s, \chi)$.

Theorem 8.6 *With the above notation we have :*

(a) *$f(s, \chi)$ is holomorphic in the half plane $\mathfrak{R}(s) > 0$,*
(b) *$f(s, \chi) \neq 0$ in $R_{2\alpha}(k)$,*
(c) $\left|\dfrac{f'(s, \chi)}{f(s, \chi)}\right| \leqslant A \operatorname{Log} \tau(k, t)$ *in $R_\alpha(k)$, where A is an effectively computable absolute constant.*

Proof. Parts (*a*) and (*b*) are immediate consequences of theorems 7.7 and 8.3. To prove part (*c*) we shall consider the following two cases :

$$\mathfrak{R}(s) > 1 + \frac{\alpha}{2 \operatorname{Log} \tau(k, t)} \quad \text{and} \quad \mathfrak{R}(s) \leqslant 1 + \frac{\alpha}{2 \operatorname{Log} \tau(k, t)}.$$

The first case is easily dealt with. From the definition of $f(s, \chi)$ and the Dirichlet series expansion of $L'(s, \chi)/L(s, \chi)$ we have, for $\mathfrak{R}(s) = \sigma > 1$:

$$\left|\frac{f'(s, \chi)}{f(s, \chi)}\right| \leqslant \left|\frac{L'(s, \chi)}{L(s, \chi)}\right| + \left|\frac{1}{s - b}\right|$$

$$\leqslant \left|\frac{\zeta'(\sigma)}{\zeta(\sigma)}\right| + \frac{1}{\sigma - 1} \leqslant \frac{A_1}{\sigma - 1}, \tag{8.8}$$

for a suitable effective absolute constant A_1. Thus, for $\mathfrak{R}(s) > 1 + \frac{1}{2}\alpha/\operatorname{Log} \tau(k, t)$ we certainly have

$$\left|\frac{f'(s, \chi)}{f(s, \chi)}\right| \leqslant \frac{2 A_1}{\alpha} \operatorname{Log} \tau(k, t). \tag{8.9}$$

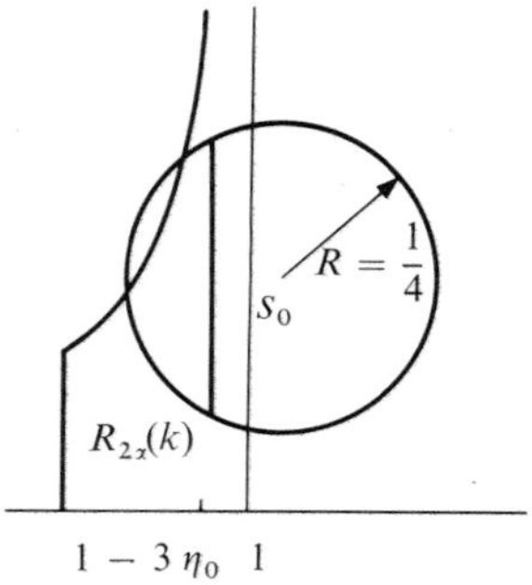

Fig. 14

It now remains to deal with the case $\mathcal{R}(s) \leqslant 1 + \frac{1}{2}\alpha/\text{Log}\,\tau(k, t)$. This will be done by applying theorem 4.2(*d*) to $f(s, \chi)$. First we choose the parameters of theorem 4.2 and then check that $f(s, \chi)$ satisfies the hypotheses.

Choose a fixed $t_0 \geqslant 0$ and define :

$$\tau_0 = \tau(k, t_0)$$

$$\eta_0 = \frac{\alpha}{2\,\text{Log}\,\tau_0}$$

$$s_0 = 1 + \eta_0 + it_0 = \sigma_0 + it_0$$

$$R = \frac{1}{4}, \qquad q = 2, \qquad r = 3\,\eta_0, \qquad d = 4\,\eta_0\,.$$

The function $f(s, \chi)$ is holomorphic in the disc $|\,s - s_0\,| \leqslant 2\,R = \frac{1}{2}$ and $f(s_0, \chi) \neq 0$. To apply theorem 4.2 we must deduce an upper bound for $|\,f(s, \chi)/f(s_0, \chi)\,|$ on the disc $|\,s - s_0\,| \leqslant \frac{1}{2}$. To do this we must consider the two possible choices for $f(s, \chi)$ separately.

In case (i) we use theorem 8.1 with $\theta = \frac{1}{2}$ to deduce that for $\mathcal{R}(s) \geqslant \frac{1}{2}$ we have

$$|\,f(s, \chi)\,| \leqslant 12(k\,|\,s\,|)^{1/2}\,|\,s - 1\,|^{E(\chi)} + E(\chi)\,.$$

and by theorem 8.2(*c*)

$$|\,f(s_0, \chi)\,|^{-1} < \frac{1 + \eta_0}{\eta_0^{1 + E(\chi)}} < \frac{2}{\eta_0^{1 + E(\chi)}}\,.$$

Thus, on the disc $|\,s - s_0\,| \leqslant \frac{1}{2}$ we certainly have

$$\left|\frac{f(s, \chi)}{f(s_0, \chi)}\right| \leqslant e^{\mathrm{U}}$$

where $\mathrm{U}_1 = A_2\,\text{Log}\,\tau_0$ and A_2 is an effectively computable absolute constant.

In case (ii) we write

$$f(s, \chi) = (s - \beta)^{-1}\int_\beta^s L'(u, \chi)\,du$$

$$\leqslant \max_{u \in \mathfrak{L}}\,|\,L'(u, \chi)\,|\,,$$

where $\mathfrak{L}$ is the line segment joining β to s. We once again appeal to theorem 8.1 with $\theta = \frac{1}{2}$ to deduce that for $\mathfrak{R}(s) \geqslant \frac{1}{2}$

$$| f(s, \chi) | \leqslant 24(km)^{1/2} \{ 2 + \operatorname{Log} (2\, km) \} ,$$

where $m = \max (| s |, \beta) \leqslant | s | + 1$. To estimate $| f(s_0, \chi) |^{-1}$ we use theorem 8.2(c):

$$| f(s_0, \chi) |^{-1} < | s_0 - \beta | \frac{\sigma_0}{\sigma_0 - 1} \leqslant (| 1 + \eta_0 - \beta | + t_0) \frac{2}{\eta_0}$$
$$\leqslant (1 + t_0) \frac{2}{\eta_0} \leqslant (1 + \tau_0) \frac{2}{\eta_0} .$$

Thus, on the disc $| s - s_0 | \leqslant \frac{1}{2}$ we again have the estimate

$$\left| \frac{f(s, \chi)}{f(s_0, \chi)} \right| \leqslant e^{\mathrm{U}}$$

where $\mathrm{U}_2 = A_3 \operatorname{Log} \tau_0$ and A_3 is an effective absolute constant.

If we take $A_4 = \max (A_2, A_3)$, then an upper bound for $| f(s, \chi)/f(s_0, \chi) |$ on the disc $| s - s_0 | \leqslant \frac{1}{2}$ is given, in both cases, by $\exp(A_4 \operatorname{Log} \tau_0)$.

We can now apply theorem 4.2(d), since $f(s, \chi)$ has no zeros in the region

$$| s - s_0 | \leqslant R ; \quad \mathfrak{R}(s) \geqslant \sigma_0 - d = 1 - 3\, \eta_0 ,$$

which is contained in $R_{2\alpha}(k)$. Thus, upon using (8.8) we conclude

$$\left| \frac{f'(s, \chi)}{f(s, \chi)} \right| \leqslant \frac{7}{1} \left\{ \frac{2\, A_4}{\alpha} \operatorname{Log} \tau_0 + \frac{\frac{1}{2} A_2 \operatorname{Log} \tau_0}{\left(\frac{1}{4} - 4\, \eta_0 \right)^2} \right\}$$
$$\leqslant A_5 \operatorname{Log} \tau_0$$

where A_5 is a suitable effectively computable absolute constant. In particular this last inequality holds on the line joining $s_0 - 3\, \eta_0$ to s_0.

Combining this result with (8.9) we deduce that for all $s = \sigma + it_0$ with $\sigma \geqslant 1 - 2\, \eta_0$ we have

$$\left| \frac{f'(s, \chi)}{f(s, \chi)} \right| \leqslant A_6 \operatorname{Log} \tau_0$$

where $A_6 = \max (A_5, 2\, A_1/\alpha)$. Using this latter inequality for each $t_0 \geqslant 0$ we obtain part (c) of the theorem. ■

The next theorem is a consequence of theorem 8.6 and is the last technical result which we require for the proof of theorem 8.5.

Theorem 8.7 (a) *If $L(s, \chi)$ has no zeros in $R_{2\alpha}(k)$, then for $s \in R_\alpha(k)$*

$$| L(s, \chi) |^{-1} < A_1 \operatorname{Log} \tau(k, t) ;$$

(b) *If $L(s, \chi)$ has a zero $\beta \in R_{2\alpha}(k)$, then for $s \in R_\alpha(k)$ and $s \neq \beta$*

$$| L(s, \chi) |^{-1} < A_1 \{ \operatorname{Log} \tau(k, t) + | s - \beta |^{-1} \}$$

where A_1 is an effectively computable absolute constant.

Proof. As in the proof of theorem 8.6 we shall consider the two cases :

$$\Re(s) \geqslant 1 + \frac{\alpha}{2 \log \tau(k, t)} \quad \text{and} \quad \Re(s) < 1 + \frac{\alpha}{2 \operatorname{Log} \tau(k, t)} .$$

As before the first case is easy, for if $\Re(s) > 1$ we have

$$| L(s, \chi) |^{\pm 1} \leqslant \zeta(\sigma) < \frac{\sigma}{\sigma - 1}$$

and so for $\sigma \geqslant 1 + \frac{1}{2} \alpha/\operatorname{Log} \tau(k, t)$ we certainly have

$$| L(s, \chi) |^{\pm 1} \leqslant \frac{4}{\alpha} \operatorname{Log} \tau(k, t) .$$

Now for the second case. Let $t_0 \geqslant 0$ be a fixed real number and as in theorem 8.6 we write $s_0 = 1 + \eta_0 + it_0$. If $s = \sigma + it_0$ where $1 - 2\,\eta_0 \leqslant \sigma \leqslant 1 + \eta_0$, then

$$\left| \operatorname{Log} \left| \frac{f(s, \chi)}{f(s_0, \chi)} \right| \right| \leqslant \left| \operatorname{Log} \frac{f(s, \chi)}{f(s_0, \chi)} \right| = \left| \int_{\sigma_0}^{\sigma} \frac{f'(u + it_0, \chi)}{f(u + it_0, \chi)} du \right|$$
$$\leqslant 3\,\eta_0\, A \operatorname{Log} \tau_0 .$$

It now follows from theorem 8.6(c) that

$$\left| \operatorname{Log} \left| \frac{f(s, \chi)}{f(s_0, \chi)} \right| \right| \leqslant 3\,\eta_0\, A \operatorname{Log} \tau_0 = \frac{3\,\alpha}{2} A = \operatorname{Log} A_2 ,$$

where A_2 is an effective absolute constant. Hence, we have

$$\left| \frac{f(s, \chi)}{f(s_0, \chi)} \right|^{\pm 1} \leqslant \exp(3\,\alpha A/2) = A_2 .$$

and in particular

$$| f(s, \chi) |^{-1} \leqslant A_2 \, | f(s_0, \chi) |^{-1} .$$

If we now define the function ρ by

$$\rho(s) = \begin{cases} \left| \dfrac{s-1}{s_0-1} \right|^{E(\chi)} & \text{in case } (a) \\ \left| \dfrac{s_0-\beta}{s-\beta} \right| & \text{in case } (b), \end{cases}$$

then by theorem 8.2(c) :

$$| L(s, \chi) |^{-1} \leqslant A_2\, \rho(s)\, | L(s_0, \chi) |^{-1},$$
$$\leqslant A_2\, \rho(s) \frac{4}{\alpha} \operatorname{Log} \tau_0 .$$

since $s = \sigma + it_0$ and $1 - 2\,\eta_0 \leqslant \sigma \leqslant 1 + \eta_0$ we have, in case (a)

$$\rho(s) \leqslant \left| \frac{s-1}{s_0-1} \right| = \left| 1 + \frac{s-s_0}{s_0-1} \right| \leqslant 1 + \frac{| \sigma - \sigma_0 |}{| s_0 - 1 |} \qquad \rho(s) \leqslant 1 + \frac{3\,\eta_0}{\eta_0} = 4 .$$

and in case (b)

$$\rho(s) = \left| 1 + \frac{s_0 - s}{s - \beta} \right| \leqslant 1 + \frac{3\,\eta_0}{| s - \beta |} .$$

The theorem now follows. ■

5.3 Proof of theorem 8.5. We are now ready to begin the proof of theorem 8.5. Let $f(s, \chi)$ be the function defined in § 5.2. From the definition of $f(s, \chi)$ and (8.7) it follows that if we define F by

$$F(s) = -\frac{1}{\varphi(k)} \sum_{\chi} \bar{\chi}(l) \frac{f'(s, \chi)}{f(s, \chi)}$$

then for $\mathfrak{R}(s) > 1$

$$F(s) = \sum_{n \equiv l(k)} \frac{\Lambda(n)}{n^s} + \frac{1}{\varphi(k)} \left(-\frac{1}{s-1} + \frac{\lambda}{s-b} \right).$$

For $x \geqslant 1$ we write

$$I(x) = \frac{1}{2\,i\pi} \int_{2-i\infty}^{2+i\infty} \frac{F(s)}{s(s+1)} x^{s+1}\, ds ,$$

and then

$$I(x) = \frac{1}{2\,i\pi}\int_{2-i\infty}^{2+i\infty} \sum_{n\equiv l(k)} \frac{\Lambda(n)}{n^s}\,\frac{x^{s+1}}{s(s+1)}\,ds$$
$$-\frac{1}{\varphi(k)\,2\,i\pi}\int_{2-i\infty}^{2+i\infty}\left(\frac{1}{s-1}-\frac{\lambda}{s-b}\right)\frac{x^{s+1}}{s(s+1)}\,ds\,. \qquad (8.10)$$

We can interchange the order of summation in the first integral to obtain

$$\sum_{n\equiv l(k)} \Lambda(n)\,\frac{x}{2\,i\pi}\int_{2-i\infty}^{2+i\infty}\frac{(x/n)^s}{s(s+1)}\,ds = \sum_{\substack{n\leqslant x\\ n\equiv l(k)}} \Lambda(n)\,x\left(1-\frac{n}{x}\right)$$
$$= \psi_1(x\,;k,l)\,.$$

The second integral in (8.10) is estimated by " moving " the line of integration to the line $\Re(s) = -\frac{1}{2}$ and then using Cauchy's residue theorem to obtain

$$\frac{1}{2\,i\pi}\int_{2-i\infty}^{2+i\infty}\left(\frac{1}{s-1}-\frac{\lambda}{s-b}\right)\frac{x^{s+1}}{s(s+1)}\,ds$$
$$=\frac{1}{2\,i\pi}\int_{-\frac{1}{2}-i\infty}^{-\frac{1}{2}+i\infty}\left(\frac{1}{s-1}-\frac{\lambda}{s-b}\right)\frac{x^{s+1}}{s(s+1)}\,ds$$
$$+\frac{x^2}{2}-\lambda\,\frac{x^{b+1}}{b(b+1)}-x\left(1-\frac{\lambda}{b}\right).$$

The integral along the line $\Re(s) = -\frac{1}{2}$ is $O(\sqrt{x})$, thus

$$\frac{1}{2\,i\pi}\int_{2-i\infty}^{2+i\infty}\left(\frac{1}{s-1}-\frac{\lambda}{s-b}\right)\frac{x^{s+1}}{s(s+1)}\,ds = \frac{x^2}{2}-\lambda\,\frac{x^{b+1}}{b(b+1)}+O(x)$$

where the implied " O " constant is absolute and effective.

Thus we have shown for $x \geqslant 1$

$$I(x) = \psi_1(x\,;k,l) - \frac{1}{\varphi(k)}\left\{\frac{x^2}{2}-\frac{\lambda x^{b+1}}{b(b+1)}\right\} + O\left(\frac{x}{\varphi(k)}\right).$$

and it remains to estimate $|\,I(x)\,|$. This will be done by " moving " the line of integration to the left of the line $\Re(s) = 1$ and then using our knowledge of $f(s, \chi)$ to estimate the integral.

Recall that

$$I(x) = \frac{1}{2\,i\pi}\int_{2-i\infty}^{2+i\infty}\frac{F(s)\,x^{s+1}}{s(s+1)}\,ds$$

It follows from theorem 8.6 that in the region $R_\alpha(k)$ the function F is holomorphic and satisfies $| F(s) | < A \operatorname{Log} \tau(k, t)$. Thus, if we define the curve $\mathcal{C}_k$ by

$$\mathcal{C}_k = \left\{ s \in \mathbb{C} \mid \sigma = 1 - \frac{\alpha}{\operatorname{Log} \tau(k, t)} \right\}.$$

then the usual Cauchy theorem argument yields

$$I(x) = \frac{1}{2\, i\pi} \int_{\mathcal{C}_k} \frac{F(s)\, x^{s+1}}{s(s+1)}\, ds\,.$$

Consequently we see that

$$\begin{aligned} | I(x) | &\leqslant \frac{1}{2\,\pi} \int_{\mathcal{C}_k} | F(s) | \frac{x^{1+\sigma}}{| s(s+1) |} | ds | \\ &\leqslant \frac{A}{2\,\pi} \int_{\mathcal{C}_k} \operatorname{Log} \tau(k, t)\, x^{1+\sigma} \frac{| ds |}{| s(1+s) |} \\ &\leqslant \frac{Ax^2}{2\,\pi\alpha} \int_{\mathcal{C}_k} x^{\sigma-1}\, \tau(k, t)^{2\alpha-1}\, \tau(k, t)^{1-\alpha} \frac{| ds |}{| s(1+s) |}\,. \end{aligned}$$

On the curve $\mathcal{C}_k$ we have

$$x^{\sigma-1}\, \tau(k, t)^{2\alpha-1} = \exp \left\{ - \frac{\alpha \operatorname{Log} x}{\operatorname{Log} \tau(k, t)} - (1 - 2\,\alpha) \operatorname{Log} \tau(k, t) \right\}$$

and the arithmetic-geometric means inequality implies that

$$x^{\sigma-1}\, \tau(k, t)^{2\alpha-1} \leqslant \exp \{ - 2 \sqrt{\alpha(1 - 2\,\alpha) \operatorname{Log} x} \}.$$

We also have $\left| \dfrac{ds}{dt} \right| = \left| \dfrac{d\sigma}{dt} + i \right| \leqslant \alpha + 1$ if $t \geqslant 0$ and $t \neq k + 2$. Thus, if we write $a = \frac{1}{2}\sqrt{\alpha(1 - 2\,\alpha)}$, then we have

$$| I(x) | \leqslant A_1\, x^2\, e^{-4a\sqrt{\operatorname{Log} x}} \int_0^\infty \frac{t^{1-\alpha} + (k+2)^{1-\alpha}}{t^2 + \frac{1}{4}}\, dt$$

$$| I(x) | < A_2\, e^{-4a\sqrt{\operatorname{Log} x}}\, k^{1-\alpha}\,.$$

There are now two cases to consider :

$$k^{1-\alpha} \leqslant e^{2a\sqrt{\operatorname{Log} x}} \quad \text{and} \quad k^{1-\alpha} > e^{2a\sqrt{\operatorname{Log} x}}\,.$$

In the first case we have

$$| I(x) | \leqslant A_2 \, x^2 \exp \{ -2a\sqrt{\operatorname{Log} x} \}$$

and consequently

$$\psi_1(x; k, l) - \frac{1}{\varphi(k)} \frac{x^2}{2} + \frac{\lambda x^{b+1}}{\varphi(k)\, b(b+1)} = O\{ x^2 \exp(-2a\sqrt{\operatorname{Log} x}) \} .$$

In the second case we can obtain the above asymptotic formula for $\psi_1(x; k, l)$ in an essentially elementary manner. The integral $I(x)$ does not occur and the argument could have been given at the beginning of the proof, at the expense of being artificial. From the fact that

$$\psi_1(x; k, l) = \int_1^x \psi(u; k, l)\, du = \sum_{\substack{n \leqslant x \\ n \equiv l(k)}} \Lambda(n)(x - n) ,$$

we have

$$\left| \psi_1(x; k, l) - \frac{x^2}{2\varphi(k)} + \frac{\lambda x^{b+1}}{\varphi(k)\, b(b+1)} \right|$$

$$\leqslant x \operatorname{Log} x \sum_{\substack{n \leqslant x \\ n \equiv l(k)}} 1 + \frac{1}{\varphi(k)} \left(\frac{x^2}{2} + \frac{4}{3} x^2 \right)$$

$$\leqslant x \operatorname{Log} x \left(\frac{x}{k} + 1 \right) + \frac{2x^2}{\varphi(k)}$$

$$= \frac{x^2}{k^{1-\alpha}} \left(\frac{\operatorname{Log} x}{k^\alpha} + \frac{2k}{k^\alpha \varphi(k)} \right) + x \operatorname{Log} x .$$

Using our hypothesis we trivially have

$$\frac{\operatorname{Log} x}{k^\alpha} \leqslant \operatorname{Log} x . \exp \left\{ -\frac{2a\alpha}{1-\alpha} \sqrt{\operatorname{Log} x} \right\} \leqslant A_3 ,$$

and

$$\frac{k}{k^\alpha \varphi(k)} \leqslant \prod_{p | k} p^{-\alpha}(1 - p^{-1})^{-1} < \prod_{p^\alpha < 2} 2p^{-\alpha} < A_4 .$$

Thus, in the second case

$$\psi_1(x; k, l) - \frac{x^2}{2\varphi(k)} + \frac{\lambda x^{b+1}}{b(b+1)\varphi(k)} = O(x^2 e^{-2a\sqrt{\operatorname{Log} x}}) .$$

This completes the proof of part (*a*) of the theorem.

(*b*) The asymptotic expression for $\psi(x; k, l)$ will be deduced from the formula for $\psi_1(x; k, l)$ by the usual differencing argument. We shall suppose that $x \geqslant 2$ and let η be a function of x, to be chosen explicitly later, which satisfies $0 < \eta \leqslant \frac{1}{2}x$ for all $x \geqslant 2$. Because $\psi_1(x; k, l)$ is monotonic increasing function we have :

$$\frac{1}{\eta}\int_{x-\eta}^{x} \psi(u; k, l)\, du \leqslant \psi(x; k, l) \leqslant \frac{1}{\eta}\int_{x}^{x+\eta} \psi(u; k, l)\, du .$$

The two outer expressions are respectively :

$$\frac{\psi_1(x; k, l) - \psi_1(x-\eta; k, l)}{\eta} \leqslant \psi(x; k, l) \leqslant \frac{\psi_1(x+\eta; k, l) - \psi_1(x; k, l)}{\eta} .$$

Upon using the asymptotic formula for $\psi_1(x; k, l)$ and noting that

$$\sqrt{\text{Log}\,(x-\eta)} \geqslant \sqrt{\text{Log}\,(x/2)} \geqslant \sqrt{\text{Log}\,x} - 1$$

we conclude

$$\frac{\psi_1(x \pm \eta; k, l) - \psi_1(x; k, l)}{\pm\,\eta} = \frac{1}{\varphi(k)}\left\{x - \frac{\lambda x^b}{b}\right\} + \text{O}(\eta) + \text{O}\left(\frac{x^2}{\eta}e^{-2a\sqrt{\text{Log}\,x}}\right)$$

Hence we now have

$$\psi(x; k, l) = \frac{1}{\varphi(k)}\left\{x - \frac{\lambda x^b}{b}\right\} + \text{O}\left(\eta + \frac{x^2}{\eta}e^{-2a\sqrt{\text{Log}\,x}}\right).$$

and choosing

$$\eta = \frac{1}{2}x\,e^{-a\sqrt{\text{Log} x}} \leqslant \frac{1}{2}x$$

we obtain part (*b*) of the theorem.

(*c*) This is an immediate consequence of part (*b*) after we note that

$$0 \leqslant \psi(x; k, l) - \theta(x; k, l) \leqslant \sum_{\substack{p^\nu \leqslant x \\ \nu > 1}} \text{Log}\,p \leqslant \sum_{p \leqslant \sqrt{x}} \text{Log}\,p\left[\frac{\text{Log}\,x}{\text{Log}\,p}\right] \leqslant \sqrt{x}\,\text{Log}\,x$$

(*d*) By the Abel summation formula we have :

$$\pi(x; k, l) = \sum_{\substack{p \leqslant x \\ p \equiv l(k)}} 1 = \sum_{\substack{p \leqslant x \\ p \equiv l(k)}} \text{Log}\,p \cdot \frac{1}{\text{Log}\,p}$$

$$= \frac{\theta(x; k, l)}{\text{Log}\,x} - \int_2^x \theta(u; k, l)\frac{d}{du}\left(\frac{1}{\text{Log}\,u}\right) du .$$

If we write

$$\theta(x; k, l) = \frac{1}{\varphi(k)}\left\{ x - \frac{\lambda}{b} x^b \right\} + Q(x),$$

then

$$\pi(x ; k, l) = \frac{1}{\varphi(k) \operatorname{Log} x}\left(x - \frac{\lambda}{b} x^b\right) - \frac{1}{\varphi(k)} \int_2^x \left(u - \frac{\lambda}{b} u^b\right) \frac{d}{du}\left(\frac{1}{\operatorname{Log} u}\right) du$$

$$+ \frac{Q(x)}{\operatorname{Log} x} - \int_2^x \frac{Q(u)}{u \operatorname{Log}^2 u} du$$

$$= \frac{1}{\varphi(k)} \int_2^x \frac{du}{\operatorname{Log} u} - \frac{\lambda}{\varphi(k)} \int_2^x \frac{u^{b-1}}{\operatorname{Log} u} du + \frac{Q(x)}{\operatorname{Log} x}$$

$$+ \int_2^x \frac{Q(u)}{u \operatorname{Log}^2 u} du + \mathrm{O}(1)$$

$$= \frac{1}{\varphi(k)} \operatorname{Li}(x) - \frac{\lambda}{\varphi(k)} \operatorname{Li}(x^b) + \frac{Q(x)}{\operatorname{Log} x} + \int_2^x \frac{Q(u)}{u \operatorname{Log}^2 u} + \mathrm{O}(1).$$

We have already proved that

$$Q(x) = \mathrm{O}\{ x \exp(- a\sqrt{\operatorname{Log} x}) \}.$$

Since $x^{\frac{1}{2}} \exp(- a\sqrt{\operatorname{Log} x})$ is a monotone, increasing function of x for $x \geqslant 2$ it follows that

$$\int_2^x \frac{Q(u)}{u \operatorname{Log}^2 u} du = \mathrm{O}\left\{ \int_2^x \frac{x \exp(- a\sqrt{\operatorname{Log} x})}{\sqrt{ux}} du \right\}$$

$$= \mathrm{O}\{ x \exp(- a\sqrt{\operatorname{Log} x}) \}.$$

Hence we have

$$\pi(x ; k, l) = \frac{1}{\varphi(k)} \operatorname{Li}(x) - \frac{\lambda}{\varphi(k)} \operatorname{Li}(x^b) + \mathrm{O}\{ x e^{-a\sqrt{\operatorname{Log} x}} \}.$$

and this completes the proof of the theorem. ■

5.4 Uniform estimates. For many purposes it is useful to have asymptotic formulae which are valid uniformly in k or at least for all values of k in some suitable range. The following theorem, which is a simple consequence of theorem 8.5, gives such results. Unfortunately the implied " O " constants will depend upon the constants of theorem 8.4. Thus, the theorem is ineffective.

Theorem 8.8 *Let H be any fixed positive number.*

(a) *If k satisfies $k \leqslant (\operatorname{Log} x)^H$, then for $(k, l) = 1$*

$$\left.\begin{array}{l}\theta(x;k,l)\\ \psi(x;k,l)\end{array}\right\} = \frac{x}{\varphi(k)} + O\{x \exp(-a\sqrt{\operatorname{Log} x})\}$$

$$\pi(x;k,l) = \frac{\operatorname{Li}(x)}{\varphi(k)} + O\{x \exp(-a\sqrt{\operatorname{Log} x})\}.$$

(b) *For any k we have, for $(k, l) = 1$:*

$$\left.\begin{array}{l}\theta(x;k,l)\\ \psi(x;k,l)\end{array}\right\} = \frac{x}{\varphi(k)} + O\{x(\operatorname{Log} x)^{-H}\}$$

$$\pi(x;k,l) = \frac{\operatorname{Li}(x)}{\varphi(k)} + O\{x(\operatorname{Log} x)^{-H}\}.$$

The above " O " constants do not depend on k or l and are explicit functions of H, but they are not effective.

Proof. Only the formulae for $\psi(x; k, l)$ will be proved, the deduction of the remaining expressions is left as an easy exercise. We must show that either of the above hypotheses implies that the term x^b/b can be absorbed into the error term.

If no $L(s, \chi_k)$ has a real zero in $R_{2\alpha}(k)$, then $\lambda(k, l) = 0$, $b(k, l) = 1$ and the theorem is an immediate consequence of theorem 8.5. So we shall suppose that some $L(s, \chi_k)$ has a real zero in $R_{2\alpha}(k)$. Let $\varepsilon > 0$ be a fixed number to be chosen explicitly later. By theorems 8.3 and 8.4 we have

$$\frac{1}{2} \leqslant 1 - 2\alpha \leqslant b \leqslant 1 - c(\varepsilon)\, k^{-\varepsilon}$$

consequently

$$\frac{x^b}{b} \leqslant 2\,x \exp\{-c(\varepsilon)\, k^{-\varepsilon} \operatorname{Log} x\}.$$

(*a*) If we suppose that $k \leqslant (\operatorname{Log} x)^H$, then the above expression is majorised by

$$2\,x \exp\{-c(\varepsilon)(\operatorname{Log} x)^{1 - H\varepsilon}\}.$$

If we choose $\varepsilon = \dfrac{1}{3} H$ we deduce that

$$\frac{x^b}{b} \leqslant x \exp(-a\sqrt{\operatorname{Log} x}) \exp H_1$$

where H_1 is defined by

$$H_1 = \max_{\xi>0} \left\{ a\xi^{1/2} - c\left(\frac{1}{3\,H}\right)\xi^{2/3} \right\}.$$

Thus the term $\lambda x^b/b\varphi(k)$ can be absorbed into the error term and we have proved part (*a*).

(*b*) From exercise 1.18 it follows that there exists an effective absolute constant K_1 such that

$$\frac{1}{\varphi(k)} \leqslant \frac{A_1}{\sqrt{k}}$$

hence

$$\frac{x^b}{b\varphi(k)} \leqslant \frac{2\,A_1}{\sqrt{k}}\, x \exp\left\{ -\, c(\varepsilon)\, k^{-\varepsilon} \operatorname{Log} x \right\}$$

$$\leqslant 2\,A_1 \frac{x}{(\operatorname{Log} x)^H} k^{-1/2} (\operatorname{Log} x)^H \exp\left\{ -\, c(\varepsilon)\, k^{-\varepsilon} \operatorname{Log} x \right\}.$$

If we choose $\varepsilon = \frac{1}{2} H$, then $k^{-\frac{1}{2}} = (k^{-\varepsilon})^H$ from which we deduce that

$$k^{-1/2}(\operatorname{Log} x)^H \exp\left\{ -\, c\left(\frac{1}{2\,H}\right) k^{-(2H)^{-1}} \operatorname{Log} x \right\}$$

$$\leqslant \max_{u>0} \left\{ u^H \exp\left(-\, c\left(\frac{1}{2\,H}\right) u \right) \right\} = H_2\,.$$

Thus we have shown that

$$\frac{x^b}{b\varphi(k)} \leqslant 2\,A_1\,H_2 \frac{x}{(\operatorname{Log} x)^H}\,.$$

This proves part (*b*) since

$$x \exp(-\,a\sqrt{\operatorname{Log} x}) = \mathrm{O}\left\{ \frac{x}{(\operatorname{Log} x)^H} \right\},$$

Notes to chapter 8

REAL ZEROS OF *L*-FUNCTIONS

If χ is a given character modulo k, then it is not very difficult to verify whether or not $L(s, \chi)$ has a zero between 0 and 1. The simple method indicated in exercice 8.3 serves to show that for many " small " values of k the functions $L(s, \chi)$ have no positive real zeros. However, as Heilbronn [1] has shown, there exist infinitely many values of k for which the method breaks down. In contrast the technique given in exercice-8.4 enables one to solve the problem of positive real zeros for any given real character. It was in this way that Rosser [1] and [2] proved that for each $k \leqslant 227$ no L function formed with a character modulo k has a positive real zero. (An unpublished manuscript of Rosser extends the range to $k \leqslant 1\,000$.) A different computational approach, based on the theory of the Epstein zeta function, has been given by Low [1]. He proves that if $k < 593\,000$, $k \neq 115\,147$ and $\chi_k(-1) = -1$, then $L(s, \chi_k) > 0$ for $s > 0$.

For theoretical purposes one would like either a proof that no L-function has a positive real zero, or a non-trivial effective upper bound for the possible real zeros of $L(s, \chi)$. It is not inconcievable that there exist Dirichlet L-functions with a zero at $s = \frac{1}{2}$, examples of L-functions of algebraic number fields with this property have been given by Armitage [1]. The known results on effective upper bounds for real zeros of L-functions are rather sparse. They are discussed below.

From (8.6) we know that if χ is a real proper character modulo k and $\beta(k)$ is a real zero of $L(s, \chi)$, then

$$1 - \beta = \frac{L(1, \chi)}{L'(s_1, \chi)}.$$

for some $s_1 \in [\beta, 1]$. Thus, in order to obtain an upper bound for $\beta(k)$ we need a lower bound for $L(1, \chi)$ and an upper bound for $L'(s, \chi)$. Theorem 8.1 gives us an upper bound for $L'(s, \chi)$, namely

$$| L'(s, \chi) | < \frac{6}{\theta(1 - \theta)} (k \mid s \mid)^{1-\theta} \left\{ \frac{1}{\theta} + \operatorname{Log}\left(\frac{k \mid s \mid}{\theta} \right) \right\}.$$

valid for $\theta \leqslant \Re(s) \leqslant 1$. If we choose $\theta = 1 - (\text{Log } k)^{-1}$ in the above formula and take s real and at most equal to 1 we obtain

$$L'(s, \chi) \leqslant \frac{6\, e \text{ Log } k}{1 - (\text{Log } k)^{-1}} \left\{ \frac{1}{1 - (\text{Log } k)^{-1}} + \text{Log} \frac{k}{1 - (\text{Log } k)^{-1}} \right\}$$
$$< 1\,500 \text{ Log}^2 k \,.$$

valid for $1 - (\text{Log } k)^{-1} \leqslant s \leqslant 1$ and with an effective " O " constant.

Thus, if $L(s, \chi)$ has a real zero satisfying

$$1 - \frac{1}{\text{Log } k} \leqslant \beta(k) < 1 \,,$$

then

$$1 - \beta(k) \geqslant \frac{L(1, \chi)}{1\,500 \text{ Log}^2 k} \,.$$

The proof of theorem 8.4, which gives a lower bound for $L(1, \chi)$, can almost be made completely effective. For example, Tatuzawa [1] proves the following result.

Let χ be a non-principal real proper character modulo k. If $\varepsilon < 0$ is a fixed real number, then with the exception of at most one value of k one has

$$L(1, \chi) > \frac{\varepsilon}{10\, k^{\varepsilon}}$$

Hence, apart from the possible exceptional $k(\varepsilon)$ we have

$$1 - \beta(k) \geqslant \frac{\varepsilon}{15.10^3\, k^{\varepsilon} \text{ Log}^2 k} \,.$$

A different method of proof of theorem 8.4 has been given by Knapowski [4], who used Turán's lower bound method. But, like all other known proofs it is ineffective.

As for completely effective upper bounds for $\beta(k)$ the known results are either very weak or not very explicit. For example, Turán [4] proves the following effective theorem.

If $k > k_0$ and $\text{P} = \exp \{ \text{Log}^2 k (\text{Log Log } k)^2 \}$, *then*

$$\frac{1}{2} \leqslant \beta(k) \leqslant 2 \frac{\text{Log Log Log } k}{\text{Log Log } k} + \frac{1}{\text{Log P}} \max_{1 \leqslant x \leqslant \text{P}} \left\{ \text{Log} \left| \sum_{n \leqslant x} \Lambda(n)\, \chi(n) \right| \right\} .$$

Naturally one would like the last expression in a more convenient form as a function of k, but this seems difficult to achieve. The only obvious completely explicit lower bound for $L(1, \chi)$ is given by Dirichlet's clars number formula, which we mentioned in the *Notes* to chapter 7. From this formula we have

$$L(1, \chi) > \begin{cases} c_1 \dfrac{h(-k)}{\sqrt{k}} & \text{if} \quad \chi(-1) = -1 \\ c_2 \dfrac{h(k)}{\sqrt{k}} \operatorname{Log} \varepsilon & \text{if} \quad \chi(-1) = +1 \end{cases}$$

where c_2 and c_3 are effective constants. Since $h(\pm k) \geqslant 1$ and

$$\operatorname{Log} \varepsilon \geqslant \operatorname{Log} \frac{1}{2}(1 + \sqrt{k})$$

we certainly have

$$\beta(k) < \begin{cases} 1 - \dfrac{c_3}{\sqrt{k} \operatorname{Log}^2 k} & \text{if} \quad \chi(-1) = -1 \\ 1 - \dfrac{c_4}{\sqrt{k} \operatorname{Log} k} & \text{if} \quad \chi(-1) = +1 \end{cases}$$

where c_4 and c_5 are effective constants.

By a careful analysis of certain character sums, Davenport [2] gave a direct proof that :

$$\beta(k) < 1 - \frac{c_6}{\sqrt{k} \operatorname{Log} \operatorname{Log} k} \quad \text{if} \quad \chi(-1) = -1$$

and

$$\beta(k) < 1 - \frac{c_7 \operatorname{Log} k}{\sqrt{k} \operatorname{Log} \operatorname{Log} k} \quad \text{if} \quad \chi(-1) = +1\,.$$

Moreover, the factor Log Log k can be omitted if k is a prime.

ZERO-FREE REGIONS AND ERROR TERMS

Let χ run over the characters modulo k and define the function $\mathfrak{L}_k(s)$ by

$$\mathfrak{L}_k(s) = \prod_{\chi} L(s, \chi)$$

Theorem 8.3 implies that for at most one real point β the function $\mathfrak{L}_k(s)$ does not vanish in the region $R_\alpha(k)$, where α is an effectively computable absolute constant.

The calculation of α is somewhat tedious. However, in certain problems it is important to have a numerical value for α and Miech [1] has partially carried out the computation of such an " almost " zero-free region for $\mathfrak{L}_k(s)$. His result is as follows :

Apart from the possibility of a unique real zero $\mathfrak{L}_k(s)$ does not vanish in the region

$$\left\{ s = \sigma + it \mid \sigma \geqslant 1 - \frac{1}{20 \operatorname{Log} k(2 + |t|)} \right\}$$

provided that $k > k_0$, where k_0 is an effectively computable constant.

As is to be expected, the relationship between almost zero-free regions for $\mathfrak{L}_k(s)$ and the error term in, says the asymptotic formula for $\psi(x, k, l)$ can be made quite explicit. Tatuzawa [2] proved that if $\mathfrak{L}_k(s) \neq 0$ for $s \neq \beta$ in the region

$$\left\{ s = \sigma + it \mid \sigma \geqslant 1 - \frac{c_1}{\max \{ \operatorname{Log} k, \operatorname{Log}^{\gamma}(|t| + 3) \}} \right\},$$

then

$$\left| \psi(x ; k, l) - \frac{x}{\varphi(k)} + \frac{\lambda(k, l)}{b\varphi(k)} x^b \right| \leqslant c \frac{x}{\varphi(k)} \exp \left\{ - \frac{c_2 \operatorname{Log} x}{\max \{ \operatorname{Log} k, \operatorname{Log}^{\delta} x \}} \right\}$$

for k satisfying $1 \leqslant k \leqslant \exp\left(\dfrac{c_3 \operatorname{Log} x}{\operatorname{Log} \operatorname{Log} x} \right)$.

By using Turán's method a converse implication can be proved. The details have been carried out by Wiertelak [1]. His theorem is as follows.

Suppose that

$$\left| \psi(x ; k, l) - \frac{x}{\varphi(k)} + \frac{\lambda(k, l)}{b\varphi(k)} x^b \right| \leqslant c_1 \frac{x}{\varphi(k)} \exp \left\{ - \frac{c_2 \operatorname{Log} x}{\max \{ \operatorname{Log} k \operatorname{Log}^{\delta} x \}} \right\},$$

for $x > x_0(k)$ and any l satisfying $0 < l \leqslant k$, $(l, k) = 1$, then $\mathfrak{L}_k(s)$ does not vanish in the region

$$\left\{ s = \sigma + it \mid \sigma \geqslant 1 - \frac{c_3}{\max \{ c_4 \operatorname{Log} k, \operatorname{Log}^{\gamma} |t| \}}, \; |t| \geqslant \max \{ c_5, \omega(k) \} \right\}$$

where c_3, c_4, c_5 are explicit functions of c_1, c_2, v and $\omega(k)$ is an explicit function of k.

Exercises to chapter 8

8.1 Use the lower bound for $L(1, \chi)$ given by Dirichlet's class number formula to prove the following result :

" If $\delta > 0$ is fixed and k satisfies $0 < k \leqslant (\text{Log } x)^{1-\delta}$, then

$$\psi(x ; k, l) = \frac{x}{\varphi(k)} + \mathrm{O}\{ x \exp(- c \sqrt{\text{Log } x}) \}$$

where c and the implied " O " constant are both absolute and effective. "

8.2 (*a*) Let χ_1 and χ_2 be distinct real proper characters to the moduli k_1 and k_2 respectively. Suppose that the corresponding L-functions have real zeros β_1 and β_2. Prove that there exists an absolute effective constant $c > 0$ such that

$$\min \{ \beta_1, \beta_2 \} < 1 - \frac{c}{\text{Log } (k_1 k_2)} .$$

(*b*) Suppose that $\{ k_i \}$ is a sequence of positive integers with the property that for each i there exists a real proper character χ_i to the modulus k_i for which $L(s, \chi_i)$ has a real zero satisfying

$$\beta_i > 1 - \frac{c}{\alpha \text{ Log } k_i} .$$

Prove that $k_{i+1} > k_i^{\alpha - 1}$ for $i = 1, 2, \dots$.

8.3 Let χ be a real non-principal character modulo k. For $m = 1, 2, \dots$ define the functions S_m recursively as follows :

$$\mathrm{S}_1(x) = \sum_{n \leqslant x} \chi(n) , \qquad \mathrm{S}_m(x) = \sum_{n \leqslant x} \mathrm{S}_{m-1}(n) .$$

Suppose that there exists an integer $M = M(\chi)$ such that for all $x \geqslant 1 : \mathrm{S}_M(x) \geqslant 0$.

(*a*) Prove that $L(s, \chi)$ satisfies $L(s, \chi) > 0$ for $s > 0$.

(*b*) $L(1, \chi) > (M + 1)^{-1}$.

[Hint : Use repeated partial summation and induction to prove that for any $m \geqslant 1$

$$L(s, \chi) = \sum_{n=1}^{\infty} S_m(n) \sum_{j=0}^{m} (-1)^j \frac{m!}{j!\,(m-j)!} (n+j)^{-s}$$

$$= s(s+1) \cdots (s+m-1) \sum_{n=1}^{\infty} S_m(n) \int_0^1 \cdots \int_0^1 (n + u_1 + \cdots + u_m)^{-n-s}\, du_1 \cdots du_m .]$$

8.4 Let χ be a non-principal character modulo k. Prove the following representation of $L(s, \chi)$, valid for all s

(*a*) If $\chi(-1) = 1$, then

$$L(s, \chi) = 2 \sum_{n=1}^{\infty} \frac{s(s+1) \cdots (s+2n-1)}{4^n (2n)!\, k^{s+2n}} (2^{s+2n} - 1)\, \zeta(s+2n)\, S_{2n}$$

where

$$S_n = \sum_{m \leqslant k/2} \chi(m)(k - 2m)^n .$$

(*b*) If $\chi(-1) = -1$, then

$$L(s, \chi) = \sum_{n=0}^{\infty} \frac{s(s+2) \cdots (s+2n)}{4^n (2n+1)!\, k^{s+2n+1}} (2^{s+2n+1} - 1)\, \zeta(s+2n+1)\, T_n .$$

where

$$T_n = \sum_{m \leqslant k/2} \chi(m)(k - 2m)^{2n+1} .$$

8.5 If χ is a non-principal character modulo k and x satisfies $x \geqslant \max\{2k, k|t|\}$, then

$$L(s, \chi) = \sum_{n \leqslant x} \frac{\chi(n)}{n^s} + O(kx^{-\sigma}) .$$

[Hint : Use exercises 5.17 and 7.4.]

8.6 Let χ be a real non-principal character. From an effective upper bound for $L'(1, \chi)$ deduce an effective lower bound for $L(1, \chi)$ in the following way. Write

$$F(s) = \zeta(s)\, L(s, \chi) = \sum_{n=1}^{\infty} \frac{b_n}{n^s} .$$

and in the disc $| s - 2 | < 1$ expand $F(s)$ as a power series :

$$F(s) = \sum_{m=0}^{\infty} a_m(2 - s)^m$$

noting that $a_n \geqslant 0$ and $a_{n^2} > 1$ for all n. Now consider the function

$$g(s) = F(s) - \frac{L(1, \chi)}{s - 1}.$$

and show that

$$g(1) = \gamma L(1, \chi) + L'(1, \chi) = \sum_{m=0}^{\infty} (a_m - L(1, \chi)).$$

Use theorem 4.1, with $R = \frac{3}{2}$, $r = 1$ to obtain the estimate

$$\left| \sum_{m>r} (a_m - L(1, \chi)) \right| < f(k) \left(\frac{2}{3}\right)^r$$

where f is an explicit, effective function of k. Choose M such that

$$f(k) \left(\frac{2}{3}\right)^{M+1} \leqslant a_{M+1} \qquad \text{and} \qquad a_1 + \cdots + a_M > L'(1, \chi)$$

and conclude that

$$L(1, \chi) > \frac{a_1 + \cdots + a_M - L'(1, \chi)}{M + \gamma}.$$

CHAPTER 9

Sums of prime numbers

§1 INTRODUCTION

1.1 So far in this book we have been concerned with the density of the primes in the positive integers. In this chapter we shall study a new type of problem, namely the relationship between the positive integers and the *additive* semigroup generated by the primes. This branch of number theory abounds with unsolved problems and conjectures. The two most famous problems are the conjecture of Goldbach, which asserts that every even integer greater than 2 can be written as a sum of two primes, and the " twin prime " conjecture, which asserts that there is an infinity of primes p such that $p + 2$ is also a prime.

In contrast to the conjectures, the theorems in additive prime number theory are either of a qualitative nature or are numerically far from the conjectured best possible result. For example, Brun proved that the sum of the reciprocals of the " twin primes " is convergent and Schnirelmann proved that there exists an absolute constant k such that every positive integer can be written as a sum of at most k primes. It is known that the constant in Schnirelmann's theorem satisfies $k \leqslant 73$ (see Deshouillers [2]) but one is inclined to conjecture that k can be taken to be 3. It is a theorem of Vinogradov that every *sufficiently large* odd integer can be written as a sum of three primes and one suspects that " sufficiently large " can be taken to mean " greater than 5 ".

Roughly speaking, the methods of proof for theorems in additive number theory are of two types :

(i) The analytic method initiated by Hardy, Ramanujan, Littlewood and Vinogradov.

(ii) The " sieve methods " introduced by Brun and extensively developed by Selberg.

This latter technique lies outside the scope of this book as does the study of the additive properties of general sequences of integers. For an introduction to this fascinating topic we refer the reader to *Sequences* by Halberstam and Roth. We shall confine our attention to an application of the analytic method to prove the theorem of Vinogradov mentioned above. The basic technique is one of great versatility and is frequently used to prove the deeper theorems of additive number theory.

For some other applications of the method and variations on the theme see Birch [1], Cassels [3], Davenport [3] and Hua [4].

1.2 Since we wish to study the representation of integers as sums of primes, a natural function to introduce is

$$v_r(n) = \sum_{p_1 + \cdots + p_r = n} 1 .$$

Goldbach's conjecture can now be formulated as :

If N *is even and* $N > 3$, *then* $v_2(N) > 0$,

and Vinogradov's theorem can be expressed as :

If N *is odd and sufficiently large, then* $v_3(N) > 0$.

Our main result in this chapter will be an asymptotic formula for $v_r(N)$ when $r \geqslant 3$ and $N \equiv r \pmod 2$, namely

$$v_r(N) = \mathfrak{S}_r(N) \frac{N^{r-1}}{(\operatorname{Log} N)^r} + \mathrm{O}\left(\frac{N^{r-1} \operatorname{Log} \operatorname{Log} N}{(\operatorname{Log} N)^{r+1}}\right),$$

where $\mathfrak{S}_r(N)$ is a rather complicated arithmetic function of N, but is such that $0 < c_1(r) < \mathfrak{S}_r(N) < c_2(r) < \infty$ for all N. Clearly, Vinogradov's three primes theorem follows from the above formula with $r = 3$.

Just as in the proof of the Prime Number Theorem it was better to work with $\psi(x)$ rather than $\pi(x)$ it will be technically easier to work with a function $\lambda_r(N)$ which is closely related to $v_r(N)$, but is a less " natural " function to consider. The function λ_r is defined by

$$\lambda_r(N) = \sum_{p_1 + \cdots + p_r = N} (\operatorname{Log} p_1) \cdots (\operatorname{Log} p_r) .$$

1.3 The relationship between $\lambda_r(N)$ and $v_r(N)$. Since we trivially have

$$\lambda_r(N) > 0 \Leftrightarrow v_r(N) > 0 ,$$

it is clear that in order to establish the three primes theorem it suffices to show $\lambda_r(N) > 0$. We do this by showing that for any fixed $\Delta > 0$,

$$\lambda_r(N) = \mathfrak{S}_r(N)\, N^{r-1} + \mathrm{O}\left(\frac{N^{r-1}}{\operatorname{Log}^{\Delta} N}\right)$$

The asymptotic formula for $v_r(N)$ is a direct consequence of that for $\lambda_r(N)$ as follows.

Let θ be a real number, to be chosen explicitly later, satisfying $\frac{1}{2} < \theta < 1$. Then we see

$$\lambda_r(N) \geqslant \sum_{\substack{p_1+\cdots+p_r=N \\ \forall i\ p_i>N^\theta}} (\operatorname{Log} p_1) \cdots (\operatorname{Log} p_r) \geqslant (\theta \operatorname{Log} N)^r \{ \nu_r(N) - \rho_r(N, \theta) \}$$

where

$$\rho_r(N, \theta) = \sum_{i=1}^{r} \sum_{\substack{p_i \leqslant N^\theta \\ p_1+\cdots+p_r=N}} 1 = r \sum_{\substack{p_1 \leqslant N^\theta \\ p_1+\cdots+p_r=N}} 1 = \sum_{p_1 \leqslant N^\theta} \sum_{p_2+\cdots+p_r=N-p_1} 1$$

$$\leqslant \sum_{p_1 \leqslant N^\theta} \sum_{p_3+\cdots+p_r<N-p_1} 1 \leqslant \Big(\sum_{p_1 \leqslant N^\theta} 1\Big) \Big(\sum_{p_3+\cdots+p_r<N} 1\Big)$$

$$\leqslant \Big(\sum_{p_1 \leqslant N^\theta} 1\Big) \Big(\sum_{p<N} 1\Big)^{r-2},$$

and by a repeated application of Tchebycheff's theorem we have

$$\rho_r(N, \theta) = \mathrm{O}\left\{\frac{N^\theta}{\operatorname{Log} N}\left(\frac{N}{\operatorname{Log} N}\right)^{r-2}\right\}.$$

Thus we obtain

$$\nu_r(N)(\operatorname{Log} N)^r \leqslant \theta^{-r} \lambda_r(N) + \mathrm{O}(N^{r-2+\theta} \operatorname{Log} N),$$

and since $\lambda_r(N) \leqslant \nu_r(N)(\operatorname{Log} N)^r$, it follows that

$$0 \leqslant \nu_r(N)(\operatorname{Log} N)^r - \lambda_r(N) = \mathrm{O}\{(1-\theta)\lambda_r(N) + N^{r-1} N^{\theta-1} \operatorname{Log} N\}.$$

Upon choosing θ so that

$$N^{\theta-1} \operatorname{Log} N = 1,$$

the above formula becomes

$$0 \leqslant \nu_r(N)(\operatorname{Log} N)^r - \lambda_r(N) = \mathrm{O}\left\{\frac{\operatorname{Log}\operatorname{Log} N}{\operatorname{Log} N}\lambda_r(N) + N^{r-1}\right\}. \tag{9.1}$$

Since $\lambda_r(N) = \mathrm{O}(N^{r-1})$, we have the asymptotic formula for $\nu_r(N)$ after choosing $\Delta \geqslant 1$.

§ 2 THE ASYMPTOTIC FORMULA FOR $\lambda_r(N)$

This section is devoted to the proof of the following theorem due to Vinogradov.

Theorem 9.1 *If $r \geqslant 3$ is a fixed integer and $N \equiv r \pmod 2$, then for any fixed $\Delta > 0$ we have*

$$\lambda_r(N) = N^{r-1}\,\mathfrak{S}_r(N) + O\left\{\frac{N^{r-1}}{(\operatorname{Log} N)^{\Delta}}\right\}$$

where

$$\mathfrak{S}_r(N) = \frac{2}{(r-1)\,!}\prod_{p>2}\left(1-\left(\frac{-1}{p-1}\right)^r\right)\prod_{\substack{p\mid N\\ p>2}}\frac{(p-1)^r+(-1)^r(p-1)}{(p-1)^r-(-1)^r}\,.$$

and there exist constants $c_1(r)$, $c_2(r)$ such that

$$0 < c_1(r) < \mathfrak{S}_r(N) < c_2(r) < \infty\,.$$

The implied " O " constant is not effective.

The proof of theorem 9.1, although long, is not at all difficult. It provides a classic example of techniques introduced by Hardy, Littlewood and Vinogradov for the estimation of arithmetic functions. Briefly the method is as follows.

2.1 Outline of the Hardy-Littlewood-Vinogradov method. One has a pair of functions G and g which are related by

$$G(\alpha) = \sum_{m=1}^{N} g(m)\, e^{2i\pi m\alpha}\,,$$

and one wants information about $g(m)$, usually an asymptotic formula. It is easy to express g as a function of G, for we have

$$G(\alpha)\, e^{-2\pi in\alpha} = \sum_{m=1}^{N} g(m)\, e^{2\pi i\alpha(m-n)}\,.$$

Integrating over any unit interval I and observing that

$$\int_I e^{2\pi i\alpha(m-n)}\, d\alpha = \begin{cases} 1 & \text{if} \quad m = n\,, \\ 0 & \text{if} \quad m \neq n\,, \end{cases}$$

we obtain

$$g(n) = \int_I G(\alpha)\, e^{-2i\pi n\alpha}\, d\alpha\,.$$

If one possesses information about G, then one can hope to deduce some information about $g(n)$ from the above integral.

In most situations the function G is rather complicated and one cannot deal directly with the integral. What one does at this point is to divide the interval I into disjoint subintervals of two sorts, called, for historical reasons, the major arcs and the minor arcs respectively. On each of the major arcs one finds a simpler function which is a " good " approximation to the integrand and so obtains a " good " approximation to the integral on the subinterval. On the minor arcs one shows that the integrand is " small ". The actual choice of these subintervals is a matter of experience, intelligence and some experimentation.

While the Hardy-Littlewood-Vinogradov method is conceptually straightforward, the details of most applications are quite long.

Proof of theorem 9.1. We set up the situation described above for our particular problem and introduce some notation.

2.2 For typographical convenience we write

$$e(z) = \exp(2\, i\pi z).$$

Let x be a fixed real number; define the function f_x by

$$f_x(\alpha) = \sum_{n \leqslant x} \lambda_1(n)\, e(\alpha n)$$

$$= \sum_{p \leqslant x} (\log p)\, e(\alpha p)\,.$$

By multiplication, we have for any positive integers r and N :

$$f_N^r(\alpha) = \sum_{n \leqslant rN} \lambda_r(n, N)\, e(\alpha n)\,,$$

where the coefficient $\lambda_r(n, N)$ is equal to $\lambda_r(n)$ for each $n \leqslant N$. Since primes in the range $N < p \leqslant rN$ do not contribute to $\lambda_r(n, N)$ we do not necessarily have $\lambda_r(n, N) = \lambda_r(n)$ when $n > N$. On multiplying by $e(-\alpha N)$ and integrating over any unit interval I we obtain

$$\lambda_r(N) = \int_I f_N^r(\alpha)\, e(-\alpha N)\, d\alpha\,.$$

2.3 The major and minor arcs. We shall now divide the interval I into two sets; M, the union of the major arcs, and $m = I \setminus M$, which is in effect the union of the minor arcs. The above equation will then be written

$$\lambda_r(N) = \int_M f_N^r(\alpha)\, e(-\alpha N)\, d\alpha + \int_m f_N^r(\alpha)\, e(-\alpha N)\, d\alpha$$
$$= \mathfrak{J}_M + \mathfrak{J}_m$$

As we mentioned above, the actual choice for the major and minor arcs can be a rather delicate matter. Our choice will, at first sight, appear arbitrary.

It is plausible that for a/q rational we might be able to estimate $f_x\left(\frac{a}{q}\right)$ by using information about the density of primes in arithmetic progressions. Thus if M consists of real numbers which are in some sense well approximable by rationals, we ought to be able to estimate $\mathfrak{J}_M$. A classical result of Dirichlet tells us that for any real numbers α and Q there exist a rational a/q such that

$$(a, q) = 1, \quad 1 \leqslant q \leqslant Q \quad \text{and} \quad \left|\alpha - \frac{a}{q}\right| < \frac{1}{qQ} \tag{9.2}$$

For some α we can find an approximation with q quite " small ". Such α are well approximable. For other α, q must be close to Q.

Let H be a fixed positive real number. We carry H as a parameter throughout the proof of theorem 9.1. At the end of the proof it will be chosen in a manner which gives us a tidy final result. We define η and Q as follows :

$$\eta = (\operatorname{Log} N)^H \quad \text{and} \quad Q = N(\operatorname{Log} N)^{-H} = N\eta^{-1}.$$

Consider the set of rational numbers a/q which satisfy

$$(a, q) = 1 \quad \text{and} \quad 0 \leqslant a < q \leqslant \eta. \tag{9.3}$$

There are only a finite number of such rationals. For each one of them we define an interval $I_{a/q}$ by

$$I_{a/q} = \left]\frac{a}{q} - \frac{1}{Q}, \frac{a}{q} + \frac{1}{Q}\right].$$

We observe that for N sufficiently large we have

$$\frac{1}{2}Q > \eta^2 > 1,$$

thus if a_1/q_1 and a_2/q_2 satisfy (9.3),

$$\left| \frac{a_1}{q_1} - \frac{a_2}{q_2} \right| \geqslant \frac{1}{q_1 q_2} \geqslant \frac{1}{\eta^2} > \frac{2}{Q}$$

and so the intervals $I_{a/q}$ are disjoint.

We shall take as our unit interval I the interval

$$I = \left] - \frac{1}{Q}, 1 - \frac{1}{Q} \right[$$

and define the major set M to be

$$M = \bigcup_{a,q} I_{a/q}$$

The minor set m is defined by

$$m = I - M$$

By the definition of m, it follows that all rational approximations to $\alpha \in m$ which satisfy the conditions (9.2) also satisfy

$$\eta < q \leqslant Q = N\eta^{-1} .$$

2.4 The integral $\mathfrak{J}_m$. A vital tool which we shall need in order to estimate $\mathfrak{J}_m$ is an estimate for the absolute value of the sum

$$S_x(\alpha) = \sum_{p \leqslant x} e(\alpha p) .$$

The required estimate, due to Vinogradov, is given in § 3. It is :

If $\left| \alpha - \frac{a}{q} \right| < \frac{1}{q^2}$, *where* $(a, q) = 1$ *and* $q > 0$, *then*

$$| S_x(\alpha) | = \left| \sum_{p \leqslant x} e(\alpha p) \right| \leqslant A \left\{ \frac{1}{q} + \frac{q}{x} + \exp(-\sqrt{\operatorname{Log} x}) \right\}^{1/2} x(\operatorname{Log} x)^3 ,$$

where A is an effectively computable absolute constant.

We now proceed to obtain an upper bound for $| \mathfrak{J}_m |$ using the above theorem. The following inequalities are quite trivial :

$$| \mathfrak{J}_m | = \left| \int_m f_N^r(\alpha)\, e(-\alpha N)\, d\alpha \right| \leqslant \int_m | f_N^r(\alpha) |\, d\alpha$$

$$\leqslant \max_{\alpha \in m} | f_N(\alpha) |^{r-2} \int_I | f_N^2(\alpha) |\, d\alpha .$$

It is clear that

$$\int_I |f_N(\alpha)|^2 \, d\alpha = \sum_{p \leqslant N} (\operatorname{Log} p)^2 \leqslant \theta(N) \operatorname{Log} N = O(N \operatorname{Log} N)$$

and consequently we have

$$\mathfrak{J}_m = O\left\{ \max_{\alpha \in m} |f_N(\alpha)|^{r-2} N \operatorname{Log} N \right\}.$$

Now we shall consider the expression $\max_{\alpha \in m} |f_N(\alpha)|$. By an application of theorem 1.8 we see that

$$|f_N(\alpha)| = \left| \sum_{p \leqslant N} (\operatorname{Log} p)\, e(\alpha p) \right|$$

$$\leqslant 2 \operatorname{Log} N \max_{1 \leqslant x \leqslant N} |S_x(\alpha)|.$$

By Vinogradov's estimate for $S_x(\alpha)$ we see

$$\max_{1 \leqslant x \leqslant N} |S_x(\alpha)| \leqslant A \left\{ \frac{1}{q} + \frac{q}{N} + \exp(-\sqrt{\operatorname{Log} N}) \right\}^{1/2} N(\operatorname{Log} N)^3.$$

We know that if $\alpha \in m$ and $|\alpha - a/q| < 1/q^2$ where a/q satisfies (9.2), then q satisfies $\eta < q \leqslant Q = N\eta^{-1}$. Thus we certainly have

$$\max_{\alpha \in m} \max_{1 \leqslant x \leqslant N} |S_x(\alpha)| \leqslant A\{\eta^{-1} + \eta^{-1} + \exp(-\sqrt{\operatorname{Log} N})\}^{1/2} N(\operatorname{Log} N)^3$$

$$\leqslant B \frac{N}{\sqrt{\eta}} (\operatorname{Log} N)^3.$$

It now follows that

$$\mathfrak{J}_m = O\left\{ N^{r-1} \left(\frac{\operatorname{Log}^4 N}{\sqrt{\eta}} \right)^{r-2} \operatorname{Log} N \right\}. \tag{9.4}$$

2.5 An asymptotic formula for $\mathfrak{J}_M$. Since the details in the approximation of $\mathfrak{J}_M$ are rather long, we give here an outline of the method without proof.

Our first objective will be to find a relatively simple approximating function to $f_N(\alpha)$. This will be done in § 2.6 and is the following.

For $\alpha \in I_{a/q}$ we have, writing $t = \alpha - a/q$,

$$f_N(\alpha) = \frac{\mu(q)}{\varphi(q)} g_N\left(\alpha - \frac{a}{q}\right) + O(N\eta^{-3}) \tag{9.5}$$

where $g_N(t) = \sum_{n \leqslant N} e(nt)$. The "O" constant is ineffective.

Using this estimate we will show in § 2.7 that

$$\int_{I_{a/q}} f_N^r(\alpha)\, e(-\alpha N)\, d\alpha = \left(\frac{\mu(q)}{\varphi(q)}\right)^r e\left(-\frac{Na}{q}\right) \int_{-1/Q}^{+1/Q} g_N^r(t)\, e(-Nt)\, dt + \\ + \mathrm{O}\left(\frac{N^r}{\eta^3\, Q\varphi^{r-1}(q)}\right),$$

then summing this expression over all the major arcs and performing a long but routine computation we arrive at

$$\mathfrak{J}_M = \frac{N^{r-1}}{(r-1)\,!} \sum_{q=1}^{\infty} \left(\frac{\mu(q)}{\varphi(q)}\right)^r c_q(-N) + \mathrm{O}\left(\frac{N^{r-1}}{\sqrt{\eta}}\right)$$

where

$$c_q(-N) = \sum_{\substack{a=1\\(a,q)=1}}^{q} e\left(-\frac{Na}{q}\right).$$

We denote by $\mathfrak{S}_r(N)$ the sum

$$\mathfrak{S}_r(N) = \sum_{q=1}^{\infty} \left(\frac{\mu(q)}{\varphi(q)}\right)^r c_q(-N)\,.$$

It will now remain to simplify $\mathfrak{S}_r(N)$ and to show that there exist absolute constants $c_1(r)$, $c_2(r)$ so that for all N, $0 < c_1(r) < \mathfrak{S}_r(N) < c_2(r) < \infty$. This will imply, after choosing H, that the " main " term of the asymptotic formula is in fact the dominant term.

2.6 We begin the programme outlined in § 2.5 establishing the asymptotic formula for $f_N(\alpha)$ given by (9.5).

For $x \geqslant 2$ we have

$$\begin{aligned} f_x\left(\frac{a}{q}\right) &= \sum_{p \leqslant x} (\operatorname{Log} p)\, e\left(\frac{ap}{q}\right) \\ &= \sum_{\substack{p \leqslant x\\(p,q)=1}} (\operatorname{Log} p)\, e\left(\frac{ap}{q}\right) + \sum_{\substack{p \leqslant x\\ p \mid q}} (\operatorname{Log} p)\, e\left(\frac{ap}{q}\right) \\ &= \sum_{l(q)}{}' \sum_{\substack{p \leqslant x\\ p \equiv l(q)}} (\operatorname{Log} p)\, e\left(\frac{ap}{q}\right) + \mathrm{O}\left(\sum_{p \mid q} \operatorname{Log} p\right) \\ &= \sum_{l(q)}{}' \sum_{\substack{p \leqslant x\\ p \equiv l(q)}} (\operatorname{Log} p)\, e\left(\frac{al}{q}\right) + \mathrm{O}\left(\sum_{p \mid q} \operatorname{Log} p\right) \\ &= \sum_{l(q)}{}' \theta(x\,;q,l)\, e\left(\frac{al}{q}\right) + \mathrm{O}\left(\sum_{p \mid q} \operatorname{Log} p\right) \end{aligned}$$

where the symbol $\sum'_{l(q)}$ denotes a summation of a complete set of reduced residues modulo q. We now apply theorem 8.8(*b*), which gives an asymptotic formula for $\theta(x; q, l)$, with $5\,H$ in place of H, to obtain :

$$f_x\left(\frac{a}{q}\right) = \sum_{l(q)}{}' \left\{ e\left(\frac{la}{q}\right) \frac{x}{\varphi(q)} + \mathrm{O}\left(\frac{x}{\mathrm{Log}^{5H} x}\right) \right\} + \mathrm{O}(\mathrm{Log}\, q)\,. \qquad (9.6)$$

We recall that the implied " O " constant here is absolute but not effective.

Using simple properties of the Möbius function

$$E = \sum_{l(q)}{}' e\left(\frac{la}{q}\right)$$

$$E = \sum_{l=1}^{q} e\left(\frac{la}{q}\right) \sum_{\delta \mid (l,q)} \mu(\delta)$$

$$= \sum_{\delta \mid q} \mu(\delta) \sum_{\substack{l=1 \\ \delta \mid l}}^{q} e\left(\frac{la}{q}\right).$$

If we now write $q = \delta d$ and $l = \delta s$, then the inner sum is

$$\sum_{s=1}^{d} e\left(\frac{sa}{q}\right) = \begin{cases} d & \text{if } \ d \mid a \\ 0 & \text{if } \ d \nmid a\,, \end{cases}$$

Thus we have

$$\sum_{l(q)}{}' e\left(\frac{la}{q}\right) = \sum_{\substack{\delta d = q \\ d \mid a}} \mu(\delta)\, d = \sum_{d \mid (a,q)} \mu\left(\frac{q}{d}\right) d = \mu(q)\,.$$

Since $(a, q) = 1$. Applying this to (9.6) we see that

$$f_x\left(\frac{a}{q}\right) = \frac{\mu(q)}{\varphi(q)}\, x + \mathrm{O}\left(\frac{xq}{\mathrm{Log}^{5H} x}\right).$$

We now make use of the Abel summation formula to relate $f_x(\alpha)$ to $f_x\left(\frac{a}{q}\right)$. If $\alpha \in I_{a/q}$, then

$$\alpha = \frac{a}{q} + t \quad \text{where} \quad |\, t \,| \leqslant Q^{-1}$$

hence

$$f_x(\alpha) - \frac{\mu(q)}{\varphi(q)}\, g_x(t) = \sum_{n \leqslant x} \left\{ \lambda_1(n)\, e\left(\frac{na}{q}\right) - \frac{\mu(q)}{\varphi(q)} \right\} e(nt)\,.$$

Denoting by $D(x, t)$ the sum

$$D(x, t) = \sum_{n \leqslant x} \left\{ \lambda_1(n)\, e\left(\frac{na}{q}\right) - \frac{\mu(q)}{\varphi(q)} \right\} e(nt),$$

and using the Abel summation formula we obtain

$$D(N, t) = D(N, 0)\, e(nt) - 2\, i\pi t \int_0^N D(x, 0)\, e(xt)\, dx,$$

from which we infer that

$$| D(N, t) | \leqslant | D(N, 0) | + 2\,\pi\, | t |\, N \sup_{0 \leqslant x \leqslant N} | D(x, 0) |.$$

Since

$$D(x, 0) = f_x\left(\frac{a}{q}\right) - \frac{\mu(q)}{\varphi(q)}\, x$$

we conclude

$$| D(N, t) | \leqslant c_1 \frac{Nq}{(\mathrm{Log}\, N)^{5H}} (1 + 2\,\pi\, | t |\, N)$$

and recalling that $| t | \leqslant Q^{-1} = N^{-1}(\mathrm{Log}\, N)^H$ and $q \leqslant \eta = (\mathrm{Log}\, N)^H$ we have

$$| D(N, t) | \leqslant c_2 \frac{N}{(\mathrm{Log}\, N)^{3H}} = c_2 \frac{N}{\eta^3}.$$

The asymptotic formula for $f_N(\alpha)$ is now established.

2.7 The integral over a major arc. Since $| g_N(t) | \leqslant N$, an immediate consequence of § 2.6 is that

$$f_N^r(\alpha) = \left\{ \frac{\mu(q)}{\varphi(q)}\, g_N(t) \right\}^r + \mathrm{O}\left(\frac{N}{\eta^3} \max \left\{ \frac{N}{\varphi(q)}, \frac{N}{\eta^3} \right\}^{r-1} \right)$$

and since $\eta^3 > \varphi(q) \geqslant 1$ the error term is in fact

$$\mathrm{O}\left(\frac{N^r}{\eta^3\, \varphi(q)^{r-1}} \right).$$

It now follows that

$$\int_{I_{a/q}} f_N^r(\alpha)\, e(-\,\alpha N)\, d\alpha = \left(\frac{\mu(q)}{\varphi(q)} \right)^r e\left(-\frac{Na}{q} \right) \int_{-1/Q}^{+1/Q} g_N^r(t)\, e(-\,Nt)\, dt +$$
$$\mathrm{O}(N^r\, \eta^{-3}\, Q^{-1}\, \varphi(r)^{1-r}).$$

2.8 The integral over the major set. Because M is the union of the disjoint intervals $\{ I_{a/q} \}$ it is clear that

$$\mathfrak{J}_M = \sum_{q \leqslant \eta} \sum_{\substack{a=1 \\ (a,q)=1}}^{q} \int_{I_{a/q}} f_N^r(\alpha)\, e(-\alpha N)\, d\alpha$$

$$= \sum_{q \leqslant \eta} \left(\frac{\mu(q)}{\varphi(q)}\right)^r c_q(-N) \int_{-1/Q}^{+1/Q} g_N^r(t)\, e(-Nt)\, dt$$

$$+ \mathrm{O}\left(N^r \eta^{-3} Q^{-1} \sum_{q \leqslant \eta} \varphi^{2-r}(q)\right)$$

where we have written $c_q(-N)$ for the sum :

$$\sum_{\substack{a=1 \\ (a,q)=1}}^{q} e\left(-\frac{Na}{q}\right).$$

(The sums $c_q(-N)$ are known as Ramanujan sums in honour of the mathematician S. Ramanujan who investigated many of their arithmetic properties.)

Since $\varphi^{r-2}(q) \geqslant 1$ it is trivial that the sum

$$\sum_{q \leqslant \eta} \varphi^{2-r}(q) = \mathrm{O}(\eta)$$

and so the error term in the expression for $\mathfrak{J}_M$ is

$$\mathrm{O}(N^{r-1} \eta^{-1}),$$

thus we have

$$\mathfrak{J}_M = \sum_{q \leqslant \eta} \left(\frac{\mu(q)}{\varphi(q)}\right)^r c_q(-N) \int_{-1/Q}^{+1/Q} g_N^r(t)\, e(-Nt)\, dt + \mathrm{O}(N^{r-1} \eta^{-1}).$$

Now we consider the " main " term and try to simplify it. First of all, if we replace the range of integration $[-1/Q, 1/Q]$ by $\left[-\frac{1}{2}, \frac{1}{2}\right]$ then the resulting error which we introduce in our expression for $\mathfrak{J}_M$ is

$$\mathrm{O}\left(\sum_{q \leqslant \eta} \frac{| c_q(-N) |}{\varphi^r(q)} \int_{1/Q}^{1/2} | g_N^r(t) |\, dt\right).$$

For t satisfying $0 < | t | < \frac{1}{2}$ it is easy to see that

$$| g_N(t) | = \left| \sum_{n \leqslant N} e(nt) \right| = \left| \frac{e(t) - e(Nt + t)}{1 - e(t)} \right|$$

$$\leqslant \left| \frac{2}{1 - e(t)} \right| \leqslant | \sin \pi t |^{-1} \leqslant 2 | t |^{-1}. \qquad (9.7)$$

Furthermore, since $c_q(-N)$ contains only $\varphi(q)$ terms, $|c_q(-N)| \leqslant \varphi(q)$. Thus the above error term is at most

$$\mathrm{O}\left(\sum_{q \leqslant \eta} \left(\frac{Q}{\varphi(q)}\right)^{r-1}\right) = \mathrm{O}\left(Q^{r-1} \sum_{q \leqslant \eta} \varphi^{1-r}(q)\right)$$
$$= \mathrm{O}(Q^{r-1}\,\eta) = \mathrm{O}(N^{r-1}\,\eta^{2-r})$$

since $r \geqslant 3$. Our expression for $\mathfrak{J}_M$ is now

$$\mathfrak{J}_M = \sum_{q \leqslant \eta} \left(\frac{\mu(q)}{\varphi(q)}\right)^r c_q(-N) \int_{-1/2}^{+1/2} g_N^r(t)\, e(-Nt)\, dt + \mathrm{O}(N^{r-1}\,\eta^{-1}).$$

The integral $\int_{-1/2}^{+1/2} g_N^r(t)\, e(-Nt)$ can be evaluated explicitly, for

$$\int_{-1/2}^{+1/2} g_N^r(t)\, e(-Nt)\, dt$$

$$\int_{-1/2}^{+1/2} \{ b_0 + b_1\, e(t) + \cdots + b_N\, e(Nt) + \cdots + b_{Nr}\, e(Nrt) \}\, e(-Nt)\, dt = b_N$$

where b_i is the coefficient of x^i in the expansion of $(x + \cdots + x^N)^r$. This coefficient is

$$b_N = C_{N-1}^{r-1} = \frac{N^{r-1}}{(r-1)!} + \mathrm{O}(N^{r-2}).$$

If we replace the integral in our expression for $\mathfrak{J}_M$ by the above asymptotic formula, then the error which we introduce is

$$\mathrm{O}\left(\sum_{q \leqslant \eta} \frac{|c_q(-N)|}{\varphi^r(q)} N^{r-2}\right) = \mathrm{O}\left(\sum_{q \leqslant \eta} \frac{N^{r-2}}{\varphi^{r-1}(q)}\right)$$
$$= \mathrm{O}(\eta N^{r-2}) = \mathrm{O}\left(\frac{N^{r-1}}{\eta}\right)$$

and the resulting expression for $\mathfrak{J}_M$ now becomes

$$\mathfrak{J}_M = \frac{N^{r-1}}{(r-1)!} \sum_{q \leqslant \eta} \left(\frac{\mu(q)}{\varphi(q)}\right)^r c_q(-N) + \mathrm{O}(N^{r-1}\,\eta^{-1}).$$

Extending the summation over q to infinity, we have

$$\sum_{q=1}^{\infty} \left|\frac{\mu(q)}{\varphi(q)}\right|^r |c_q(-N)| \leqslant \sum_{q=1}^{\infty} \frac{1}{\varphi(q)^{r-1}} \tag{9.8}$$

where the terms of the latter series are multiplicative functions of q. Since $r \geqslant 3$ we can apply theorem 1.1 using condition (ii) to deduce that

$$\varphi(q) = q \prod_{p \mid q} \left(1 - \frac{1}{p}\right) \geqslant q \prod_{p \leqslant q} \left(1 - \frac{1}{p}\right) \geqslant c_1 \frac{q}{\operatorname{Log} q}$$

$$\sum_{q=1}^{\infty} \left| \frac{\mu(q)}{\varphi(q)} \right|^r | c_q(-N) | \leqslant \frac{1}{c_1^{r-1}} \sum_{q=1}^{\infty} \left(\frac{\operatorname{Log} q}{q}\right)^{r-1} < \infty .$$

and so the first series of (9.8) is absolutely convergent.

We now examine the error introduced in the expression for $\mathfrak{J}_M$ by replacing the finite series by the infinite series. The error is

$$\mathrm{O}\left(N^{r-1} \sum_{q > \eta} \frac{| \mu(q) |}{\varphi^{r-1}(q)}\right) .$$

From exercise 1.18 we know that

$$\varphi(q) = q \prod_{p \mid q} \left(1 - \frac{1}{p}\right) \geqslant q \prod_{p \leqslant q} \left(1 - \frac{1}{p}\right) \geqslant \frac{c_1 q}{\operatorname{Log} q},$$

consequently

$$\frac{| \mu(q) |}{\varphi^{r-1}(q)} < c_2 \left(\frac{\operatorname{Log} q}{q}\right)^{r-1} < c_3 q^{r - \frac{3}{2}} .$$

Thus we have

$$\sum_{q > \eta} \frac{| \mu(q) |}{\varphi^{r-1}(q)} < c_3 \sum_{q > \eta} q^{r - \frac{3}{2}} = \mathrm{O}(\eta^{-1/2})$$

To sum up, we have now shown that

$$\mathfrak{J}_M = \frac{N^{r-1}}{(r-1)\,!} \sum_{q=1}^{\infty} \left(\frac{\mu(q)}{\varphi(q)}\right)^r c_q(-N) + \mathrm{O}\left(\frac{N^{r-1}}{\sqrt{\eta}}\right)$$

and if we define the function $\mathfrak{S}_r$ by

$$\mathfrak{S}_r(N) = \frac{1}{(r-1)\,!} \sum_{q=1}^{\infty} \left(\frac{\mu(q)}{\varphi(q)}\right) c_q(-N) ,$$

we can write our expression for the integral over the major set as

$$\mathfrak{J}_M = N^{r-1} \mathfrak{S}_r(N) + \mathrm{O}\left(\frac{N^{r-1}}{\sqrt{\eta}}\right) \qquad (9.9)$$

2.9 The function $\mathfrak{S}_r(N)$. We now investigate the function $\mathfrak{S}_r(N)$ in a little more detail and show that it can be transformed into an Euler product. First we show that the terms of the series

$$\sum_{q=1}^{\infty} \left(\frac{\mu(q)}{\varphi(q)}\right)^r c_q(-N)$$

are multiplicative functions of q. The proof is not difficult. It is well known that φ and μ are multiplicative functions and we will now give a proof that $c_q(-N)$ is a multiplicative function of q.

If a, b are integers such that $(a, b) = 1$, then

$$c_a(n)\, c_b(n) = \sum_{r(a)}' e\left(\frac{rn}{a}\right) \sum_{s(b)}' e\left(\frac{sn}{b}\right)$$

$$c_a(n)\, c_b(n) = \sum_{r(a)}' \sum_{s(b)}' e\left(\frac{br + as}{ab} n\right) .$$

and since $\{ br + as \}$ runs over a complete set of reduced residues modulo ab if r and s run over complete sets of reduced residues modulo a and b, we certainly have

$$c_a(-N)\, c_b(-N) = c_{ab}(-N) .$$

Because the series $\mathfrak{S}_r(N)$ is absolutely convergent and the terms of the series are multiplicative we can apply theorem 1.1 and write the function $\mathfrak{S}_r(N)$ as

$$\mathfrak{S}_r(N) = \frac{1}{(r-1)!} \prod_p \left\{ 1 + \left(\frac{\mu(p)}{\varphi(p)}\right) c_p(-N) \right\} .$$

The factor corresponding to $p = 2$ is $(1 + (-1)^r (-1)^N)$. Hence we have

$$\mathfrak{S}_r(N) = \begin{cases} 0 & \text{if } N \not\equiv r (\text{mod } 2) \\ \dfrac{2}{(r-1)!} \mathrm{P}_r(N) & \text{if } N \equiv r (\text{mod } 2) . \end{cases}$$

where $\mathrm{P}_r(N)$ denotes the product

$$\mathrm{P}_r(N) = \prod_{p>2} \left\{ 1 + \left(\frac{-1}{p-1}\right)^r c_p(-N) \right\} ,$$

One can easily see that if $p \mid N$, then $c_p(-N) = p - 1$ and that if $p \times N$, then $c_p(-N) = -1$. Thus, we can write

$$\mathrm{P}_r(N) = \prod_{\substack{p>2 \\ p\nmid N}} \left\{1 - \frac{1}{(1-p)^r}\right\} \prod_{\substack{p>2 \\ p\mid N}} \left\{1 + \frac{p-1}{(1-p)^r}\right\}$$

$$= \prod_{p>2} \left\{1 - \left(\frac{-1}{p-1}\right)^r\right\} \prod_{\substack{p>2 \\ p\mid N}} \left\{1 - \left(\frac{-1}{p-1}\right)^r\right\}^{-1} \left\{1 - \left(\frac{-1}{p-1}\right)^{r-1}\right\}$$

$$= H(r)\, Q_r(N)\,.$$

We look first at $Q_r(N)$. If r is even then each term of $Q_r(N)$ is greater than 1, hence

$$\liminf_{N\to\infty} Q_r(N) = 1 \leqslant Q_r(N)$$

$$\prod_{p>2} \left\{1 - \left(\frac{1}{p-1}\right)^r\right\}^{-1} \left\{1 + \left(\frac{1}{p-1}\right)^{r-1}\right\} = \limsup_{N\to\infty} Q_r(N)\,.$$

Similarly, if r is odd

$$\limsup_{N\to\infty} Q_r(N) = 1 \geqslant Q_r(N)$$

$$\prod_{p>2} \left\{1 - \left(\frac{1}{p-1}\right)^{r-1}\right\} \left\{1 + \left(\frac{1}{p-1}\right)^r\right\}^{-1} = \liminf_{N\to\infty} Q_r(N)\,.$$

Since $r \geqslant 3$, each of the above infinite products is convergent. Consequently, for all $N \equiv r(\mathrm{mod}\ 2)$, there exist $c_1(r)$, $c_2(r)$ such that

$$0 < c_1(r) < \mathfrak{S}_r(N) < c_2(r) < \infty\,.$$

This completes our study of the function $\mathfrak{S}_r(N)$.

2.10 The choice of H. Recall that $\eta = (\mathrm{Log}\, N)^H$. If we now set $H = 2\Delta + 10$ then from (9.4) we have

$$\mathfrak{J}_m = \mathrm{O}\left(N^{r-1}\left(\frac{\mathrm{Log}^4 N}{\sqrt{\eta}}\right)^{r-2} \mathrm{Log}\, N\right)$$

$$= \mathrm{O}\left(\frac{N^{r-1}}{\mathrm{Log}^{\Delta} N}\right),$$

and from (9.9),

$$\mathfrak{J}_M = N^{r-1}\,\mathfrak{S}_r(N) + \mathrm{O}\left(\frac{N^{r-1}}{\sqrt{\eta}}\right)$$

$$= N^{r-1}\,\mathfrak{S}_r(N) + \mathrm{O}\left(\frac{N^{r-1}}{\operatorname{Log}^A N}\right).$$

Thus we now have

$$\lambda_r(N) = \mathfrak{J}_M + \mathfrak{J}_m$$

$$= N^{r-1}\,\mathfrak{S}_r(N) + \mathrm{O}\left(\frac{N^{r-1}}{\operatorname{Log}^A N}\right)$$

and the proof of theorem 9.1 is complete. ■

§ 3 THE SUM $\sum_{p \leqslant x} e(\alpha p)$

3.1 We shall ultimately estimate the sum $\sum_{p \leqslant x} e(\alpha p)$ in terms of x and a rational approximation to α. This will be given in theorem 9.4. Before proving this theorem we shall prove two general results which will be needed in the course of our estimations. The first theorem gives an estimate for a type of sum which often arises in the study of trigonometric sums and the second gives an estimate for the absolute value of a bilinear form. This latter result and its generalisations have numerous other applications in number theory, see for example Montgomery [1].

Sums which are often going to occur in the proof of theorem 9.4 are of the following type. Let $\mathcal{N}$ be a finite set of positive integers, U a fixed positive real number, and α a fixed real number, then

$$\mathrm{S}(\mathcal{N}, \mathrm{U}) = \sum_{n \in \mathcal{N}} \sum_{r \leqslant \mathrm{U}} e(\alpha r n)\,.$$

We shall need estimates for $|\mathrm{S}(\mathcal{N}, \mathrm{U})|$. Trivially we have

$$|\mathrm{S}(\mathcal{N}, \mathrm{U})| \leqslant \sum_{n \in \mathcal{N}} \left| \sum_{r \leqslant \mathrm{U}} e(\alpha r n) \right|.$$

The inner sum is a geometric progression which is easily computed :

$$\left| \sum_{r \leqslant \mathrm{U}} e(\alpha r n) \right| = \left| \frac{e(\alpha n) - e(\alpha n[\mathrm{U}] + \alpha n)}{1 - e(\alpha n)} \right| \leqslant \frac{2}{|1 - e(\alpha n)|} = \frac{1}{|\sin \pi \alpha n|}\,.$$

This estimate is good provided sin $\pi\alpha n$ is not " small ". When sin $\pi\alpha n$ is, in fact, small we cannot, in general, deduce anything better than the trivial estimate

$$\left| \sum_{r \leqslant U} e(\alpha r n) \right| \leqslant U .$$

Thus we certainly have

$$| S(\mathcal{N}, U) | \leqslant \sum_{n \in \mathcal{N}} \min \{ U, | \sin \pi\alpha n |^{-1} \}$$

and the problem of estimating $| S(\mathcal{N}, U) |$ is reduced to the study of sums of the type

$$\sigma = \sum_{n \in \mathcal{N}} \min \{ U, | \sin \pi\alpha n |^{-1} \}$$

where the minimum is defined to be U when sin $\pi\alpha n = 0$.

Theorem 9.2 *Let a, q be integers such that $(a, q) = 1$, $q > 0$ and $| \alpha - a/q | < q^{-2}$. If $\mathcal{N} \subseteq I(\beta) = \left[\beta - \frac{1}{2} q, \beta + \frac{1}{2} q \right]$, then*

$$\sigma < 3\,U + q(1 + \operatorname{Log} q)$$

whilst if $\mathcal{N} \subseteq I(0) \setminus \{ 0 \}$ we have

$$\sigma < q(2 + \operatorname{Log} q) .$$

Proof. From our rational approximation to α we deduce that

$$n\alpha = \frac{na}{q} + \theta \frac{n}{q^2}$$

where $| \theta | < 1$ and as n varies in the interval $I(\beta)$ the number $n\theta/q$ is confined to an interval of length at most $q(\theta)/q < 1$. Hence we certainly have

$$\frac{n\theta}{q} = b + \theta_n .$$

where b is an integer *independent of* n and $| \theta_n | < 1$. Thus we see that

$$n\alpha = \frac{na + b}{q} + \frac{\theta_n}{q} .$$

Since $(a, q) = 1$ we see that $na + b$ describes with n the whole or part of a complete set of residues modulo q. Replacing $(na + b)$ by its absolutely least residue l, modulo q we obtain

$$\sigma \leqslant \sum_{-\frac{1}{2}q < l \leqslant \frac{1}{2}q} \min \left\{ U, \left| \sin \pi \left(\frac{l}{q} + \frac{\tau_l}{q} \right) \right|^{-1} \right\}$$

where $|\theta_l| < 1$.

If $2 \leqslant |l| \leqslant \frac{1}{2} q$, then we have

$$\left| \sin \pi \left(\frac{l}{q} + \frac{\tau_l}{q} \right) \right| = \sin \pi \left(\frac{|l|}{q} \pm \frac{\tau_l}{q} \right) \geqslant \frac{2(|l| - 1)}{q},$$

since $\sin \pi x \geqslant 2 \min(x, 1 - x)$ for $0 \leqslant x \leqslant 1$.

Hence we certainly have

$$\begin{aligned} \sigma &\leqslant 3\,\mathrm{U} + 2 \sum_{l=2}^{[q/2]} \frac{q}{2(l-1)} \\ &< 3\,\mathrm{U} + q(1 + \operatorname{Log} q) . \end{aligned}$$

In the second case we have $\beta = 0$, $b = 0$, $|\theta_n| \leqslant \frac{1}{2}$ and hence

$$\sigma \leqslant 2 \sum_{l=1}^{[q/2]} \frac{q}{2\left(l - \frac{1}{2}\right)} < q \left(2 + \int_1^{\frac{1}{2}q + \frac{1}{4}} \frac{du}{u - \frac{1}{2}} \right)$$

$$< q(2 + \operatorname{Log} q) .$$

This completes the proof of the theorem. ■

Theorem 9.3 *Let $B(\mathbf{x}, \mathbf{y})$ be the bilinear form*

$$B(\mathbf{x}, \mathbf{y}) = \sum_{m,n} b_{m,n} x_m y_n .$$

Then we have

$$|B(\mathbf{x}, \mathbf{y})| \leqslant \|\mathbf{x}\| \cdot \|\mathbf{y}\| \cdot B$$

where $\|\mathbf{x}\| = \left\{ \sum_m |x_m|^2 \right\}^{1/2}$, $\|\mathbf{y}\| = \left\{ \sum_n |y_n|^2 \right\}^{1/2}$ *and*

$$B = \max_n \left(\sum_r \left| \sum_m b_{m,r} \bar{b}_{m,n} \right| \right)^{1/2} .$$

Proof. Let $\eta_m = \sum_n b_{m,n} y_n$, then we have

$$B(\mathbf{x}, \mathbf{y}) = \sum_m x_m \eta_m ,$$

and consequently by Cauchy's inequality we have

$$\begin{aligned}
|B(\mathbf{x},\mathbf{y})|^2 &\leqslant \sum_m |x_m|^2 \sum_m |\eta_m|^2 \\
&= \|\mathbf{x}\|^2 \sum_m \sum_n b_{m,n}\, y_n \sum_r \overline{b}_{m,r}\, \overline{y}_r \\
&\leqslant \|\mathbf{x}\|^2 \sum_{n,r} \frac{1}{2}(|y_n|^2 + |y_r|^2) \left| \sum_m b_{m,n}\, \overline{b}_{m,r} \right| \\
&= \|\mathbf{x}\|^2 \sum_{n,r} |y_n|^2 \left| \sum_m b_{m,n}\, \overline{b}_{m,r} \right| \\
&\leqslant \|\mathbf{x}\|^2\, \|\mathbf{y}\|^2\, B^2 .
\end{aligned}$$

3.2 We are now in a position to estimate the sum $\mathrm{S}_N(\alpha) = \sum_{p \leqslant N} e(\alpha p)$. During the proof of the theorem, $\mathrm{S}_N(\alpha)$ will be split up into several subsums, each of which is handled differently.

Theorem 9.4 *Let α be a real number and a/q a rational approximation to α which satisfies*

$$(a, q) = 1\,, \quad q > 0 \quad \textit{and} \quad |\alpha - a/q| < q^{-2} .$$

Then for all $N > 1$ we have

$$|\mathrm{S}_N(\alpha)| = \left| \sum_{p \leqslant N} e(\alpha p) \right| < A \left\{ \frac{1}{q} + \frac{q}{N} + \exp(-\sqrt{\operatorname{Log} N}) \right\}^{1/2} N(\operatorname{Log} N)^3 .$$

where A is an effectively computable absolute constant.

Proof. Since we obviously have $|\mathrm{S}_N(\alpha)| \leqslant N$ the result is trivial if $q \geqslant N$. So we shall suppose that $q < N$.

We begin our estimation of $\mathrm{S}_N(\alpha)$ by writing

$$\begin{aligned}
\mathrm{S}_N(\alpha) &= \sum_{p \leqslant \sqrt{N}} e(\alpha p) + \sum_{\sqrt{N} < p \leqslant N} e(\alpha p) \\
&= \mathrm{O}(\sqrt{N}) + \sum_{\sqrt{N} < p \leqslant N} e(\alpha p) .
\end{aligned}$$

Define the integer D as

$$D = \prod_{p < \sqrt{N}} p$$

then the set of integers $\{ n \leqslant N : (n, D) = 1 \}$ consists of the primes in the interval $[\sqrt{N}, N]$ together with 1. Thus we have

$$\begin{aligned} S_N(\alpha) &= O(\sqrt{N}) + \sum_{\substack{n \leqslant N \\ (n,D)=1}} e(\alpha n) \\ &= O(\sqrt{N}) + S \,. \end{aligned}$$

We shall estimate the sum S. From the elementary properties of the Möbius function

$$\begin{aligned} S &= \sum_{n \leqslant N} e(\alpha n) \sum_{l \mid (D,n)} \mu(l) = \sum_{\substack{n \leqslant N \\ l \mid D, l \mid n}} e(\alpha n)\, \mu(l) \\ &= \sum_{\substack{lm \leqslant N \\ l \mid D}} e(\alpha l m)\, \mu(l) \,. \end{aligned}$$

Let $h \geqslant 2$ be a fixed integer, to be chosen explicitly later. For any $l \mid D$ we define the function v_h by :

$$v_h(l) = \operatorname{card}\{ p \mid p > h \,;\, p \mid l \} \,.$$

Furthermore we shall write

$$H_0 = \{ l \mid l \mid D;\ v_h(l) = 0 \}$$

and

$$H_+ = \{ l \mid l \mid D;\ v_h(l) > 0 \} \,.$$

We now split the sum (9.10) into two parts as follows :

$$\begin{aligned} S &= \sum_{\substack{lm \leqslant N \\ l \in H_0}} e(\alpha l m)\, \mu(l) + \sum_{\substack{lm \leqslant N \\ l \in H_+}} e(\alpha l m)\, \mu(l) \\ &= S_0 + S_+ \,. \end{aligned}$$

The sums S_0 and S_+ will be considered separately.

3.3 The sum S_0.

Summing over m we obtain, by trivial estimation,

$$\left| \sum_{m \leqslant N/l} e(\alpha l m) \right| \leqslant \min \left\{ \frac{N}{l}, \frac{1}{|\sin \pi \alpha l|} \right\}$$

$$\begin{aligned} |S_0| &\leqslant \sum_{\substack{l \leqslant N \\ l \in H_0}} \min \left\{ \frac{N}{l}, \frac{1}{|\sin \pi \alpha l|} \right\} \\ &\leqslant \sum_{l \leqslant L} \min \left\{ \frac{N}{l}, |\sin \pi \alpha l|^{-1} \right\} + \sum_{\substack{L < l \leqslant N \\ l \in H_0}} \min \left\{ \frac{N}{l}, |\sin \pi \alpha l|^{-1} \right\} \\ &= S_{01} + S_{02} \,, \end{aligned}$$

where the real number L satisfies $1 < L < N$ and will be chosen explicitly later.

We first consider the sum $S_{0,1}$. For integers in the range $0 \leqslant r \leqslant \left[L/q + \frac{1}{2}\right]$ we define $I_r = \left(rq - \frac{1}{2}q, rq + \frac{1}{2}q\right]$. The union of the intervals I_r contains $[1, L]$. If $l \in I_r \cap [1, L]$, we have for $r > 0$

$$\frac{N}{l} \leqslant \frac{N}{rq - \frac{1}{2}q} < \frac{2N}{rq}$$

and if σ_r denotes the sum $\sum_{\substack{l \in I_r \\ 1 \leqslant l \leqslant L}} \min\{U_r, |\sin \pi l\alpha|^{-1}\}$, where $U_r = 2N/rq$ if $r \geqslant 1$ and $U_0 = N$, then we certainly have

$$S_{01} \leqslant \sum_{r=0}^{R} \sigma_r .$$

Applying theorem 9.2 to each of the sums σ_r we deduce that

$$\begin{aligned} S_{01} &\leqslant q(2 + \operatorname{Log} q) + \sum_{r=1}^{R} \left\{\frac{6N}{rq} + q(1 + \operatorname{Log} q)\right\}, \\ &= O\left(\frac{N}{q}\log N + \left(\frac{L}{q} + 1\right) q \log q\right). \end{aligned}$$

We now consider the sum $S_{0.2}$. If $l \in H_0$ has λ prime factors, then $l \leqslant h^\lambda$. Denoting, as usual, the number of divisors of l by $d(l)$, we have $d(l) = 2^\lambda \geqslant l^{1/\gamma}$ where γ is defined by $h = 2^\gamma$ and consequently,

$$\begin{aligned} S_{02} &\leqslant \sum_{\substack{L < l \leqslant N \\ l \in H_0}} \frac{N}{l} \leqslant \sum_{L < l \leqslant N} \frac{N}{l} \frac{d(l)}{l^{1/\gamma}} \\ &\leqslant \frac{N}{L^{1/\gamma}} \sum_{l \leqslant N} \frac{d(l)}{l} . \end{aligned}$$

Quite trivially we have

$$\sum_{l \leqslant N} \frac{d(l)}{l} = \sum_{l \leqslant N} \sum_{rm = l} \frac{1}{rm} = \sum_{rm \leqslant N} \frac{1}{rm} \leqslant \left(\sum_{l \leqslant N} \frac{1}{l}\right)^2 \leqslant (\operatorname{Log} N + 1)^2 , \qquad (9.11)$$

thus

$$S_{02} \leqslant \frac{N}{L^{1/\gamma}} (\operatorname{Log} N + 1)^2 .$$

If we now choose L to satisfy

$$L = N^{\gamma/(\gamma+1)}$$

then, since $q \leqslant N$,

$$S_{02} \leqslant N^{1-1/(1+\gamma)} (\operatorname{Log} N + 1)^2 ,$$

where $\delta = 1/(\gamma + 1)$ and $\gamma = \operatorname{Log} h/\operatorname{Log} 2$.

We wish to choose h so that δ is maximised while at the same time S_+ is minimised. If $h \geqslant \sqrt{N}$ then S_+ is empty, thus minimal. Unfortunately this choice of h yields worse than the trivial estimate, N, for $S_0 = S$. So we must choose h small enough so that S_+ is non-empty.

3.4 The sum S_+. Recall that the sum S_+ is defined by

$$S_+ = \sum_{\substack{lm \leqslant N \\ l \in H_+}} e(lm\alpha)\, \mu(l)$$

where $H_+ = \{ l \mid D \text{ with at least one prime factor } p > h \}$. Thus $l \in H_+$ if and only if $l = pr$, where p and r satisfy

$$h < p \leqslant \sqrt{N}, \quad r \mid D, \quad (r, p) = 1 . \tag{9.12}$$

By the definition of $v_h(l)$, any $l \in H_+$ has $v_h(l)$ such representations and for each of these representations we have

$$\mu(l) = -\mu(r) \quad \text{et} \quad v_h(l) = v_h(r) + 1 .$$

Consequently,

$$\mu(l) = v_h(l)\left(\frac{\mu(l)}{v_h(l)}\right) = \frac{\mu(l)}{v_h(l)} \sum_{pr=l}' 1 = \sum_{pr=l}' \frac{-\mu(r)}{v_h(r) + 1} ,$$

where $\sum'$ implies the restrictions (9.12) on the summation variables p and r. We can now write the sum S_+ as

$$\begin{aligned} S_+ &= \sum_{prm \leqslant N}' e(\alpha prm)\, \varepsilon(r) \\ &= T - R \end{aligned} \tag{9.13}$$

where $\varepsilon(r) = -\mu(r)/(v_h(r) + 1)$, T is the sum (9.13) with the condition $(r, p) = 1$ omitted and R is the sum (9.13) with the condition $(r, p) = 1$ replaced by $(r, p) = p$. The sums R and T will be considered separately.

The sum R is easy to estimate, for we have

$$\begin{aligned} | R | &\leqslant \sum_{\substack{p^2 sm \leqslant N \\ p > h}} 1 \leqslant \sum_{\substack{p^2 s \leqslant N \\ p > h}} \frac{N}{p^2 s} = N \sum_{\substack{p^2 \leqslant N \\ p > h}} \frac{1}{p^2} \sum_{s \leqslant N/p^2} \frac{1}{s} \\ &= O\left(\frac{N}{h} \operatorname{Log} N\right) . \end{aligned}$$

To estimate the sum T we are going to use theorem 9.3, so we express T as a bilinear form. For positive integers n, u and v, we write

$$T = \sum_{u,v} b_{u,v}\, x_u\, y_v \qquad \left(h < u < \sqrt{N},\, 0 < v < \frac{N}{h}\right)$$

where

$$b_{u,v} = \begin{cases} e(\alpha uv) & \text{if } 1 \leqslant uv \leqslant N \\ 0 & \text{if } uv > N \end{cases}$$

and

$$x_u = \begin{cases} 1 & \text{if } u \text{ is prime} \\ 0 & \text{otherwise}, \end{cases}$$

$$y_v = \sum_{\substack{rm=v \\ r|D}} \varepsilon(r) = \sum_{r|(v,D)} \varepsilon(r)\,.$$

for the summation ranges

$$h < u < \sqrt{N} \quad \text{and} \quad 0 < v < \frac{N}{h}.$$

We would like to apply theorem 9.3 directly, but this would not give a good result. For this reason, we split the sum T into subsums and apply theorem 9.3 to each of these subsums. The sum T is split into O(Log N) subsums $T(\mathrm{U})$ with summation ranges for u and v defined by

$$\mathrm{U} < u \leqslant \mathrm{U}', \qquad 0 < v \leqslant \frac{N}{\mathrm{U}}$$

where $\mathrm{U} = h, 2h, 2^2 h, \ldots, < \sqrt{N}$

$$\mathrm{U}' = \min(2\,\mathrm{U}, \sqrt{N})\,.$$

Theorem 9.3 is now applied to each of the sums $T(\mathrm{U})$. First we estimate the quantities $\| x \|$, $\| y \|$ and B :

$$\| x \|^2 = \sum_u x_u^2 \leqslant \sum_{u=\mathrm{U}+1}^{2\mathrm{U}} 1 = \mathrm{U}\,,$$

$$\| y \|^2 = \sum_v y_v^2 \leqslant \sum_{0<v<N/\mathrm{U}} (d(v))^2\,.$$

We observe that

$$\begin{aligned}
\sum_{n \leqslant x} d(n)^2 &= \sum_{n \leqslant x} d(n) \sum_{lm=n} 1 = \sum_{lm \leqslant x} d(lm) \\
&\leqslant \sum_{lm \leqslant x} d(l)\, d(m) = \sum_{l \leqslant x} d(l) \sum_{rs \leqslant x/l} 1 \\
&= \sum_{l \leqslant x} d(l) \sum_{s \leqslant x'} \sum_{r \leqslant x/sl} 1 \leqslant \sum_{l \leqslant x} d(l) \sum_{s \leqslant x/l} \frac{x}{sl} \\
&\leqslant \sum_{l \leqslant x} \frac{d(l)}{l} x \left(\log\left(\frac{x}{l}\right) + 1 \right) \\
&\leqslant x(\log x + 1) \sum_{l \leqslant x} \frac{d(l)}{l} \leqslant x(\log x + 1)^3
\end{aligned}$$

and as in (9.11) this expression is bounded above by

$$x(\mathrm{Log}\, x + 1)\, (\mathrm{Log}\, x + 1)^2 .$$

So we have

$$\| y \|^2 \leqslant \frac{N}{\mathrm{U}} \mathrm{Log}^3 N .$$

Obtaining an upper bound for B is a little harder. We write

$$\begin{aligned}
| B_{v,v'} | &= \sum_{\mathrm{U} < u < \min\{\mathrm{U}', N/v, N/v'\}} e(\alpha u(v - v')) \\
&\leqslant \min \{ \mathrm{U}, | \sin \pi\alpha(v - v') |^{-1} \}
\end{aligned}$$

whence

$$\sum_u b_{u,v} \overline{b}_{u,v'} \leqslant \min \{ \mathrm{U}, | \sin \pi\alpha(v - v') |^{-1} \} .$$

Writing $t = v' - v$ and applying theorem 9.2 to the $\left[\frac{N}{\mathrm{U}q}\right] + 1$ intervals of the type $I(\beta)$ with $\beta = -v + \frac{1}{2}q,\ -v + \frac{3}{2}q, \ldots$ etc., we obtain

$$\begin{aligned}
\sum_{v'} | B_{v,v'} | &\leqslant \sum_{-v < t \leqslant \frac{N}{\mathrm{U}} - v} \min \{ \mathrm{U}, | \sin \pi\alpha t |^{-1} \} \\
&\leqslant \left(\frac{N}{\mathrm{U}q} + 1 \right) \{ 3\,\mathrm{U} + q(1 + \mathrm{Log}\, q) \} .
\end{aligned}$$

Thus we have

$$B^2 = \max_v \sum_{v'} | B_{v,v'} | \leqslant \left(\frac{3\,N}{q} + 3\,\mathrm{U} \right) + \left(\frac{N}{\mathrm{U}} + q \right) \mathrm{Log}\, q .$$

We now apply theorem 9.3 to deduce that

$$| T(\mathrm{U}) |^2 \leqslant \left\{ \left(\frac{3N}{q} + 3\,\mathrm{U} \right) + \left(\frac{N}{\mathrm{U}} + q \right) \operatorname{Log} q \right\} \mathrm{U} \cdot \frac{N}{\mathrm{U}} \cdot \operatorname{Log}^3 N$$

$$\leqslant \left(\frac{1}{q} + \frac{\mathrm{U}}{N} + \frac{1}{\mathrm{U}} + \frac{q}{N} \right) N^2 \operatorname{Log}^4 N$$

and so

$$T(\mathrm{U}) = \mathrm{O} \left\{ \left(\frac{1}{q} + \frac{q}{N} + \frac{\mathrm{U}}{N} + \frac{1}{\mathrm{U}} \right)^{1/2} N \operatorname{Log}^2 N \right\}.$$

For each of the sums $T(\mathrm{U})$ we have

$$\frac{\mathrm{U}}{N} < \frac{1}{\mathrm{U}} \leqslant \frac{1}{h}$$

and T is the sum of $\mathrm{O}(\operatorname{Log} N)$ sums of the type $T(\mathrm{U})$. Hence as an upper bound for T we obtain

$$T = \mathrm{O} \left\{ \left(\frac{1}{q} + \frac{1}{h} + \frac{q}{N} \right)^{1/2} N \operatorname{Log}^3 N \right\}.$$

The estimate for S_+ is therefore

$$\mathrm{S}_+ = T - R = \mathrm{O} \left\{ \left(\frac{1}{q} + \frac{1}{h} + \frac{q}{N} \right)^{1/2} N \operatorname{Log}^3 N \right\} + \mathrm{O} \left(\frac{N}{h} \operatorname{Log} N \right).$$

3.5 The choice of *h*. We recall that

$$\begin{aligned} \mathrm{S}_N(\alpha) &= \mathrm{S} + \mathrm{O}(\sqrt{N}) \\ &= \mathrm{S}_0 + \mathrm{S}_+ + \mathrm{O}(\sqrt{N}) \\ &= \mathrm{O} \left\{ \left(\frac{1}{q} + \frac{1}{N^\delta} + \frac{q}{N} \right) N \operatorname{Log}^2 N \right\} \\ &\quad + \mathrm{O} \left\{ \left(\frac{1}{q} + \frac{1}{h} + \frac{q}{N} \right)^{1/2} N \operatorname{Log}^3 N \right\} + \mathrm{O} \left(\frac{N}{h} \operatorname{Log} N \right) + \mathrm{O}(\sqrt{N}). \end{aligned}$$

Since $q < N$,

$$\left(\frac{1}{q} + \frac{q}{N} + \frac{1}{N^\delta} \right) N \operatorname{Log}^2 N = \mathrm{O} \left\{ \left(\frac{1}{q} + \frac{q}{N} + \frac{1}{N^\delta} \right)^{1/2} N \operatorname{Log}^3 N \right\}.$$

and defining $\xi = \min(N^{2\delta}, h) = \min(N^{2/(\gamma+1)}, 2^\gamma)$ we have

$$\mathrm{S}_N(\alpha) = \mathrm{O} \left\{ \left(\frac{1}{q} + \frac{q}{N} + \frac{1}{\xi} \right)^{1/2} N \operatorname{Log}^3 N \right\}.$$

If we choose γ such that $\gamma + 1 = [2\sqrt{\operatorname{Log} N}]$, then, for $N \geqslant 3$, $h = 2^\gamma$ is an integer greater than 1 and $N^{2/(\gamma+1)} \geqslant \exp(\sqrt{\operatorname{Log} N})$, $2^\gamma \geqslant \exp\left(\frac{1}{2}\gamma\right) > \exp(\sqrt{\operatorname{Log} N} - 1)$, so $\xi > \exp(\sqrt{\operatorname{Log} N} - 1)$. The cases $N < 3$ are trivial, thus the theorem is now proved. ■

Remark. An interesting method for estimating sums of the type $\sum_{p \leqslant x} e(\alpha p)$ has been given in chapter 16 of Montgomery [1]. The method of proof is to relate the sum to the explicit formula for $\psi(x; k, l)$ and to the " density " of zeros of L-functions in the critical strip. A similar technique for the estimation of the sum

$$\sum_{p \leqslant x} e(f(p)),$$

for certain special functions f is given in Čudakov [1]. A systematic discussion of these techniques lies beyond the scope of this book.

§ 4 EFFECTIVE ERROR TERMS

4.1 In this section we shall derive an asymptotic formula for $\lambda_r(N)$ which has an error term weaker than that given in theorem 9.1, but which has the advantage that the implied " O " constant is effective. One can then use (9.1) to deduce an effective asymptotic formula for $v_r(N)$.

From now on all " O " constants are effective.

Theorem 9.5 *If $r \geqslant 3$ and $N \equiv r (\mathrm{mod}\ 2)$, then for any fixed $\varepsilon > 0$*

$$\lambda_r(N) = N^{r-1}\,\mathfrak{S}_r(N) + O\left(\frac{N^{r-1}}{(\operatorname{Log} N)^{r-2-\varepsilon}}\right)$$

where the implied " O " constant is an effective function of ε.

Proof. The proof of this theorem is basically the same as the proof of theorem 9.1. We shall assume that the reader is very familiar with the details of the proof of theorem 9.1 and consequently we only indicate how the ineffective constants can be removed.

The major and minor arcs are defined exactly as in § 2.3, in terms of $\eta = (\operatorname{Log} N)^H$ and $Q = N\eta^{-1}$, where H is a parameter to be chosen later.

The discussion of the integral $\mathfrak{J}_m$ was completely effective, hence

$$\mathfrak{J}_m = \mathrm{O}\left\{ N^{r-1}\left(\frac{\mathrm{Log}^4 N}{\sqrt{\eta}}\right)^{r-2} \mathrm{Log}\, N \right\},$$

as before.

The difference in the two proofs lies in the treatment of $\mathfrak{J}_M$. Instead of the approximation to $f_N(\alpha)$ on the major arc $I_{a/q}$ given in § 2.6, we shall prove a modified result with an effective error term.

4.2 Our principal tool will be theorem 8.3 which asserts the existence of regions $R_{2a}(K)$ of the complex plane with the following properties :

(i) The set $\{ L(s, \chi_q) : 1 \leqslant q \leqslant K, \chi_q \text{ proper} \}$ contains at most one L-function, $L(s, \chi_{q(K)})$ which has a zero, $\beta(a, K)$, in $R_{2a}(K)$.

(ii) If $\beta(a, K)$ exists, it is a real, simple zero of $L(s, \chi_{q(K)})$ and $\chi_{q(K)}$ is a real character.

(iii) The L-function $L(s, \chi_{q(K)})$ has no other zero in $R_{2a}(K)$.

We apply the theorem with $K = [\eta]$. If $\beta(a, \eta)$ exists, then of course $q(\eta)$ and all of its multiples less than or equal to η give rise to L-functions having $\beta(a, \eta)$ as a zero. We call $q(\eta)$ and its multiples less than or equal to η *exceptional relative to* η. Clearly, no non-exceptional q can give rise to an L-function having a zero in the region $R_{2a}([\eta])$.

4.3 Approximation to $f_N(\alpha)$ on a major arc. Before we are in a position to state the new approximation to $f_N(\alpha)$, we need to introduce some notation.

If χ_q is a character modulo q, then

$$\omega(\chi_q) = \sum_{r(q)}{}' \chi_q(r)\, e\left(\frac{r}{q}\right)$$

and setting $\beta = \beta(a, \eta)$, when it exists, define

$$g_N(t, \beta) = \sum_{n \leqslant N} n^{\beta - 1}\, e(nt)\,.$$

Recall that in § 2.5 we defined

$$g_N(t) = \sum_{n \leqslant N} e(nt)\,.$$

The required approximation is :

(a) *If* $\alpha \in I_{a/q}$ *and* q *is exceptional, then for* $t = \alpha - \dfrac{a}{q}$

$$f_N(\alpha) = \frac{\mu(q)}{\varphi(q)} g_N(t) - \chi_q(a) \frac{\omega(\chi_q)}{\varphi(q)} g_N(t, \beta) + O\left(\frac{N}{\eta^3}\right)$$

(b) *If* $\alpha \in I_{a/q}$ *and* q *is not exceptional, then for* $t = \alpha - \dfrac{a}{q}$

$$f_N(\alpha) = \frac{\mu(q)}{\varphi(q)} g_N(t) + O\left(\frac{N}{\eta^3}\right). \tag{9.14}$$

As in the proof of (9.5) we may write

$$f_x\left(\frac{a}{q}\right) = \sum_{l(q)}' e\left(\frac{al}{q}\right) \theta(x; q, l) + O\left(\sum_{p \mid q} \operatorname{Log} p\right).$$

Instead of using theorem 8.8 to obtain (9.6) we employ theorem 8.5(*c*), which implies that for $3 \leqslant x \leqslant N$

$$\theta(x; q, l) = \frac{x}{\varphi(q)} - \frac{\chi_q(l)\, x^\gamma}{\varphi(q)\, \gamma} + O\left(\frac{N}{\operatorname{Log}^{5H} N}\right), \tag{9.15}$$

with an effective " O " constant. The second term on the right hand side only occurs if $L(s, \chi_q)$ has a real zero $\gamma \in R_{2a}(q)$. If γ does *not* occur, then we can continue as in the treatment of (9.6) to obtain (9.14). On the other hand, if γ does occur, we distinguish two cases :

$$\gamma \notin R_{2a}([\eta]) \quad \text{and} \quad \gamma = \beta \in R_{2a}([\eta]).$$

If $\gamma \notin R_{2a}([\eta])$, then by the definition of $R_{2a}([\eta])$,

$$1 - \gamma > \frac{2\,a}{\operatorname{Log}([\eta] + 2)},$$

which implies that

$$N^\gamma < N \exp\left(-\frac{2\,a \operatorname{Log} N}{\operatorname{Log}(\eta + 2)}\right).$$

Thus, the second term in (9.15) can be absorbed into the error term $O(N/(\operatorname{Log} N)^{5H})$ and we again obtain (9.14).

Hence it remains to consider the case when $\gamma = \beta(a, \eta) = \beta \in R_{2a}([\eta])$, that is, when q is exceptional relative to η. In this case, the new term is added to the expression for $\theta(x; q, l)$, and this contributes to $f_x\left(\frac{a}{q}\right)$ the quantity

$$-\sum_{l(q)}' e\left(\frac{la}{q}\right)\frac{\chi_q(l)}{\varphi(q)}\frac{x^\beta}{\beta}.$$

By the definition of $R_{2a}([\eta])$ we know that β is bounded away from zero, and since

$$\int_1^x u^{\beta-1}\,du \leqslant \sum_{n\leqslant x} n^{\beta-1} \leqslant 1 + \int_1^x u^{\beta-1}\,du\,,$$

we have

$$\frac{x^\beta}{\beta} = \sum_{n\leqslant x} n^{\beta-1} + \mathrm{O}(1)\,.$$

Observing that la runs over a reduced set of residues modulo q when l does, we can write the new term as

$$-\chi_q(a)\frac{\omega(\chi_q)}{\varphi(q)}\sum_{n\leqslant x} n^{\beta-1} + \mathrm{O}(1)$$

and we have

$$f_x\left(\frac{a}{q}\right) = \frac{\mu(q)}{\varphi(q)}x - \chi_q(a)\frac{\omega(\chi_q)}{\varphi(q)}\sum_{n\leqslant x} n^{\beta-1} + \mathrm{O}\left(\frac{xq}{\mathrm{Log}^{5H} x}\right).$$

We now use this to approximate $f_x(\alpha)$. Writing

$$\begin{aligned} f_x(\alpha) - \frac{\mu(q)}{\varphi(q)}g_N(t) + \chi_q(a)\frac{\omega(\chi_q)}{\varphi(q)}g_N(t, \beta) & \\ &= \sum_{n\leqslant x}\left\{\lambda_1(n)\,e\left(\frac{na}{q}\right) - \frac{\mu(q)}{\varphi(q)} + \chi_q(a)\frac{\omega(\chi_q)}{\varphi(q)}n^{\beta-1}\right\} e(nt) \\ &= H(x, t)\,, \end{aligned}$$

we apply the Abel summation formula as in the proof of (9.5) to obtain the required approximation to $f_x(\alpha)$.

4.4 The integral over a major arc. If $\alpha \in I_{a/q}$ we write, for simplicity

$$f_N(\alpha) = \mathrm{U} - \mathrm{V} + E$$

where

$$U = \frac{\mu(q)}{\varphi(q)} g_N(t)\,, \qquad t = \alpha - \frac{a}{q}\,,$$

$$V = \begin{cases} \chi_q(a)\dfrac{\omega(\chi_q)}{\varphi(q)} g_N(t, \beta) & \text{if } q \text{ is exceptional} \\ 0 \quad \text{otherwise}\,, \end{cases}$$

$$E = O(N\eta^{-3})\,.$$

We consider three cases separately.

(i) q is not exceptional.

Let $1 > \delta > 0$ be a fixed real number to be chosen explicitly later and recall that by the definition of $R_{2a}([\eta])$, $\beta = \beta(a, \eta) < 1$.

(ii) q is exceptional and $1 - \beta > (\operatorname{Log} N)^{\delta - 1}$.

(iii) q is exceptional and $1 - \beta \leqslant (\operatorname{Log} N)^{\delta - 1}$.

Case (i). If q is not exceptional the term V does not occur and we integrate over a major arc as in § 2.7 to obtain

$$\int_{I_{a/q}} f_N^r(\alpha)\, e(-\alpha N)\, d\alpha = \left(\frac{\mu(q)}{\varphi(q)}\right)^r e\left(-\frac{Na}{q}\right) \int_{-1/Q}^{+1/Q} g_N^r(t)\, e(-Nt)\, dt + O\left(\frac{N^r}{\eta^3\, Q\varphi(q)^{r-1}}\right). \qquad (9.16)$$

Case (ii). This case is easily dealt with, for by the trivial estimate for $g_N(t, \beta)$ we have

$$V = O(N^\beta) = O\{N \exp(-(\operatorname{Log} N)^\delta)\} = O(E)\,.$$

Thus the term V can be absorbed into the error term and we obtain the same result as case (i).

Case (iii). This case is a little more troublesome than the other two. Besides the error term of (9.16) we shall see that an additional error is contributed to $\mathfrak{J}_M$ by the exceptional q when $1 - \beta \leqslant (\operatorname{Log} N)^{\delta - 1}$. It will not be until we do the final estimation of the error term of $\mathfrak{J}_M$ in § 4.6 that we will actually use the hypothesis on β.

We trivially have

$$f_N^r(\alpha) = U^r + O\{|U|^{r-1}(|V| + |E|) + |V|^r + |E|^r\}\,.$$

First we investigate the terms which involve V. Since $\beta < 1$, an application of the Abel summation formula gives us

$$|g_N(t, \beta)| \leqslant |g_N(t)|\,.$$

then by (9.7)

$$| g_N(t) | \leqslant \min \{ N, 2 | t |^{-1} \}$$

So it is clear that

$$| \mathrm{V} | \, | \mathrm{U} |^{r-1} + | \mathrm{V} |^r \leqslant \frac{| \omega(\chi_q) | + | \omega(\chi_q) |^r}{\varphi^r(q)} \min \{ N^r, 2 | t |^{-r} \} .$$

Integrating over the major arc $I_{a/q}$ we obtain

$$\mathfrak{J}_{a/q} = \int_{I_{a/q}} f_N^r(\alpha)\, e(-N\alpha)\, d\alpha = \int_{-1/Q}^{+1/Q} \mathrm{U}^r\, dt$$
$$+ \mathrm{O}\left\{ \int_{-1/Q}^{+1/Q} (| \mathrm{U} |^{r-1} | E | + | E |^r)\, dt \right\}$$
$$+ \mathrm{O}\left\{ \frac{| \omega(\chi_q) | + | \omega(\chi_q) |^r}{\varphi^r(q)} \int_{-1/Q}^{+1/Q} \min \{ N^r, 2 | t |^{-r} \}\, dt \right\} .$$

The first error term is dealt with exactly as in § 2.7 and is

$$\mathrm{O}\left(\frac{N^r}{\eta^3\, Q \varphi(q)^{r-1}} \right) .$$

Since

$$\min \{ N^r, 2 | t |^{-r} \} = \begin{cases} N^r & \text{if} \quad | t | \leqslant 2^{1/r} N^{-1} \\ 2 | t |^{-r} & \text{if} \quad | t | \geqslant 2^{1/r} N^{-1} , \end{cases}$$

the second error term is

$$\mathrm{O}\left\{ \frac{| \omega(\chi_q) | + | \omega(\chi_q) |^r}{\varphi^r(q)} N^{r-1} \right\} .$$

Let us now sumarise what we have done so far. We have shown that if q is exceptional and $1 - \beta \leqslant (\operatorname{Log} N)^{\delta - 1}$, then

$$\mathfrak{J}_{a/q} = \left(\frac{\mu(q)}{\varphi(q)} \right)^r e\left(-\frac{Na}{q} \right) \int_{-1/Q}^{+1/Q} g_N^r(t)\, e(-Nt)\, dt + \mathrm{O}\left(\frac{N^r}{\eta^3\, \varphi^{r-1}(q)\, Q} \right)$$
$$+ \mathrm{O}\left(\frac{| \omega(\chi_q) | + | \omega(\chi_q) |^r}{\varphi^r(q)} N^{r-1} \right) .$$

Otherwise

$$\mathfrak{J}_{a/q} = \left(\frac{\mu(q)}{\varphi(q)} \right)^r e\left(-\frac{Na}{q} \right) \int_{-1/Q}^{+1/Q} g_N^r(t)\, e(-Nt)\, dt + \mathrm{O}\left(\frac{N^r}{\eta^3\, \varphi^{r-1}(q)\, Q} \right) .$$

4.5 The integral over the major set. As in § 2.8 we have

$$\mathfrak{J}_M = \sum_{q \leqslant \eta} \sum_{\substack{a \leqslant q \\ (a,q)=1}} \mathfrak{J}_{a/q} \,.$$

The terms which involve the exceptional q when $1 - \beta \leqslant (\operatorname{Log} N)^{\delta - 1}$ introduce an error into the expression for $\mathfrak{J}_M$ which is of order

$$O\left(N^{r-1} \sum_{q \leqslant \eta}^{*} \frac{|\,\omega(\chi_q)\,| + |\,\omega(\chi_q)\,|^r}{\varphi^{r-1}(q)}\right),$$

the summation being over the exceptional q.

The other terms in the expressions for $\mathfrak{J}_{a/q}$ are dealt with in exactly the same way as in §§ 2.8 and 2.9. Thus we obtain

$$\mathfrak{J}_M = N^{r-1}\, \mathfrak{S}_r(N) + O\left(\frac{N^{r-1}}{\sqrt{\eta}}\right) + O\left(N^{r-1} \sum_{q \leqslant \eta}^{*} \frac{|\,\omega(\chi_q)\,| + |\,\omega(\chi_q)\,|^r}{\varphi^{r-1}(q)}\right).$$

4.6 The sum over exceptional *q*. In order to estimate the last term in the above asymptotic formula for $\mathfrak{J}_M$ we must study the sum

$$S = \sum_{q \leqslant \eta}^{*} \frac{|\,\omega(\chi_q)\,| + |\,\omega(\chi_q)\,|^r}{\varphi^{r-1}(q)}\,.$$

As we remarked in § 4.2, all the characters χ_q associated with exceptional q are equivalent to a unique proper character $\chi_{q(\eta)}$. The exceptional q are all multiples of $q(\eta)$ which we write as q_1. A classical result (see *Notes*, theorem 9.6) then implies that $|\,\omega(\chi_q)\,| \leqslant \sqrt{q_1}$.

If δ is the parameter introduced in § 4.4, then by exercise 1.18 there are constants c, $c(\delta) > 0$ such that

$$\varphi(q) \geqslant c\,\frac{q}{\operatorname{Log} q} \geqslant c(\delta)\, q^{1-\delta}\,.$$

Using the above facts we have the following relatively crude upper bound for S

$$S \leqslant c(\delta)^{1-r} \sum_{m \leqslant \eta/q_1} (q_1^{r/2} + q_1^{1/2})\,(mq_1)^{-(r-1)(1-\delta)}$$

$$= O\left(q_1^{\frac{1}{2}r-(r-1)(1-\delta)} \sum_{m \leqslant \eta/q_1} m^{-(r-1)(1-\delta)}\right).$$

If we take $\delta < (r-2)(r-1)^{-1}$, then the exponent of m is less than -1 and so the above series, when extended over all m, is convergent. Hence we now have

$$S = O\left(q_1^{\frac{1}{2}r-(r-1)(1-\delta)}\right).$$

Now supposing that δ is chosen less than $\frac{1}{2}(r-2)(r-1)^{-1}$ we see that the exponent of q_1 is negative. Thus, in order to obtain an upper bound for S in terms of N we must obtain a *lower* bound for q_1 in terms of N. This will be done by studying the function $L(s, \chi_{q_1})$, which we write as $L(s, \chi)$.

By hypothesis we know that $L(s, \chi)$ has a real zero β satisfying

$$1 - \beta \leqslant (\operatorname{Log} N)^{\delta - 1}.$$

For use later we shall also suppose that N is so large that

$$\frac{1}{2}\delta > (\operatorname{Log} N)^{\delta - 1}$$

This latter inequality implies that $\left[1 - \frac{1}{2}\delta, 1\right] \supset [\beta, 1]$.

By the mean-value theorem it follows that for some $s_1 \in [\beta, 1]$

$$L(1, \chi) - L(\beta, \chi) = (1 - \beta)\, L'(s_1, \chi)$$

hence

$$(\operatorname{Log} N)^{\delta - 1} \geqslant \frac{L(1, \chi)}{L'(s_1, \chi)}.$$

We shall now find a lower bound for $L(1, \chi)$ and an upper bound for $L'(s_1, \chi)$ in terms of q_1. By the Polya-Vinogradov inequality (see *Notes*, theorem 9.7)

$$\left| \sum_{n \leqslant x} \chi(n) \right| \leqslant \sqrt{q_1}\, \operatorname{Log} q_1 .$$

We use this bound in theorem 7.12 to obtain

$$L(1, \chi) \geqslant \frac{1}{16\sqrt{q_1}\, \operatorname{Log} q_1}.$$

To establish an upper bound for $L'(s_1, \chi)$ we shall find an upper bound for $|L'(s, \chi)|$ when $s \in \left[1 - \frac{1}{2}\delta, 1\right] \supset [\beta, 1]$.

This is done by appealing to theorem 8.1(b) with $\theta = 1 - \frac{1}{2}\delta$ and concluding

$$| L'(s, \chi) | < \frac{16}{\delta} q_1^{\frac{1}{2}\delta}(2 + \operatorname{Log}(2 q_1)).$$

$$| L'(s_1, \chi) | < \frac{16}{\delta} q_1^{\frac{1}{2}\delta}(2 + \operatorname{Log}(2 q_1)) < \frac{16}{\delta} q_1^{\frac{1}{2}\delta} \frac{(2 q_1)^{\frac{1}{2}\delta}}{\frac{1}{2}\delta}$$

$$< \frac{40}{\delta^2} q_1^{\delta} .$$

Combining our estimates for $L(1, \chi)$ and $L'(s, \chi)$ we see that

$$(\operatorname{Log} N)^{\delta - 1} \geqslant \frac{\delta^2}{640\, q_1^{\delta + \frac{1}{2}} \operatorname{Log} q_1} \geqslant \frac{c(\delta)}{q_1^{\frac{1}{2} + 2\delta}}$$

where $c(\delta)$ is a positive constant depending only on δ. Thus, we have shown

$$q_1 \geqslant (c(\delta))^{\frac{2}{1+4\delta}} (\operatorname{Log} N)^{2\frac{1-\delta}{1+4\delta}}$$

Upon substituting this lower bound for q_1 into our upper bound for S we obtain

$$\mathrm{S} = \mathrm{O}\left\{ (\operatorname{Log} N)^{2\frac{1-\delta}{1+4\delta}\left(\frac{1}{2}r - (r-1)(1-\delta)\right)} \right\}.$$

Now given any $\varepsilon > 0$ we choose δ so that

$$\delta < \frac{1}{2} \cdot \frac{r-2}{r-1}$$

and

$$2\frac{1-\delta}{1+4\delta}\left(\frac{1}{2}r - (r-1)(1-\delta)\right) < 2 - r + \varepsilon .$$

With this choice of δ we certainly have

$$\mathrm{S} = \mathrm{O}\{ (\operatorname{Log} N)^{2-r+\varepsilon} \},$$

4.7 The choice of *H*. We know that

$$\lambda_r(N) = \mathfrak{J}_M + \mathfrak{J}_m$$

and we have shown that

$$\mathfrak{J}_M = N^{r-1} \mathfrak{S}_r(N) + \mathrm{O}\left(\frac{N^{r-1}}{\sqrt{\eta}}\right) + \mathrm{O}\left(\frac{N^{r-1}}{(\operatorname{Log} N)^{r-2-\varepsilon}}\right)$$

$$\mathfrak{J}_m = \mathrm{O}\left(N^{r-1}\left(\frac{\operatorname{Log}^4 N}{\sqrt{\eta}}\right) \operatorname{Log} N\right)$$

where $\eta = (\operatorname{Log} N)^H$. If we now choose $H = 2(r-2) + 10$, then we easily see that

$$\lambda_r(N) = N^{r-1} \mathfrak{S}_r(N) + \mathrm{O}\left(\frac{N^{r-1}}{(\operatorname{Log} N)^{r-2+\varepsilon}}\right). \qquad \blacksquare$$

Notes to chapter 9

GAUSSIAN SUMS

If χ is a character modulo k, then sums of the type

$$\omega(\chi) = \sum_{n=1}^{k} \chi(n)\, e\left(\frac{n}{k}\right)$$

often occur in number theory. They are known as gaussian sums in honour of C. F. Gauss who investigated the case when χ is a quadratic character. (Several elementary properties of gaussian sums were given in the exercises to chapter 7.) Except in the special case when χ is a quadratic character virtually nothing is known about the value of $\omega(\chi)$. However, the precise value of $|\,\omega(\chi)\,|$ is known and is given by the following theorem.

Theorem 9.6 *If χ is a non-principal character modulo q with conductor f and $\chi \sim \chi_f$, then*

$$|\,\omega(\chi)\,| = \left|\,\chi_f\left(\frac{q}{f}\right)\right| \cdot \left|\,\mu\left(\frac{q}{f}\right)\right| \sqrt{f}\,.$$

Proof. The two cases χ a proper character and χ an improper character will be considered separately.

We will show that for any proper character χ and any positive integer n :

$$\chi(n)\,\omega(\bar{\chi}) = \sum_{h=1}^{q} \bar{\chi}(h)\, e\left(\frac{nh}{q}\right) \qquad (9.17)$$

Assuming, for the moment, the truth of this assertion we have

$$|\,\chi(n)\,|^2\,|\,\omega(\bar{\chi})\,|^2 = \sum_{h_1=1}^{h} \sum_{h_2=1}^{h} \bar{\chi}(h_1)\,\chi(h_2)\, e\left(\frac{n(h_1 - h_2)}{q}\right),$$

and summing over n

$$|\,\omega(\bar{\chi})\,|^2 \sum_{n=1}^{q} |\,\chi(n)\,|^2 = \sum_{h_1=1}^{q} \sum_{h_2=1}^{q} \bar{\chi}(h_1)\,\chi(h_2) \sum_{n=1}^{q} e\left(\frac{n(h_1 - h_2)}{q}\right).$$

The final sum is zero unless $h_1 = h_2$, in which case it is q, so we have

$$| \omega(\overline{\chi}) |^2 \sum_{n=1}^{q} | \chi(n) |^2 = q \sum_{h=1}^{q} | \chi(h) |^2 .$$

The sum $\sum | \chi(n) |^2$ is non-zero (in fact it is equal to $\varphi(q)$) so we have

$$| \omega(\chi) | = | \omega(\overline{\chi}) | = \sqrt{q} .$$

Now we prove (9.17). There are two cases to consider :

$$(n, q) = 1 \quad \text{and} \quad (n, q) > 1 .$$

The first case is easily dealt with, for

$$\chi(n)\, \omega(\chi) = \sum_{m=1}^{q} \chi(m)\, \chi(n)\, e\left(\frac{m}{q}\right)$$

and writing $m \equiv hn (\bmod q)$ we have

$$\chi(n)\, \omega(\overline{\chi}) = \sum_{h=1}^{q} \overline{\chi}(h)\, e\left(\frac{hn}{q}\right) .$$

The second case is more troublesome. Since $(n, q) > 1$ we have $\chi(n) = 0$, so we must prove that

$$0 = \sum_{m=1}^{q} \overline{\chi}(m)\, e\left(\frac{nm}{q}\right) . \tag{9.18}$$

Let us write

$$\frac{n}{q} = \frac{n_1}{q_1}, \quad \text{where} \quad (n_1, q_1) = 1 \quad \text{and} \quad q = q_1 q_2 ,$$

We can also suppose that $q_1 > 1$, since (9.18) is trivially true when $q \mid n$. On writing $h = uq_1 + v$, where $0 \leqslant u < q_2$, $1 \leqslant v \leqslant q_1$, we find

$$\begin{aligned}\sum_{m=1}^{q} \overline{\chi}(m)\, e\left(\frac{mn}{q}\right) &= \sum_{v=1}^{q_1} \sum_{u=0}^{q_2-1} \overline{\chi}(uq_1 + v)\, e\left(\frac{n_1}{q_1}(uq_1 + v)\right) \\ &= \sum_{v=1}^{q_1} e\left(\frac{vn_1}{q_1}\right) \sum_{u=0}^{q_2-1} \overline{\chi}(uq_1 + v) .\end{aligned}$$

Thus, to prove (9.18) it will suffice to show that for every positive integer v

$$S(v) = \sum_{u=0}^{q_2-1} \overline{\chi}(uq_1 + v) = 0 .$$

To prove this we shall use the trivial fact that $S(v)$ is periodic with period q_1 and the hypothesis that χ is a proper character. Since χ is proper it follows that for any $q_1 \mid q$ there exist c_1 and c_2 such that

$$(c_1, q) = (c_2, q) = 1\,, \qquad c_1 \equiv c_2 (\operatorname{mod} q_1)\,, \qquad \text{and} \qquad \chi(c_1) \neq \chi(c_2)\,.$$

If we take $c = c_1 c_2^{-1} (\operatorname{mod} q)$, then $\chi(c) \neq 1$ and $c \equiv 1 (\operatorname{mod} q_1)$. Now consider the expression

$$\begin{aligned}\bar{\chi}(c)\, S(v) &= \sum_{u=0}^{q_2-1} \bar{\chi}(c)\, \bar{\chi}(uq_1 + v) \\ &= \sum_{u=0}^{q_2-1} \bar{\chi}(cuq_1 + cv) \\ &= \sum_{u=0}^{q_2-1} \bar{\chi}(uq_1 + v) = S(v)\end{aligned}$$

upon using the periodicity of $S(v)$. Since $\chi(c) \neq 1$ it follows that $S(v) = 0$ and (9.18) is proved.

Finally we consider the case when χ is an improper character. If $\chi \sim \chi_f$, where χ_f is a proper character modulo f, then we are going to prove that

$$\omega(\chi) = \chi_f\left(\frac{q}{f}\right) \mu\left(\frac{q}{f}\right) \omega(\chi_f)\,; \tag{9.19}$$

from which the theorem is an immediate consequence. By the definition of χ_f we know that if $(n, q) = 1$, then $\chi(n) = \chi_f(n)$ and that if $(n, q) > 1$, $\chi(n) = 0$, so

$$\omega(\chi) = \sum_{n=1}^{q} \chi(n)\, e\left(\frac{n}{q}\right) = \sum_{n(q)}' \chi_f(n)\, e\left(\frac{n}{q}\right).$$

If we write $q = mf$, then there are two cases to consider :

$$(m, f) = 1 \quad \text{and} \quad (m, f) > 1\,.$$

As usual the first case is easy. It is clear that $rf + ms$ runs over a reduced set of residues modulo q when r, s run over reduced sets of residues modulo m and f respectively, so

$$\begin{aligned}\omega(\chi) &= \sum_{r(m)}' \sum_{s(f)}' \chi_f(fr + ms)\, e\left(\frac{fr + ms}{mf}\right) \\ &= \sum_{r(m)}' \sum_{s(f)}' \chi_f(ms)\, e\left(\frac{r}{m}\right) e\left(\frac{s}{f}\right) \\ &= \chi_f(m) \sum_{r(m)}' e\left(\frac{r}{m}\right) \sum_{s(f)}' \chi_f(s)\, e\left(\frac{s}{f}\right) \\ &= \chi_f(m)\, \mu(m)\, \omega(\chi_f)\,.\end{aligned}$$

If $(m, f) > 1$, let p be a fixed prime dividing (m, f). We write $q = pq'$, thus $f \mid q'$. A reduced set of residues modulo q is given by $\{ rq' + s \}$ when r runs over the set $\{ 0, \ldots, p-1 \}$ and s runs over the set $\{ 0 \leqslant s < q', (s, q') = 1 \}$. Consequently, we have

$$\omega(\chi) = \sum_{r\,s} \chi_f(fr + s)\, e\left(\frac{rq' + s}{pq'}\right)$$

$$= \sum_{r,s} \chi_f(s)\, e\left(\frac{r}{p}\right) e\left(\frac{s}{pq'}\right) = 0$$

since $\sum_r e\left(\frac{r}{p}\right) = 0$. This is in accord with (9.19), because $\chi_f(m) = 0$ in this case. This completes the proof of the theorem. ■

THE POLYA-VINOGRADOV INEQUALITY

Let χ be a non-principal character modulo k. As we have seen it is often very important to have estimates for the absolute value of the function

$$S(x) = \sum_{n \leqslant x} \chi(n) .$$

Let us write $\varepsilon(\chi) = 1$ if χ is proper and $\varepsilon(\chi) = 2$ if χ is improper. The following result was found independently by Polya and by Vinogradov.

Theorem 9.7 *With the above notation*

$$| S(x) | \leqslant \varepsilon(\chi) \sqrt{k} \operatorname{Log} k .$$

Proof. We shall consider the following two cases :

(i) χ a proper character and (ii) χ an improper character.

Case (i). Let us write $\rho = \exp(2\pi i/k)$ and define the " generalised Gauss sum " $\omega(a, \chi)$ by

$$\omega(a, \chi) = \sum_{n=1}^{k} \chi(n)\, \rho^{an} .$$

It is clear that if $(m, k) = 1$, then

$$\chi(m)\, \omega(m, \chi) = \omega(1, \chi)$$

and so

$$\omega(m, \chi) = \bar{\chi}(m)\, \omega(1, \chi) . \tag{9.20}$$

Thus, by summing the above equation over integers in which are coprime to k and noting that $\chi(n) = 0$ if $(n, k) \neq 1$ we obtain

$$\begin{aligned}\omega(1, \chi)\,\overline{S(N)} &= \omega(1, \chi) \sum_{m \leqslant N} \bar{\chi}(m) = \sum_{m \leqslant N} \omega(m, \chi) \\ &= \sum_{m=1}^{N} \sum_{r=1}^{k-1} \chi(r)\, \rho^{rm} \\ &= \sum_{r=1}^{k-1} \chi(r)\, \rho^{r} \frac{1 - \rho^{Nr}}{1 - \rho^{r}}.\end{aligned}$$

In order to estimate the last expression we use the following obvious facts :

$$\chi(k - r) = \chi(-1)\,\chi(r)\,, \quad \rho^{k} = 1\,, \quad \chi\left(\frac{k}{2}\right) = 0 \quad \text{if } k \text{ is even}\,,$$

to write the above equation in the form

$$\begin{aligned}\omega(1, \chi)\,\overline{S(N)} &= \sum_{r<k/2} \chi(r) \left\{ \rho^{r} \frac{1 - \rho^{Nr}}{1 - \rho^{r}} + \chi(-1)\,\rho^{-r} \frac{1 - \rho^{-Nr}}{1 - \rho^{-r}} \right\} \\ &= \sum_{r<k/2} \chi(r) \frac{\rho^{r} - \rho^{(N+1)r} - \chi(-1)\,(1 - \rho^{-rN})}{1 - \rho^{r}}.\end{aligned}$$

By taking absolute values we now conclude

$$|\,\omega(1, \chi)\,|\;|\,\overline{S(N)}\,| \leqslant \sum_{r<k/2} \frac{4}{|\,1 - \rho^{r}\,|}. \tag{9.21}$$

From theorem 9.6 we know that $|\,\omega(1, \chi)\,| = \sqrt{k}$, also the following inequality is completely trivial :

$$|\,1 - \rho^{r}\,| = 2 \sin \frac{\pi r}{k} \geqslant \frac{4\,r}{k} \qquad \text{if} \qquad r \leqslant \frac{k}{2}.$$

Using these two facts it now follows from (9.21) that

$$\begin{aligned}\sqrt{k}\,|\,S(N)\,| &\leqslant \sum_{r<k/2} \frac{k}{r} = k \sum_{r<k/2} \frac{1}{r} \\ &< k \operatorname{Log} k\,,\end{aligned}$$

$$|\,S(N)\,| < \sqrt{k} \operatorname{Log} k\,.$$

This completes the discussion of case (i).

Case (ii). Let f be the conductor of χ and suppose that $\chi \sim \chi^*$, where χ^* is a proper character. For the moment we assume that

$$\sum_{n \leqslant N} \chi(n) = \sum_{\substack{r \leqslant N \\ r | k}} \mu(r)\, \chi^*(r) \sum_{s \leqslant N/r} \chi^*(s)\,;$$

the proof of which is given later. Since χ^* is a proper character modulo f we certainly have

$$\left| \sum_{s \leqslant N/r} \chi^*(s) \right| \leqslant \sqrt{f} \operatorname{Log} f$$

and consequently

$$\left| \sum_{n \leqslant N} \chi(n) \right| \leqslant m(k) \sqrt{f} \operatorname{Log} f$$

where $m(k)$ denotes the number of square free integers of which every prime factor divides k but not f. Trivially $m(k)$ does not exceed the number of divisors of k/f and this is at most equal to $2\sqrt{(k/f)}$. Hence we certainly have the estimate

$$\left| \sum_{k \leqslant N} \chi(n) \right| \leqslant 2\sqrt{k/f}\sqrt{f} \operatorname{Log} f = 2\sqrt{k} \operatorname{Log} f \leqslant 2\sqrt{k} \operatorname{Log} k .$$

We now prove the assertion made above. For $\Re(s) > 1$ we know from chapter 7, § 3.3 that

$$\sum_{n=1}^{\infty}{}' \frac{\chi(n)}{n^s} = \prod_{p|k} \left(1 - \frac{\chi^*(p)}{p^s}\right) \sum_{m=1}^{\infty} \frac{\chi^*(m)}{m^s}$$
$$= \sum_{d|k} \frac{\mu(d)\,\chi^*(d)}{d^s} \sum_{m=1}^{\infty} \frac{\chi^*(m)}{m^s} .$$

By multiplication and equaling coefficients we obtain

$$\chi(n) = \sum_{\substack{rb=n \\ r|k}} \mu(r)\,\chi^*(r)\,\chi^*(b) .$$

and summing over n gives the stated result. ■

UNIFORM DISTRIBUTION MODULO 1 AND THE SEQUENCE $[\alpha n]$

First we recall some of the basic notions from the theory of uniform distribution modulo 1. For a much more complete account the reader is referred to Cassels [1].

A sequence $\{ x_k \}$ of real numbers is *uniformly distributed modulo* 1 if, for each $\alpha \in [0, 1]$

$$\lim_{n \to \infty} \frac{1}{n} \operatorname{card} \{ k \mid k \leqslant n ;\ x_k - [x_k] \leqslant \alpha \} = \alpha .$$

A very simple criterion for the uniform distribution modulo 1 of the sequence $\{ x_k \}$ has been given by Weyl [1], namely :

The sequence $\{ x_k \}$ *is uniformly distributed modulo* 1 *if and only if for each non-zero integer m*

$$\lim_{n \to \infty} \frac{1}{n} \sum_{k \leqslant n} e(mx_k) = 0 .$$

If α is an irrational real number, then it is a " well known " folklore theorem that the sequence $\{ \alpha p \}$ is uniformly distributed modulo 1. However, it seems to be difficult to locate a convincing proof of this assertion in the literature. Hence, as a fairly straightforward application of theorems 8.5(*d*) and 9.4 we shall give a complete proof of the result. A simple consequence of the uniform distribution of $\{ \alpha p \}$ is an asymptotic formula for the number of primes in the sequence $[\alpha n]$. This will be given as theorem 9.9. By a considerable elaboration of the methods of Vinogradov [1], Rhin [1] has shown that the sequence $\{ f(p) \}$ is uniformly distributed, where $f(x)$ is a real polynomial with at least one irrational coefficient other than the constant term.

Theorem 9.8 (a) *If* α *is an irrational number, then*

$$\lim_{N \to \infty} \frac{1}{\pi(N)} \sum_{p \leqslant N} e(\alpha p) = 0 .$$

(b) *The sequence* $\{ \alpha p \}$ is uniformly distributed modulo 1.

Proof. The second part of the theorem follows from part (*a*) by appealing to Weyl's criterion. Part (*a*) of the theorem will follow from an estimate for the sum

$$S_N(\alpha) = \sum_{p \leqslant N} e(\alpha p)$$

which will be obtained in terms of N and the denominators of a sequence of rational approximations to α. First we recall an elementary theorem, due to Dirichlet, which will be used to construct the desired approximations to α.

If α *is an irrational real number, then for any* $\tau > 1$ *there exist integers a and q such that*

$$(a, q) = 1 , \quad 1 \leqslant q \leqslant \tau , \qquad \text{and} \qquad \left| \alpha - \frac{a}{q} \right| < \frac{1}{q\tau} .$$

The particular approximations to α which will be used in the estimation of $S_N(\alpha)$ are going to be obtained by choosing $\tau = \tau(N)$, a very specific function of N, to be given later and then defining q_N to be an integer satisfying the conclusions of Dirichlet's theorem. It will be clear that $\tau(N)$, and hence q_N, tends to infinity with N. The precise result which will be proved is

$$\frac{1}{\pi(N)} \left| \sum_{p \leqslant N} e(\alpha p) \right| \leqslant A \left\{ \frac{\sqrt{q_N}}{\varphi(q_N)} + e^{-\delta \sqrt{\operatorname{Log} N}} \right\},$$

where γ and the implied " O " constant are absolute and effective. Clearly this estimate implies the truth of part (*a*).

In order to estimate $S_N(\alpha)$ it seems to be necessary to consider two cases, namely when q_N is " large " compared to N and when q_N is " small " compared to N. The precise division between " large " and " small " is fairly arbitrary. We shall consider the following two cases :

$$\text{(i)} \quad \frac{N}{\tau(N)} < q_N \leqslant \tau(N) \quad \text{and} \quad \text{(ii)} \ 1 \leqslant q_N \leqslant \frac{N}{\tau(N)},$$

where the function $\tau(N)$ will be chosen later.

Case (i). We are going to use theorem 9.4 to give an upper bound for $| S_N(\alpha) |$. The hypothesis of this theorem is satisfied since

$$\left| \alpha - \frac{a_N}{q_N} \right| < \frac{1}{q_N \, \tau(N)} \leqslant \frac{1}{q_N^2}.$$

Thus we can conclude that

$$S_N(\alpha) = O \left\{ \left(\frac{1}{q_N} + \frac{q_N}{N} + e^{-\sqrt{\operatorname{Log} N}} \right)^{1/2} N \operatorname{Log}^3 N \right\},$$

where the implied " O " constant is absolute and effective. Using our hypothesis on the size of q_N and noting that there exists an absolute, effective constant A_1 such that

$$\pi(N) > \frac{A_1 N}{\operatorname{Log} N} > 0$$

we can conclude that

$$\frac{S_N(\alpha)}{\pi(N)} = O \left\{ \left(\frac{\tau(N)}{N} + e^{-\sqrt{\operatorname{Log} N}} \right)^{1/2} \operatorname{Log}^4 N \right\}.$$

This estimate will be simplified later, when we choose $\tau(N)$ explicitly.

Case (ii). Write $\alpha = \dfrac{a}{q} + \beta$, where $1 \leqslant q \leqslant N/\tau$ and $| \beta | \leqslant \dfrac{1}{q} \tau$. In order to estimate $S_N(\alpha)$ we shall first estimate $S_n(a/q)$ for $1 \leqslant n \leqslant N$ and then use a partial summation argument to relate these sums to $S_N(\alpha)$.

We have

$$\mathrm{S}_n\left(\frac{a}{q}\right) = \sum_{p \leqslant n} e\left(a\frac{p}{q}\right) = \sum_{\substack{p \leqslant n \\ (p,q)=1}} e\left(a\frac{p}{q}\right) + \sum_{\substack{p \leqslant n \\ p \mid q}} e\left(a\frac{p}{q}\right)$$

$$= \sum_{\substack{l=1 \\ (l,q)=1}}^{q} \sum_{\substack{p \leqslant n \\ p \equiv l(q)}} e\left(\frac{al}{q}\right) + \mathrm{O}(q)$$

$$= \sum_{\substack{l=1 \\ (l,q)=1}}^{q} \pi(n\,;q,l)\, e\left(\frac{al}{q}\right) + \mathrm{O}(q)\,. \tag{9.22}$$

With the notation of theorem 8.3(d)

$$\pi(n\,;q,l) = \frac{1}{\varphi(q)}\,\mathrm{Li}\,(n) + \frac{\lambda(l)}{\varphi(q)}\,\mathrm{Li}\,(n^b) + \mathrm{O}(n\,e^{-c\sqrt{\mathrm{Log}\,n}})\,.$$

Substituting this expression into (9.22) we obtain :

$$\mathrm{S}_n\left(\frac{a}{q}\right) = \frac{\mathrm{Li}\,(n)}{\varphi(q)} \sum_{\substack{l=1 \\ (l,q)=1}}^{q} e\left(\frac{al}{q}\right) + \frac{\mathrm{Li}\,(n^b)}{\varphi(q)} \sum_{\substack{l=1 \\ (l,q)=1}}^{q} \lambda(l)\, e\left(\frac{al}{q}\right) + \mathrm{O}(qn\,e^{-c\sqrt{\mathrm{Log}\,n}})$$

$$= \frac{\mu(q)}{\varphi(q)}\,\mathrm{Li}\,(n) + \frac{\omega(a,\chi_q)}{\varphi(q)}\,\mathrm{Li}\,(n^b) + \rho(n)$$

Now for the general case :

$$\mathrm{S}_N(\alpha) = \sum_{p \leqslant N} e\left(\frac{ap}{q}\right) e(\beta p) = \sum_{n=3}^{N} \left\{ \mathrm{S}_n\left(\frac{a}{q}\right) - \mathrm{S}_{n-1}\left(\frac{a}{q}\right) \right\} e(\beta n) + \mathrm{O}(1)$$

$$= \frac{\mu(q)}{\varphi(q)} \sum_{n=3}^{N} \{\,\mathrm{Li}\,(n) - \mathrm{Li}\,(n-1)\,\}\, e(\beta n)$$

$$+ \frac{\omega(a,\chi_q)}{\varphi(q)} \sum_{n=3}^{N} \{\,\mathrm{Li}\,(n^b) - \mathrm{Li}\,((n-1)^b)\,\}\, e(\beta n) + R + \mathrm{O}(1)\,.$$

To simplify the " O " term we use the following inequality, valid for all real x and y :

$$|\,e(x) - e(y)\,| \leqslant 2\,\pi\,|\,x - y\,|$$

The " O " term then becomes

$$R = \mathrm{O}\,\{\,qN\,e^{-c\sqrt{\mathrm{Log}\,N}}(1 + N\,|\,\beta\,|)\,\}\,.$$

Thus, absorbing some initial terms into the error term we have :

$$| S_N(\alpha) | \leqslant \frac{| \mu(q) |}{\varphi(q)} \mathrm{Li}(N) + \frac{| \omega(a, \chi_q) |}{\varphi(q)} \mathrm{Li}(N^b) + \mathrm{O}\{ qN e^{-c\sqrt{\mathrm{Log}\, N}}(1 + N | \beta |) \}.$$

Taking absolute values and noting that $\mathrm{Li}(x)$ is a monotone increasing function we see

$$| S_N(\alpha) | \leqslant \frac{| \mu(q) |}{\varphi(q)} \mathrm{Li}(N) + \frac{| \omega(a, \chi_q) |}{\varphi(q)} \mathrm{Li}(N^b) + \mathrm{O}\{ qN e^{-c\sqrt{\mathrm{Log}\, N}}(1 + N | \beta |) \}.$$

From theorem 9.6 and (9.20) we have $| \omega(a, \chi_q) | \leqslant \sqrt{q}$, and by Tchebycheff's theorem there exist absolute effective constants A_1 and A_2 such that for all $N \geqslant 2$

$$\pi(N) \geqslant A_2 \mathrm{Li}(N) \geqslant A_1 \frac{N}{\mathrm{Log}\, N}.$$

Hence, since $b < 1$,

$$\frac{S_N(\alpha)}{\pi(N)} = \mathrm{O}\left\{ \frac{\sqrt{q}}{\varphi(q)} + q \,\mathrm{Log}\, N\, e^{-c\sqrt{\mathrm{Log}\, N}} + Nq | \beta | \mathrm{Log}\, N\, e^{-c\sqrt{\mathrm{Log}\, N}} \right\}.$$

Now we use our hypothesis on q, namely

$$1 \leqslant q \leqslant \frac{N}{\tau} \quad \text{and} \quad \left| \alpha - \frac{a}{q} \right| = | \beta | < \frac{1}{q\tau}.$$

The " O " term now becomes

$$\mathrm{O}\left\{ \frac{\sqrt{q}}{\varphi(q)} + \frac{N}{\tau} \mathrm{Log}\, N\, e^{-c\sqrt{\mathrm{Log}\, N}} \right\}.$$

Combining cases (i) and (ii) we always have

$$\frac{S_N(\alpha)}{\pi(N)} = \mathrm{O}\left\{ \frac{\sqrt{q}}{\varphi(q)} + \frac{N}{\tau} \mathrm{Log}\, N\, e^{-c\sqrt{\mathrm{Log}\, N}} + \left(\frac{\tau}{N} + e^{-\sqrt{\mathrm{Log}\, N}} \right)^{1/2} \mathrm{Log}^4 N \right\}.$$

All that remains now is to choose τ as a function of N. We choose

$$\tau = N e^{-\gamma\sqrt{\mathrm{Log}\, N}}$$

where $0 < \gamma < \alpha$. If δ satisfies

$$0 < \delta < \min\left\{ \frac{1}{2}, \frac{\gamma}{2}, c - \gamma \right\}$$

then the " O " term is certainly

$$O\left\{\frac{\sqrt{q}}{\varphi(q)} + e^{-\delta\sqrt{\operatorname{Log} N}}\right\}.$$

and the theorem is proved. ■

The preceding theorem allows us to deduce information about the asymptotic distribution of primes in the sequence $[\alpha n]$ when α is irrational.

Theorem 9.9 *Let $\alpha > 1$ be a fixed irrational number. The number of primes less than x in the sequence $[\alpha n]$ is*

$$\pi_\alpha(x) \sim \frac{x}{\alpha \operatorname{Log} x}.$$

Proof. The following assertions are trivialities : The statement :

" $p \leqslant x$ and $p \in \{[\alpha n]\}$ "

is equivalent to

" There exists an integer $m \in \mathbb{N}$ such that $\alpha m - 1 < p < \alpha m$ and $p \leqslant x$. "

which is equivalent to

" $p \leqslant x$ and $1 - \frac{1}{\alpha} < \frac{1}{\alpha}p - \left[\frac{1}{\alpha}p\right] < 1$ "

Thus we see

$$\pi_\alpha(x) = \operatorname{card}\left\{p \leqslant x \quad \text{and} \quad 1 - \frac{1}{\alpha} < \frac{1}{\alpha}p - \left[\frac{1}{\alpha}p\right]\right\}.$$

Since

$$1 = \frac{1}{\pi(x)}\operatorname{card}\left\{p \leqslant x \quad \text{and} \quad \frac{1}{\alpha}p - \left[\frac{1}{\alpha}p\right] \leqslant 1 - \frac{1}{\alpha}\right\} + \frac{\pi_\alpha(x)}{\pi(x)},$$

and since the sequence $\left\{\frac{1}{\alpha}p\right\}$ is uniformly distributed modulo 1, we have

$$\lim_{x\to\infty} \frac{\pi_\alpha(x)}{\pi(x)} = \frac{1}{\alpha}.$$ ■

Remark. The above theorem is purely qualitative and one cannot use it to deduce numerical information about the behaviour of $\pi_\alpha(x)$. However, by using more refined techniques one can deduce an asymptotic formula for $\pi_\alpha(x)$ with a completely explicit and effective error term.

GOLDBACH'S CONJECTURE

As mentioned at the beginning of the chapter, one of the most famous unsolved problems in additive prime number theory is Goldbach's conjecture, which asserts that every even integer $n \geqslant 4$ can be written as the sum of two primes.

By using the technique of this chapter one can obtain an upper bound for $E(x)$, the number of even integers less than x which are *not* the sum of two prime numbers. For example, van der Corput [1] showed that $E(x) = o(x)$ and later Vaughan [1] improved this to $E(x) = O\{x \exp(-\alpha\sqrt{\operatorname{Log} x})\}$.

Heuristic arguments leads one to conjecture that $R_2(n)$, the number of representations of the even integer n as a sum of two primes satisfies

$$R_2(n) \sim A_0 \frac{n}{(\operatorname{Log} n)^2} \prod_{\substack{p>2 \\ p \mid n}} \frac{p-1}{p-2},$$

where

$$A_0 = 2 \prod_{p>2} \left\{1 - \frac{1}{(p-1)^2}\right\}.$$

Turán [5] has given an " explicit formula " for $R_2(n)$ in terms of n and the complex zeros of certain L-functions. The quantity

$$A_0 \frac{n}{(\operatorname{Log} n)^2} \prod_{\substack{p>2 \\ p \mid n}} \frac{p-1}{p-2}$$

does occur in Turáns equation, but one cannot, as yet, show that the other terms in the expression for $R_2(n)$ are of a lower order of magnitude.

Exercises to chapter 9

9.1 At what point in the proof of theorem 9.1 was it absolutely necessary to suppose that $r > 2$?

9.2 Let $E(x)$ denote the number of even integers less than x which cannot be written as a sum of two primes. Prove that $E(x) = o(x)$.

9.3 Compute an explicit integer N_0 such that every odd integer $n > N_0$ can be written as a sum of three primes.

9.4 Let f be a real valued function with a continuous derivative. By using theorems 5.8(*b*) and 5.9 prove that

$$\frac{1}{\pi(x)} \sum_{p \leqslant x} e(f(p)) = \frac{1}{x} \int_1^x e(f(u))\, du - \lim_{T \to \infty} \sum_{|\rho| \leqslant T} \frac{1}{x} \int_1^x u^{\rho - 1} e(f(u))\, du + O\left(\frac{1}{\operatorname{Log} x}\right),$$

where the summation is over the complex zeros of $\zeta(s)$.

9.5 Prove that the following two statements are equivalent :

(i) Every even integer $n > 6$ can be written as a sum of two distinct prime numbers.

(ii) Every integer $n > 17$ can be written as a sum of three distinct prime numbers.

9.6 Denote by G the statement of Goldbach's conjecture, namely : G : Every even integer $n \geqslant 6$ can be written as the sum of two odd primes.

(i) Prove that G implies that every odd integer $n > 7$ can be written as a sum of three primes.
(ii) Prove the following result, due to Schinzel [1] : " G implies that every odd integer $n > 17$ can be written as a sum of three distinct primes. "

[Hint : Observe that every odd integer lies in one of the following ten progressions : $6k+3$, $10k+5$, $30k+r$, where $r = 7, 11, 13, 17, 19, 23, 29$. Then consider the following ten cases :

$$6k+3 = 3 + p_1 + p_2, \qquad 10k+5 = 5 + p_1 + p_2,$$
$$30k+1 = 13 + p_1 + p_2, \qquad 30k+7 = 17 + p_1 + p_2,$$
$$30k+11 = 23 + p_1 + p_2, \qquad 30k+13 = 23 + p_1 + p_2,$$
$$30k+17 = 7 + p_1 + p_2, \qquad 30k+19 = 7 + p_1 + p_2,$$
$$30k+23 = 13 + p_1 + p_2, \qquad 30k+29 = 17 + p_1 + p_2 .]$$

9.7 Consider the following two conjectures :

C_1 : There exist an infinity of primes p such that $p+2$ is also a prime.

C_2 : If $\{\varepsilon_n\} \in \{-1, +1\}$, then the sequence $\{4n + \varepsilon_n\}$ contains an infinity of prime numbers.

Prove that :

(i) $C_1 \Leftrightarrow C_2$.

(ii) There exists a sequence $\{\varepsilon_n\}$ such that

$$S(\{\varepsilon_n\}) = \sum_{p \in \{4n+\varepsilon_n\}} \frac{1}{p} < \infty .$$

(iii) For almost all sequences $\{\varepsilon_n\}$ the series $S(\{\varepsilon_n\})$ is divergent.

9.8* Let α be a fixed real irrational number. Erdös conjectures that as N tends to infinity

$$\frac{1}{\pi(N)} \sum_{v < p \leqslant N+v} e(\alpha p)$$

does not tend to zero uniformly with respect to v.

CHAPTER 10

Trigonometric sums

§ 1 INTRODUCTION

1.1 If $\{ a_n \}$ is a sequence of real numbers, then a study of sums of the type

$$S = \sum_{n=N+1}^{N+M} \exp(2\, i\pi a_n) \tag{10.1}$$

has important applications in number theory. For example, if $a_n = \alpha p_n$, then we saw in chapter 9 that an estimate of the corresponding sum was needed in the proof of Vinogradov's " three primes " theorem. Whilst if $a_n = \alpha n^k$, then information regarding the sum is needed in the study of Waring's problem. Also, one is obliged to study sums of the type (10.1) when investigating the distribution modulo 1 of the sequence $\{ a_n \}$. For a multitude of other applications of trigonometric sums we refer the reader to Hua [4], [5], Vinogradov [1] and Walfisz [1], [2].

It is beyond the scope of this book to give a comprehensive account of all the known results concerning trigonometric sums. We shall restrict our attention to the very important special case when the sequence $\{ a_n \}$ is defined by $a_n = f(n)$, where f is a real valued, " sufficiently regular " function. (Precise conditions on f will emerge later.) For historical reasons the corresponding trigonometric sums are known as *Weyl sums.* The chapter will be devoted to an exposition of a technique for estimating the absolute value of a Weyl sum which was initiated by Vinogradov and later much simplified by Hua.

1.2 What do we mean by a non-trivial estimate for the absolute value of the sum (10.1) ? In general one cannot assert anything more than $| S | \leqslant N$, since this upper bound is attained when the $\{ a_n \}$ are all integers. Apart from this rather special case one would expect to have a better upper bound and it would seem to be natural to take as a non-trivial estimate for $| S |$ an upper bound of the form KN, where $0 < K < 1$. However, in applications one usually needs a stronger result, at least $S = o(N)$ and often $| S | < AN^{1-\lambda}$, where $A > 0$, $\lambda > 0$ are independent of N. Because of its importance we shall always mean by a non-trivial estimate of S an upper bound of this latter kind.

1.3 Contents of the chapter. In § 2 we shall describe rather informally the skeletal structure of the Hua-Vinogradov method and indicate how upper bounds for $| S |$ can be related to the number of solutions of certain systems of diophantine equations. Then in § 3 we state the classical inequalities of Hölder and Cauchy upon which the whole method hinges. Section § 4 is devoted to a preliminary study of the diophantine equations which arise in the study of Weyl sums. We shall return to this topic again in § 6 when we prove a very important result known as Vinogradov's Mean Value theorem. The results of § 3 and § 4 will be used in § 5 to give detailed proofs of the theorems outlined in § 2. These results will be in the form of " parametric " upper bounds. Choosing specific values for the parameters will be done in § 7.

§ 2 OUTLINE OF THE HUA-VINOGRADOV METHOD

2.1 Because the Hua-Vinogradov technique is applicable in a great variety of situations one cannot produce a simple upper bound for the absolute value of the sum $\sum e(f(n))$ which is the best that one is able to obtain under all possible circumstances.

In this section we shall describe, without detailed proof, some basic principles of the method. As we shall see, at various points in the discussion one has a choice amongst various courses of action. Each choice leads to *some* upper bound for $| S |$. In appropriate circumstances it seems that one can usually find a sequence of choices which ultimately leads to a good final result. *However, for a specific problem it is often better to start at the beginning and follow the general principles we shall describe, taking care to exploit any special features of the problem at hand, rather than trying to modify the problem so that one of the following theorems is applicable* ! !

We have chosen to illustrate the method by giving three variations upon the same theme. The first two are given because they are simple and clearly show how the process works. The third example is a little more subtle but the result will be needed in chapter 11.

In § 2.3 we shall describe the Skeletal structure of our first two examples which differ only slightly. The detailed proofs will be given in § 5. In § 2.4 we shall discuss our third example which will also be proven completely in § 5.

2.2 Our three techniques eventually yield the following three theorems which will be proved in § 7.

Since the Hua-Vinogradov method is applicable to the estimation of $| \sum e(f(n)) |$ when f can be " well approximated " by a polynomial, we shall consider first the special case when f is a polynomial. The result we shall obtain is important in the study of Waring's problem.

Theorem 10.1 *Suppose* $N \geqslant 1$ *and* $f(x)$ *is the polynomial*

$$f(x) = \alpha_{k+1} x^{k+1} + \cdots + \alpha_1 x,$$

where $k \geqslant 8$. *If* a/q *is a rational approximation to* α_{k+1} *which satisfies* $(a, q) = 1$, $| \alpha_{k+1} - a/q | < q^{-2}$ *and* $N \leqslant q \leqslant N^{k-1}$, *then we have*

$$\left| \sum_{n=1}^{N} e(f(n)) \right| \leqslant c(k) N^{1-\delta}$$

where $c(k) \leqslant k^{10k}$ *and* $\rho^{-1} = 4 k^2 \left(\operatorname{Log} k + \frac{1}{2} \operatorname{Log} \operatorname{Log} x + 3 \right)$.

The next theorem is a variation of theorem 10.1 but with slightly stronger hypotheses and a correspondingly stronger conclusion.

Theorem 10.2 *Let* $N \geqslant 1$ *and* f *the polynomial* $f(x) = \alpha_{k+1} x^{k+1} + \cdots + \alpha_1 x$. *If* a/q *is a rational approximation to* α_{k+1} *which satisfies :*

$$(a, q) = 1, \quad \left| \alpha_{k+1} - \frac{a}{q} \right| < q^{-2} \quad \text{and} \quad q = N^r$$

where r *satisfies* $\sqrt{k} \operatorname{Log} k < r < k - \sqrt{k} \operatorname{Log} k$, *then we have*

$$\left| \sum_{n=1}^{N} e(f(n)) \right| \leqslant c N^{1-\delta}$$

where $\rho = \gamma / k^2 \operatorname{Log} k$ *and* c *and* γ *are absolute constants.*

The above two theorems give estimates for the corresponding trigonometric sum in terms of just *one* of the coefficients $\alpha_1, \ldots, \alpha_{k+1}$. It is reasonable to suppose that by using information about the other coefficients one could derive better estimates. In a sense this is so, but one pays a price. Because in order to be in a position to use the properties of $\alpha_1, \ldots, \alpha_k$ it seems that one is forced to employ a rather wasteful argument. However, in certain circumstances the additional gain from the other coefficients more than makes up for the wasteful part of the argument and the final result is better than than the result obtained by considering just one coefficient. Such an upper bound is given as theorem 10.9 and we shall use it to prove the following result.

Theorem 10.3 *Let* N, M *be integers which satisfy* $1 < N < M \leqslant 2N$ *and* f *a real, infinitely differentiable function defined on* $[N, 2N]$ *with the property that for some fixed* $\lambda \geqslant \frac{1}{2}$ *we have*

$$\frac{N^{\lambda - r}}{4^r} \leqslant \frac{| f^{(r)}(u) |}{r!} \leqslant N^{\lambda - r}.$$

for all $u \in (N, 2N)$ *and all* $r \geqslant 1$. *Then we have*

$$\left| \sum_{n=N+1}^{M} e(f(n)) \right| < AN^{1-\frac{\alpha}{\lambda^2}}.$$

where A *and* α *are absolute constants, possible values of which are* $A = 3$ *and* $\alpha = (49\,152)^{-1}$.

2.3 The first two methods. For simplicity we shall consider the special case when f is a polynomial, but these methods are also applicable when f is sufficiently differentiable. We treat the Weyl sum

$$S = \sum_{n=1}^{N} e(f(n)).$$

Let M be an integer satisfying $1 \leqslant M \leqslant N$, to be chosen judiciously later. Denote by $S_Q(x)$ the sum

$$S_Q(x) = \sum_{n=1}^{Q} e(f(n+x)).$$

It is quite easy to see that

$$S = \sum_{z=1}^{N} e(f(z)) = M^{-1} \sum_{n=1}^{M} \sum_{z=1}^{N} e(f(z))$$

$$= M^{-1} \sum_{n=1}^{M} \sum_{y=1-n}^{N-n} e(f(n+y))$$

$$= M^{-1} \sum_{n=1}^{M} \sum_{y=1}^{N} e(f(n+y)) + \theta(M+1) \tag{10.2}$$

where $|\theta| \leqslant 1$. Now we have a choice. Either we take absolute values in the above sum and deduce that

$$|S| \leqslant M^{-1} \sum_{n=1}^{M} |S_N(n)| + M + 1; \tag{10.3}$$

or, interchanging the summations in (10.2) and then taking absolute values we deduce that

$$|S| \leqslant M^{-1} \sum_{y=1}^{N} |S_M(y)| + M + 1. \tag{10.4}$$

We can now work with either (10.3) or (10.4). We shall work first with (10.4), while our second method deals with (10.3). Thus we must now consider the problem of finding an upper bound for the sum

$$\sum_{y=1}^{N} | S_M(y) | .$$

Upon using Hölder's inequality we deduce that for any integer $b \geqslant 1$

$$\left\{ N^{-1} \sum_{y=1}^{N} | S_M(y) | \right\}^{2b} \leqslant N^{-1} \sum_{y=1}^{N} | S_M(y) |^{2b} \tag{10.5}$$

Consequently, if we can find a " good " upper bound for the sum

$$\sum_{y=1}^{N} | S_M(y) |^{2b} . \tag{10.6}$$

then we can use (10.4) and (10.5) to deduce an upper bound for $| S |$.

One of Vinogradov's key contributions to the subject was to relate the sum (10.6) to the number of integral solutions $\mathbf{x}$, $\mathbf{y}$ of the system of diophantine equations

$$x_1^r + \cdots + x_b^r = y_1^r + \cdots + y_b^r + m_r ; \qquad r = 1, 2, \ldots, k$$

which lie in the set $\mathcal{M} = \{ \mathbf{l} \in \mathbb{Z}^{(b)} \text{ with } 1 \leqslant l_i \leqslant M \}$, where $m_1, \ldots, m_k$ are fixed integers. If we denote the number of such solutions by $\mathcal{N}(\mathcal{M}, b, k, \mathbf{m})$, then we shall see that

$$\sum_{y=1}^{N} | S_M(y) |^{2b} \leqslant \sum_{\mathbf{m}} \mathcal{N}(\mathcal{M}, b, k, \mathbf{m}) \left| \sum_{y=1}^{N} e(A_1 m_1 + \cdots + A_k m_k) \right|$$

where the first summation is over all integral vectors $\mathbf{m} = (m_1, \ldots, m_k)$ which satisfy $| m_r | \leqslant b(M^r - 1)$ for $1 \leqslant r \leqslant k$. The coefficients A_r are given by

$$A_r = \frac{f^{(r)}(y)}{r\,!} .$$

The problem is now reduced to the consideration of the sum

$$\sum_{\mathbf{m}} \mathcal{N}(\mathcal{M}, b, k, \mathbf{m}) \left| \sum_{y=1}^{N} e\left(\sum_{r=1}^{k} A_r m_r \right) \right| . \tag{10.7}$$

This is the point where ingenuity is called for. Most of the essential variations of the Hua-Vinogradov method arise in the treatment of a sum of the above type, because the choice of technique depends upon the particular properties of the function f.

The most natural approach is to notice that if $\alpha \geqslant 1$ and $\beta \geqslant 1$ with $1/\alpha + 1/\beta = 1$ then we may apply Hölder's inequality to obtain the majorization

$$\left\{ \sum_{\mathbf{m}} \mathcal{N}(\mathcal{M}, b, k, \mathbf{m})^{\alpha} \right\}^{\frac{1}{\alpha}} \cdot \left\{ \sum_{\mathbf{m}} \left| \sum_{y=1}^{N} e\left(\sum_{r=1}^{k} A_r m_r \right) \right|^{\beta} \right\}^{\frac{1}{\beta}},$$

In our case we take $\alpha = \beta = 2$. Then we shall see that the above expression is quite easy to estimate fairly precisely. Consequently we obtain an upper bound for $|\mathrm{S}|$ in terms of N, $A_1, \ldots, A_k$ and the parametres M and b. This will be given explicitly as theorem 10.7. It will then remain to make a specific choice of the parametres in order to minimize the upper bound for $|\mathrm{S}|$. We will do this as theorem 10.1.

Another way of treating a sum of the form (10.7) will be given during the course of the proof of theorem 10.8. As any explanation of this treatment demands some familiarity with the results which are to be proved in § 4, we shall not discuss it now. Suffice to say that we shall obtain, as theorem 10.8 an upper bound for $|\mathrm{S}|$ of the same general character as that given by theorem 10.7, but with additional parameters at our disposal. Specific choices for the parameters will be made in § 7 and will yield theorem 10.2.

2.4 Now we shall describe a slightly more complicated variation of the Hua-Vinogradov method. The result which we shall obtain as theorem 10.9 will be used to prove theorem 10.3. This latter result will be of vital importance to our estimation of $|\zeta(\sigma + it)|$ in chapter 11, when we return to the problem of estimating the error term in the Prime Number Theorem.

For given integers M and N satisfying $1 \leqslant M < N$, we define

$$\mathrm{S} = \sum_{n=N+1}^{N+M} e(f(n)) .$$

Let q be an integer, to be chosen judiciously later, which satisfies $1 \leqslant q < \sqrt{N}$. Consider the sum

$$\mathfrak{S} = \sum_{m} \sum_{x=0}^{q-1} \sum_{y=0}^{q-1} e(f(m + xy)) \tag{10.8}$$

where m runs over the integers in the interval

$$N < m \leqslant L = \min\{ N + M, 2N - q^2 \} .$$

If we define the sum S_m by

$$\mathrm{S}_m = \sum_{x=0}^{q-1} \sum_{y=0}^{q-1} e(f(m + xy))$$

then we obviously have

$$\mathfrak{S} = \sum_m S_m . \tag{10.9}$$

Note that $N + xy$ and $L + xy$ differ from N and $N + M$ by less than q^2 and that $| e(f(m + xy)) | = 1$. Consequently, after interchanging the order of summation in (10.8) we also have

$$\mathfrak{S} = \sum_{x=0}^{q-1} \sum_{y=0}^{q-1} \sum_m e(f(m + xy))$$

$$= \sum_{x,y} \{ S + 2\,\theta(x, y)\, q^2 \} \tag{10.10}$$

where $| \theta(x, y) | \leqslant 1$. Thus, from (10.9) and (10.10) we have for a suitable θ satisfying $| \theta | \leqslant 1$:

$$| S . q^2 + 2\,\theta q^4 | = \left| \sum_m S_m \right| \leqslant N \max_m | S_m |$$

which implies that

$$\frac{| S |}{N} \leqslant \max_m | q^{-2} S_m | + \frac{2\,q^2}{N} . \tag{10.11}$$

2.5 Now we look a little more closely at the sum S_m. If we assume that f possesses $(k + 1)$ derivatives in the interval $(N, 2\,N)$, then using Taylor's theorem with Lagrange's form of the remainder we have

$$f(m + xy) - f(m) = \sum_{r=1}^{k} \alpha_r\, x^r\, y^r + \rho_k(xy)$$

where

$$\alpha_r = \frac{f^{(r)}\,(m)}{r\,!} , \quad \rho_k(xy) = \frac{f^{(k+1)}\,(m + \theta xy)}{(k + 1)\,!}\,(xy)^{k+1}$$

and θ depends on m, xy and satisfies $| \theta | \leqslant 1$. It follows that

$$S_m = \sum_{x,y} e(f(m + xy)) = e(f(m)) \sum_{x,y} e\left(\sum_{r=1}^{k} \alpha_r\, x^r\, y^r \right) e(\rho_k(xy)) .$$

Because for any real number ξ we have

$$| e(\xi) - 1 | \leqslant 2\,\pi\, | \xi |$$

we can replace the term $e(\rho_k(xy))$ by $1 + 2\pi\varepsilon_m(xy) \mid \rho_k(xy) \mid$, where $\mid \varepsilon_m(xy) \mid \leqslant 1$. We deduce that

$$S_m = e(f(m)) \sum_{x,y} \left\{ e\left(\sum_{r=1}^{k} \alpha_r x^r y^r \right) + 2\pi\varepsilon_m(xy) \mid \rho_k(xy) \mid e\left(\sum_{r=1}^{k} \alpha_r x^r y^r \right) \right\}$$

and hence

$$\mid S_m \mid \leqslant \left| \sum_{x,y} e\left(\sum_{r=1}^{k} \alpha_r x^r y^r \right) \right| + 2\pi \sum_{x,y} \mid \rho_k(xy) \mid . \qquad (10.12)$$

If we denote by J_m the sum

$$J_m = \sum_{x=0}^{q-1} \sum_{y=0}^{q-1} e\left(\sum_{r=1}^{k} \alpha_r x^r y^r \right).$$

then from (10.11) and (10.12) we have

$$\frac{\mid S \mid}{N} \leqslant \max_m \mid q^{-2} J_m \mid + \frac{2q^2}{N} + 2\pi \max_m \left\{ q^{-2} \sum_{x,y} \mid \rho_k(xy) \mid \right\}. \qquad (10.13)$$

2.6 Now we estimate the absolute value of the sum J_m. This is done by relating $\mid J_m \mid$ to $\mathcal{N}(q, b, k)$, the number of solutions to the system of diophantine equations :

$$x_1^r + \cdots + x_b^r = y_1^r + \cdots + y_b^r, \qquad 1 \leqslant r \leqslant k$$

which satisfy $0 \leqslant x_i < q$, $0 \leqslant y_i < q$ for $1 \leqslant i \leqslant b$. The result, which we shall obtain in theorem 10.9, is that for any $b \geqslant 1$ we have

$$\mid q^{-2} J_m \mid^{4b^2} \leqslant (2b)^{2k} q^{k(k+1)-4b} \mathcal{N}(q, b, k)^2 \theta_1(\alpha_1) \dots \theta_k(\alpha_k) \qquad (10.14)$$

where the function θ_r is defined for any real α by

$$\theta_i(\alpha) = \frac{1}{2bq^i} \sum_{\mid n \mid < bq^i} \min \{ 1, (2bq^i \mid \sin \pi n\alpha \mid)^{-1} \}.$$

2.7 Combining (10.13) and (10.14) we obtain an upper bound for $\mid S \mid$ in terms of the first $(k + 1)$ derivatives of f, the integers q, b and $\mathcal{N}(q, b, k)$. An upper bound for $\mathcal{N}(q, b, k)$ will be given as theorem 10.10. Estimates for the sums $\theta_i(\alpha_i)$ will depend on the particular properties of the numbers α_i and on the integers q and b. It is at this point in the argument that the special hypotheses on the function f become relevant. One must impose conditions on the derivatives of f which are sufficient to ensure that the sum $\sum |\rho_k(xy)|$ is " small " and that several of the sums $\theta_i(\alpha_i)$ are " small ", then one chooses the integers q and b in such a way as to minimize the upper bound for $|S|/N$.

§ 3 FUNDAMENTAL INEQUALITIES

Vinogradov's method uses nothing deeper than the classical inequalities of Cauchy, Hölder and the arithmetic-geometric mean inequality. For completeness, we state them here without proof.

(a) *Hölder's inequality.* For x_j and y_j complex numbers, we have

$$\left| \sum_{j=1}^{n} x_j y_j \right| \leqslant \left(\sum_{j=1}^{n} |x_j|^\alpha \right)^{\frac{1}{\alpha}} \left(\sum_{j=1}^{n} |y_j|^\beta \right)^{\frac{1}{\beta}}$$

where $\alpha > 1$, $\beta > 1$ satisfy $\dfrac{1}{\alpha} + \dfrac{1}{\beta} = 1$. The inequality becomes an equality if and only if the vector $(x_1^\alpha, ..., x_n^\alpha)$ is proportional to the vector $(y_1^\beta, ..., y_n^\beta)$.

(b) Let $x_1, ..., x_n$ be positive real numbers and $\omega_1, ..., \omega_n$ positive integers. Then for $b > 1$,

$$\left(\frac{\omega_1 x_1 + \cdots + \omega_n x_n}{\sum \omega_i} \right)^b \leqslant \frac{\omega_1 x_1^b + \cdots + \omega_n x_n^b}{\sum \omega_i},$$

with equality if and only if $x_1 = x_2 = \cdots = x_n$.

(c) *Cauchy-Schwarz inequality.* For x_j and y_j complex numbers,

$$\left| \sum_{j=1}^{n} x_j y_j \right| \leqslant \left(\sum_{j=1}^{n} |x_j|^2 \right)^{\frac{1}{2}} \cdot \left(\sum_{j=1}^{n} |y_j|^2 \right)^{\frac{1}{2}}$$

with equality if and only if $(x_1, ..., x_n)$ is proportional to $(y_1, ..., y_n)$.

(d) *Geometric and arithmetic means inequality.* For all positive real numbers $x_1, ..., x_n$,

$$(x_1 \ldots x_n)^{\frac{1}{n}} \leqslant \frac{x_1 + \cdots + x_n}{n},$$

with equality if and only if $x_1 = x_2 = \cdots = x_n$.

§ 4 SYSTEMS OF DIOPHANTINE EQUATIONS

4.1 Vinogradov's key idea was to relate the absolute value of a trigonometric sum to the number of solutions of systems of the type

$$\text{(I)}\quad \begin{cases} x_1 + x_2 + \cdots + x_b = m_1 \\ x_1^2 + x_2^2 + \cdots + x_b^2 = m_2 \\ \cdots\cdots\cdots\cdots\cdots\cdots \\ x_1^k + x_2^k + \cdots + x_b^k = m_k \end{cases} \tag{10.15}$$

and of the type

$$\text{(II)}\quad \begin{cases} x_1 + x_2 + \cdots + x_b = y_1 + y_2 + \cdots + y_b + m_1 \\ x_1^2 + x_2^2 + \cdots + x_b^2 = y_1^2 + y_2^2 + \cdots + y_b^2 + m_2 \\ \cdots\cdots\cdots\cdots\cdots\cdots\cdots\cdots\cdots\cdots\cdots\cdots \\ x_1^k + x_2^k + \cdots + x_b^k = y_1^k + y_2^k + \cdots + y_b^k + m_k \end{cases} \tag{10.16}$$

where $m_1, \ldots, m_k$ are fixed integers and the integers x_i, y_i for $1 \leqslant i \leqslant b$ are restricted to lie in certain fixed finite sets.

A certain amount of notation will be needed. Let A and B be finite subsets of $\mathbb{Z}^{(b)}$. Denote by $\nu(A, b, k, m)$ the number of solutions of the system (10.15) with

$$\mathbf{x} = (x_1, \ldots, x_b) \in A \, .$$

Denote by $\mathcal{N}(A, B, b, k, \mathbf{m})$ the number of solutions of the system (10.16) with $\mathbf{x} \in A$ and $y \in B$. However, when $A = B$ we shall simply write $\mathcal{N}(A, b, k, \mathbf{m})$ and when $\mathbf{m} = \mathbf{0}$ we shall suppress it if there is no possible confusion.

An important special set of integers is

$$\mathfrak{L}(r) = \{\, \mathbf{x} \in \mathbb{Z}^b \quad \text{with} \quad 0 \leqslant x_i < r \,;\, 1 \leqslant i \leqslant b \,\}$$

where $r \geqslant 1$ is real. For this particular set we shall use the notation

$$\begin{aligned} &\nu(r, b, k, \mathbf{m}) && \text{in place of} && \nu(\mathfrak{L}(r), b, k, \mathbf{m}) \\ &\mathcal{N}(r, b, k, \mathbf{m}) && \text{in place of} && \mathcal{N}(\mathfrak{L}(r), \mathfrak{L}(r), b, k, \mathbf{m}) \\ &\mathcal{N}(r, b, k) && \text{in place of} && \mathcal{N}(\mathfrak{L}(r), \mathfrak{L}(r), b, k, \mathbf{0}) \, . \end{aligned}$$

Where there is no confusion we shall often suppress k.

4.2 In the following theorems we give some simple but very useful properties of the numbers $\mathcal{N}(A, B, b, k, \mathbf{m})$. The results will be employed throughout our detailed discussion of the Hua-Vinogradov method.

We reserve until § 6 the derivation of a fairly precise upper bound for $\mathcal{N}(r, b, k)$ of the form

$$\mathcal{N}(r, b, k) \leqslant c(b, k)\, r^{2b - \frac{1}{2}k(k+1) + \sigma}$$

where $c(b, k)$ and σ are explicit functions of b and k. This theorem, known as Vinogradov's mean value theorem, will not be required until § 7. By using rather elaborate analytic techniques one can improve the upper bound to

$$\mathcal{N}(r, b, k) \leqslant K(b, k)\, r^{2b - \frac{1}{2}k(k+1)}$$

provided b is sufficiently large compared to k. This latter result will be discussed in a little more detail in the *Notes* at the end of this chapter.

Theorem 10.4 *With the above notation we have :*

(a) $\mathcal{N}(A, B, b, k, \mathbf{m}) \leqslant \{ \mathcal{N}(A, A, b, \mathbf{0})\, \mathcal{N}(B, B, b, \mathbf{0}) \}^{\frac{1}{2}}$

$$\{ \mathcal{N}(A, A, b, \mathbf{0})\, \mathcal{N}(B, B, b, \mathbf{0}) \}^{\frac{1}{2}} \leqslant \frac{1}{2} \{ \mathcal{N}(A, A, b, \mathbf{0}) + \mathcal{N}(B, B, b, \mathbf{0}) \} .$$

In particular it follows that

$$\mathcal{N}(A, A, b, \mathbf{m}) \leqslant \mathcal{N}(A, A, b, \mathbf{0}) .$$

(b) *If* $A = A_1 \cup \dots \cup A_n$, *then*

$$\mathcal{N}(A, A, b, \mathbf{0}) \leqslant \left\{ \sum_{i=1}^{n} \sqrt{\mathcal{N}(A_i, A_i, b, \mathbf{0})} \right\}^2 \leqslant n \sum_{i=1}^{n} \mathcal{N}(A_i, A_i, b, \mathbf{0}) .$$

(c) *The integer* $\mathcal{N}(A, B, b, \mathbf{0})$ *is unaltered if the sets* A *and* B *both undergo a translation by a vector* $\mathbf{d} = (d, \dots, d) \in \mathbb{Z}^{(b)}$.

Proof. (*a*) We obviously have

$$\mathcal{N}(A, B, b, \mathbf{m}) = \sum_{\mathbf{u} \in \mathbb{Z}^b} \nu(A, b, \mathbf{u})\, \nu(B, b, \mathbf{u} - \mathbf{m}) .$$

where the summation is over all $\mathbf{u} \in \mathbb{Z}^{(b)}$. Cauchy's inequality gives

$$\begin{aligned} \mathcal{N}(A, B, b, \mathbf{m})^2 &\leqslant \sum_{\mathbf{u}} \nu(A, b, \mathbf{u})^2 \sum_{\mathbf{u}} \nu(B, b, \mathbf{u} - \mathbf{m})^2 \\ &= \mathcal{N}(A, A, b, \mathbf{0})\, \mathcal{N}(B, B, b, \mathbf{0}) . \end{aligned}$$

and the use of the arithmetic-geometric mean inequaliy completes the proof.

(*b*) If $\mathbf{x} \in A$, $\mathbf{y} \in A$, then we must have $\mathbf{x} \in A_i$ and $\mathbf{y} \in A_j$ for some i and j. Hence by part (*a*) it follows that

$$\begin{aligned}\mathcal{N}(A, A, b, \mathbf{0}) &\leqslant \sum_{i,j=1}^{n} \mathcal{N}(A_i, A_j, b, \mathbf{0}) \\ &\leqslant \sum_{i,j=1}^{n} \{ \mathcal{N}(A_i, A_i, b, \mathbf{0})\, \mathcal{N}(A_j, A_j, b, \mathbf{0}) \}^{\frac{1}{2}} \\ &= \left\{ \sum_{i=1}^{n} \sqrt{\mathcal{N}(A_i, A_i, b, \mathbf{0})} \right\}^2 .\end{aligned}$$

and an application of Cauchy's inequality gives the result.

(*c*) Let $u_i = x_i + d$, $v_j = y_j + d$,

$$A' = \{ \mathbf{a} + \mathbf{d} \text{ where } \mathbf{a} \in A \} \quad \text{and} \quad B' = \{ \mathbf{b} + \mathbf{d} \text{ where } \mathbf{b} \in B \} .$$

Furthermore, for $1 \leqslant r \leqslant k$ let

$$\mathrm{X}_r = x_1^r + \cdots + x_b^r , \qquad \mathrm{Y}_r = y_1^r + \cdots + y_b^r ,$$
$$\mathrm{U}_r = u_1^r + \cdots + u_b^r , \qquad \mathrm{V}_r = v_1^r + \cdots + v_b^r .$$

then we have

$$\mathrm{U}_1 = \mathrm{X}_1 + db, \qquad \mathrm{U}_2 = \mathrm{X}_2 + 2\,d\mathrm{X}_1 + d^2 b, \ldots$$
$$\mathrm{V}_1 = \mathrm{Y}_1 + db, \qquad \mathrm{V}_2 = \mathrm{Y}_2 + 2\,d\mathrm{Y}_1 + d^2 b, \ldots$$

Hence for each r, $\mathrm{X}_r = \mathrm{Y}_r$ if and only if $\mathrm{U}_r = \mathrm{V}_r$ for each r. Thus we do have

$$\mathcal{N}(A, B, b, \mathbf{0}) = \mathcal{N}(A', B', b, \mathbf{0}) .$$ ■

We now relate the integers $\mathcal{N}(A, B, b, \mathbf{m})$ etc. to integrals of certain trigonometric sums. This relationship between diophantine equations and trigonometric sums, although quite trivial to prove, is one of the cornerstones of the Hua-Vinogradov method.

The following notation will be used. If $\boldsymbol{\alpha} \in \mathbb{R}^{(k)}$, then

$$\varphi(X) = \alpha_1 \mathrm{X} + \alpha_2 \mathrm{X}^2 + \cdots + \alpha_k \mathrm{X}^k$$

and

$$\mathbf{T}_A(\boldsymbol{\alpha}) = \sum_{\mathbf{x} \in A} e(\varphi(x_1) + \varphi(x_2) + \cdots + \varphi(x_b)) .$$

Theorem 10.5 *With the above notation we have :*

(a) $\nu(A, b, \mathbf{n}) = \int_0^1 \cdots \int_0^1 \mathbf{T}_A(\boldsymbol{\alpha})\, e(-\,\mathbf{n}\,.\,\boldsymbol{\alpha})\, d\alpha_1 \ldots d\alpha_k$,

(b) $\nu(q, b, \mathbf{n}) = \int_0^1 \cdots \int_0^1 \left\{ \sum_{x=0}^{q-1} e(\varphi(x)) \right\}^b e(-\,\mathbf{n}\,.\,\boldsymbol{\alpha})\, d\alpha_1 \ldots d\alpha_k$,

(c) $\mathcal{N}(A, B, b, \mathbf{n}) = \int_0^1 \cdots \int_0^1 \mathbf{T}_A(\boldsymbol{\alpha})\, \overline{\mathbf{T}}_B(\boldsymbol{\alpha})\, e(-\,\mathbf{n}\,.\,\boldsymbol{\alpha})\, d\alpha_1 \ldots d\alpha_k$,

(d) $\mathcal{N}(q, b, \mathbf{n}) = \int_0^1 \cdots \int_0^1 \left| \sum_{x=0}^{q-1} e(\varphi(x)) \right|^{2b} e(-\,\mathbf{n}\,.\,\boldsymbol{\alpha})\, d\alpha_1 \ldots d\alpha_k$.

Proof. We shall only prove (*c*), since the proof of (*a*) is similar but easier and (*b*) and (*d*) are immediate consequences of (*a*) and (*c*). Trivially we have

$$\mathbf{T}_A(\boldsymbol{\alpha})\, \overline{\mathbf{T}}_B(\boldsymbol{\alpha})\, e(-\,\mathbf{n}\boldsymbol{\alpha}) = \sum_{\substack{\mathbf{x} \in A \\ \mathbf{y} \in B}} e(\alpha_1 z_1 + \cdots + \alpha_k z_k)$$

where $z_r = x_1^r + \cdots + x_b^r - y_1^b - \cdots - y_b^r - n_r$ for $1 \leqslant r \leqslant k$. The main result now follows since

$$\int_0^1 \cdots \int_0^1 e(\alpha_1 z_1 + \cdots + \alpha_k z_k)\, d\alpha_1 \ldots d\alpha_k = \begin{cases} 1 \text{ if } z_1 = \cdots = z_k = 0 \\ 0 \text{ otherwise}. \end{cases}$$ ■

4.3 Now we consider the important special case when $A = B = \mathfrak{L}(q)$ and deduce some elementary inequalities for $\mathcal{N}(q, b)$.

Theorem 10.6 *For fixed integers b, k and h* $\geqslant 1$ *and for fixed real q* $\geqslant 1$ *we have :*

(a) $\mathcal{N}(q, b) \leqslant k\,!\, q^{2b-k}$ *if* $b > k$,

(b) $\mathcal{N}(q, b) \leqslant b\,!\, q^b$ *if* $b \leqslant k$,

(c) $\mathcal{N}(q, b) \geqslant q^b$,

(d) $\mathcal{N}(hq, b) \leqslant h^{2b}\, \mathcal{N}(q, b)$,

(e) $\mathcal{N}(q, b) \leqslant q^{2k}\, \mathcal{N}(q, b - k)$ *if* $b > k$.

Proof. (*a*) Each x_i with $k < i \leqslant b$ and each y_j with $1 \leqslant j \leqslant b$ can be given q values. By a classical result of Newton on symmetric functions the remaining x_i

are then determined, except for their order, as the roots of a polynomial equation of degree k. Hence we certainly have

$$\mathcal{N}(q, b) \leqslant k\,!\,q^{b-k}\,q^b = k\,!\,q^{2b-k}\,.$$

(*b*) This follows as in the proof of part (*a*) when we consider only the first b equations of the system.

(*c*) The vector $\mathbf{y} = (y_1, \ldots, y_b)$ can have q^b possible values and for each value of $\mathbf{y}$ the system has at least one solution, namely $\mathbf{x} = \mathbf{y}$.

(*d*) If we denote by R the sum

$$R = \sum_{0 \leqslant x < hq} e(\varphi(x))$$

then we can write R as

$$R = R_1 + \cdots + R_n$$

where R_i denotes the sum

$$R_i = \sum_{(i-1)q \leqslant x < iq} e(\varphi(x))$$

By taking absolute values and applying Hölder's inequality we deduce

$$|\,R\,|^{2b} \leqslant h^{2b-1} \sum_{i=1}^{h} |\,R_i\,|^{2b}\,,$$

We then integrate both sides of the inequality over the unit cube in $\mathbb{R}^{(k)}$ and apply theorem 10.5(*d*) with $\mathbf{n} = \mathbf{0}$ to the left hand side. To the right hand side we apply theorem 10.5(*c*) followed by theorem 10.4(*c*) and conclude that

$$\mathcal{N}(hq, b) \leqslant h^{2b-1}.h\,\mathcal{N}(q, b)\,.$$

(*e*) We can give q values to each of the variables x_i, y_i with $1 \leqslant i \leqslant k$. For each of these choices the number of possibilities for the remaining variables is $\mathcal{N}(q, b - k, \mathbf{n})$, where $\mathbf{n}$ is a function of $x_1, \ldots, x_k$ and $y_1, \ldots, y_k$. By theorem 10.4(*a*) we have

$$\mathcal{N}(q, b - k, \mathbf{n}) \leqslant \mathcal{N}(q, b - k)$$

and consequently

$$\mathcal{N}(q, b) \leqslant q^{2k}\,\mathcal{N}(q, b - k)\,. \qquad \blacksquare$$

§ 5 THE HUA-VINOGRADOV METHOD

5.1 We are now in a position to discuss in detail the three methods outlined in § 2.

A discussion of Hua-Vinogradov techniques falls naturally into two parts. First one obtains an upper bound for the trigonometric sum in terms of certain parameters, then one makes a specific choice of the parameters to minimise the upper bound. In this section we shall derive the upper bounds in terms of the parameters. These theorems will contain all the key ideas and illustrate various aspects of the estimation process. Choosing specific values for the parameters and so arriving at theorems 10.1, 10.2 and 10.3 is tedious and not at all connected with the fundamental ideas of the method. It will be carried out in § 7.

We use throughout the notation of the previous section.

Theorem 10.7 *Let $f(x) = \alpha_{k+1} x^{k+1} + \cdots + \alpha_1 x$ and suppose that a/q is a rational approximation to α_{k+1} which satisfies $(a, q)=1, q > 0$ and $|\alpha_{k+1} - a/q| < q^{-2}$. If M and b are integers which satisfy $1 \leqslant M \leqslant N$, $b \geqslant 1$, then we have*

$$M+1+N^{1-\frac{1}{4b}}\left\{3(2b)^{k-1} M^{\frac{1}{2}k(k-1)-4b} \mathcal{N}(\mathcal{M}, 2b, k)\left(\frac{N}{q}+1\right)(2bM^k + q\log q)\right\}^{\frac{1}{4b}}.$$

where $\mathcal{M} = \{\mathbf{l} \in \mathbb{Z}^{(b)}$ with $1 \leqslant l_i \leqslant M\}$.

Proof. Let M be an integer satisfying $1 \leqslant M \leqslant N$ and denote by $S_M(y)$ the sum

$$S_M(y) = \sum_{n=1}^{M} e(f(n+y)) .$$

We trivially have the following relations :

$$\begin{aligned} S = \sum_{z=1}^{N} e(f(z)) &= \frac{1}{M}\sum_{n=1}^{M}\sum_{z=1}^{N} e(f(z)) \\ &= \frac{1}{M}\sum_{n=1}^{M}\sum_{y=1-n}^{N-n} e(f(n+y)) \\ &= \frac{1}{M}\sum_{n=1}^{M}\sum_{y=1}^{N} e(f(n+y)) + \theta(M+1) \end{aligned}$$

where θ satisfies $|\theta| \leqslant 1$. Thus we certainly have

$$|S| \leqslant \frac{1}{M}\sum_{y=1}^{N} |S_M(y)| + M + 1 . \tag{10.17}$$

We may remark that the above calculation also leads to the inequality

$$\begin{aligned} |\,\mathrm{S}\,| &\leqslant \frac{1}{M}\sum_{n=1}^{M}\left|\sum_{y=1}^{N} e(f(n+y))\right| + M + 1 \\ &= \frac{1}{M}\sum_{n=1}^{M} |\,\mathrm{S}_N(n)\,| + M + 1\,. \end{aligned} \tag{10.18}$$

We shall return to this inequality during the proof of the next theorem. For the moment we shall work with (10.17). A reasonable approach when trying to estimate a sum like $\sum |\,\mathrm{S}_M(y)\,|$ is to raise the sum to a " high " power and then hope to isolate the dominant term. Hölder's inequality which relates the nth power of the mean value of a set of positive numbers to the mean value of their nth powers is a very useful tool in this respect. Thus, taking $n = 2b$ in Hölder's inequality we deduce that

$$\left\{ N^{-1} \sum_{y=1}^{N} |\,\mathrm{S}_M(y)\,| \right\}^{2b} \leqslant N^{-1} \sum_{y=1}^{N} |\,\mathrm{S}_M(y)\,|^{2b},$$

or equivalently

$$\left\{ \sum_{y=1}^{N} |\,\mathrm{S}_M(y)\,| \right\}^{2b} \leqslant N^{2b-1} \sum_{y=1}^{N} |\,\mathrm{S}_M(y)\,|^{2b}.$$

Combining the above inequality with (10.17) we deduce that

$$|\,\mathrm{S}\,| \leqslant M^{-1} N^{1-\frac{1}{2b}} \left\{ \sum_{y=1}^{N} |\,\mathrm{S}_M(y)\,|^{2b} \right\}^{\frac{1}{2b}} + M + 1\,. \tag{10.19}$$

The problem of estimating $|\,\mathrm{S}\,|$ has now been reduced to the problem of estimating the sum

$$\sum_{y=1}^{N} |\,\mathrm{S}_M(y)\,|^{2b} \quad \text{where} \quad \mathrm{S}_M(y) = \sum_{x=1}^{M} e(f(x+y))\,. \tag{10.20}$$

5.2 It is here that we relate the problem to $\mathcal{N}(\mathcal{M}, b, k, \mathbf{m})$. The way in which a system of diophantine equations arises from the sum (10.20) is quite natural. We have

$$\sum_{y=1}^{N} |\,\mathrm{S}_M(y)\,|^{2b} = \sum_{y=1}^{N} \{\,\mathrm{S}_M(y)\,\}^b \{\,\overline{\mathrm{S}_M(y)}\,\}^b$$

and now we write the products $\{\,\mathrm{S}_M(y)\,\}^b$ and $\{\,\overline{\mathrm{S}_M(y)}\,\}^b$ as b-fold summations to obtain

$$\sum_{y=1}^{N} |\,\mathrm{S}_M(y)\,|^{2b} = \sum_{y=1}^{N} \sum_{u_1=1}^{M} \cdots \sum_{u_b=1}^{M} \sum_{v_1=1}^{M} \cdots \sum_{v_b=1}^{M} e(g(y))$$

where the function g is defined by

$$g(y) = f(u_1 + y) + \cdots + f(u_b + y) - f(v_1 + y) - \cdots - f(v_b + y) .$$

On writing $f(x + y)$ as a polynomial in x, say

$$f(x + y) = A_{k+1}(y)\, x^{k+1} + A_k(y)\, x^k + \cdots + A_1(y)\, x ,$$

we observe that $A_{k+1}(y) = \alpha_{k+1}$, $A_k(y) = \alpha_k + (k + 1)\, \alpha_{k+1}\, y$ and that $g(y)$ can be written as

$$g(y) = A_{k+1} \left\{ \sum_{i=1}^{b} (u_i^{k+1} - v_i^{k+1}) \right\} + A_k(y) \left\{ \sum_{i=1}^{b} (u_i^k - v_i^k) \right\} + \\ + \cdots + A_1(y) \left\{ \sum_{i=1}^{b} (u_i - v_i) \right\}.$$

Thus we can now write

$$\sum_{y=1}^{N} |\, S_M(y)\,|^{2b} = \sum_{\mathbf{u}} \sum_{\mathbf{v}} \sum_{y=1}^{N} e \left\{ \sum_{r=1}^{k+1} A_r(y) \sum_{i=1}^{b} (u_i^r - v_i^r) \right\},$$

where the first two summations are over all integral vectors $\mathbf{u} = (u_1, \ldots, u_b)$ and $\mathbf{v} = (v_1, \ldots, v_b)$ with coordinates satisfying $1 \leqslant u_i,\ v_i \leqslant M$. Upon taking absolute values we obtain

$$\sum_{y=1}^{N} |\, S_M(y)\,|^{2b} \leqslant \sum_{\mathbf{u}} \sum_{\mathbf{v}} \left| \sum_{y=1}^{N} e \left\{ \sum_{r=1}^{k+1} A_r(y) \sum_{i=1}^{b} (u_i^r - v_i^r) \right\} \right|.$$

Because A_{k+1} does not depend upon on y it follows that

$$\sum_{y=1}^{N} |\, S_M(y)\,|^{2b} \leqslant \sum_{\mathbf{u}} \sum_{\mathbf{v}} \left| \sum_{y=1}^{N} e \left\{ \sum_{r=1}^{k} A_r(y) \sum_{i=1}^{b} (u_i^r - v_i^r) \right\} \right|. \tag{10.21}$$

Since the integers u_i, v_i are confined to the interval $[1, M]$ it is clear that if for $1 \leqslant r \leqslant k$ we write $m_r = u_1^r + \cdots + u_b^r - v_1^r - \cdots - v_b^r$, then m_r runs over a subset of the integers in the interval $[- b(M^r - 1), b(M^r - 1)]$. Consequently we can write (10.21) as

$$\sum_{y=1}^{N} |\, S_M(y)\,|^{2b} \leqslant \sum_{\mathbf{m} \in \Omega} \mathcal{N}(\mathcal{M}, b, k, \mathbf{m}) \left| \sum_{y=1}^{N} e \left\{ \sum_{r=1}^{k} A_r(y)\, m_r \right\} \right|. \tag{10.22}$$

where Ω is defined by

$$\Omega = \{\, \mathbf{m} \in \mathbb{Z}^{(k)} : |\, m_r \,| \leqslant b(M^r - 1) \text{ for } 1 \leqslant r \leqslant k \,\},$$

and $\mathcal{M} = \{\, \mathbf{l} \in \mathbb{Z}^{(b)} \text{ with } 1 \leqslant l_i \leqslant M \,\}$.

Apart from the variation given by (10.18) we have had very little choice in our actions. But now there are several ways to proceed with the consideration of (10.22). Our first choice will lead to theorem 10.7 and in the next paragraph we shall give an alternative procedure which will lead to theorem 10.8.

The most straightforward step is to separate the coefficients $\mathcal{N}(\mathcal{M}, b, k, \mathbf{m})$ and the sums $\left|\sum_{y=1}^{N} e\left\{\sum_{r=1}^{k} A_r(y)\, m_r\right\}\right|$ by appealing to Cauchy's inequality. One then obtains.

$$\sum_{y=1}^{N} |\, S_M(y)\,|^{2b} \leqslant \left\{\sum_{\mathbf{m}\in\Omega} \mathcal{N}(\mathcal{M}, b, k, \mathbf{m})^2\right\}^{1/2} \times$$

$$\times \left\{\sum_{\mathbf{m}\in\Omega} \left|\sum_{y=1}^{N} e\left\{\sum_{r=1}^{k} A_r\, m_r\right\}\right|^2\right\}^{1/2}. \tag{10.23}$$

Now we shall consider separately the two sums

$$\left\{\sum_{\mathbf{m}\in\Omega} \mathcal{N}(\mathcal{M}, b, k, \mathbf{m})^2\right\}^{\frac{1}{2}} \quad \text{and} \quad \left\{\sum_{\mathbf{m}\in\Omega} \left|\sum_{y=1}^{N} e\left\{\sum_{r=1}^{k} A_r(y)\, m_r\right\}\right|^2\right\}^{\frac{1}{2}}.$$

The first sum is easy to deal with. Applying theorem 10.5,

$$\sum_{\mathbf{m}\in\Omega} \left\{\int_0^1 \cdots \int_0^1 \left|\sum_{x=1}^{M} e\left(\sum_{r=1}^{k} \omega_r\, x^r\right)\right|^{2b} e(-\,\mathbf{m}.\boldsymbol{\omega})\, d\boldsymbol{\omega}\right\}^2$$

$$\leqslant \int_0^1 \cdots \int_0^1 \left|\sum_{x=1}^{M} e(\omega_k\, x^k + \cdots + \omega_1\, x)\right|^{4b} d\boldsymbol{\omega}$$

$$= \mathcal{N}(\mathcal{M}, 2\,b, k)\,. \tag{10.24}$$

The second sum is slightly more difficult to estimate. We have

$$\sum_{\mathbf{m}\in\Omega} \left|\sum_{y=1}^{N} e(A_k(y)\, m_k + \cdots + A_1(y)\, m_1)\right|^2 =$$

$$\sum_{y_1=1}^{N} \sum_{y_2=1}^{N} \sum_{\mathbf{m}\in\Omega} e\,\{\,(k+1)\,\alpha_{k+1}(y_1 - y_2)\, m_k + \cdots + (A_1(y_1) - A_1(y_2))\, m_1\,\}\,.$$

The coefficients of $m_1, \ldots, m_{k-1}$ are polynomials in y_1 and y_2 of degree at least two. In general we do not know any non-trivial estimates for the corresponding sums over y_1 and y_2, thus we use the trivial upper bound 1 for all terms involving $m_1, \ldots, m_{k-1}$. In certain rather special cases it may be possible to improve upon this trivial estimate, but here we bound the preceding sum by

$$\sum_{y_1=1}^{N} \sum_{y_2=1}^{N} \sum_{m_1,\ldots,m_{k-1}} \left|\sum_{m_k} e\,\{\,(k+1)\,\alpha_{k+1}(y_1 - y_2)\, m_k\,\}\right|.$$

Performing the summations with respect to $m_1, \ldots, m_{k-1}$ and noting that

$$(M - 1)(M^2 - 1) \ldots (M^{k-1} - 1) \leqslant M^{\frac{1}{2}k(k-1)}$$

we obtain the bound

$$(2b)^{k-1} . M^{\frac{1}{2}k(k-1)} \sum_{y_1=1}^{N} \sum_{y_2=1}^{N} \left| \sum_{m_k} e\{(k+1)\alpha_{k+1}(y_1 - y_2) m_k\} \right|$$

In order to estimate this sum we shall write $Y = (k+1)(y_1 - y_2)$. For a given value of Y there are at most N choices for y_1 and y_2 so the above sum is majorised by

$$(2b)^{k-1} M^{\frac{1}{2}k(k-1)} N \sum_{Y=1}^{N} \left| \sum_{m_k} e(\alpha_{k+1} Y m_k) \right| .$$

From our initial study of trigonometric sums in § 3 of chapter 9 we know that

$$\left| \sum_{m_k} e(\alpha_{k+1} Y m_k) \right| \leqslant \min \{ 2bM^k, | \sin(\pi Y \alpha_{k+1}) |^{-1} \}$$

and that

$$\sum_{y=1}^{N} \left| \sum_{m_k} e(\alpha_{k+1} Y m_k) \right| \leqslant 3\left(\frac{N}{q} + 1\right)(2bM^k + q \operatorname{Log} q),$$

where a/q is a rational approximation to α_{k+1} which satisfies $(a, q) = 1$, $q > 0$ and $| \alpha_{k+1} - a/q | < q^{-2}$. This gives us

$$\begin{aligned} \sum_{Y=1}^{N} \left| \sum_{m_k} e(\alpha_{k-1} Y m_k) \right| &\leqslant \left(\frac{N}{q} + 1\right)(3.2bM^k + q(1 + \operatorname{Log} q)) \\ &< 3\left(\frac{N}{q} + 1\right)(2bM^k + q \operatorname{Log} q) \end{aligned}$$

We combine the above inequality with (10.19), (10.23) and (10.24) to obtain the theorem. ■

5.3 As we remarked earlier we shall choose specific values of the parameters M and b in § 7 in order to obtain theorem 10.1. We shall now go on to discuss one of the variations in the treatment of the sum (10.22) which will lead to theorem 10.2.

Theorem 10.8 *Let f denote the polynomial $f(x) = \alpha_{k+1} x^{k+1} + \cdots + \alpha_1 x$. Suppose that a/q is a rational approximation to α_{k+1} which satisfies $(a, q) = 1$, $q > 0$ and $| \alpha_{k+1} - a/q | < q^{-2}$. If N, b, b_1 are all integers greater than one and s is an* *integer satisfying* $1 \leqslant s < k$, *then for all* M *satisfying* $1 \leqslant M \leqslant N$ *an upper bound for the sum*

$$\left| \sum_{n=1}^{N} e(f(n)) \right|$$

is given by

$$c_2 N^{1-\frac{1}{2b_1}} \left\{ M^{-2b_1+\frac{1}{2}k(k+1)} \mathcal{N}(\mathcal{M}, b_1, k) \times \right.$$

$$\left. \times \mathcal{N}(\mathfrak{J}, b, s) \prod_{r=1}^{s} (N^r q^{-1} + 1)(1 + qM^{r-k-1} \operatorname{Log} q) \right\}^{\frac{1}{4bb_1}} + M + 1$$

where $\mathcal{M} = \{ \mathbf{l} \in \mathbb{Z}^{(b_1)}$ *with* $1 \leqslant l_i \leqslant M \}$ *and* $\mathfrak{J} = \{ \mathbf{l} \in \mathbb{Z}^{(b)}$ *with* $1 \leqslant l_i \leqslant N \}$ *and*

$$c_2 = (b_1^{k+s}\, 3^s\, 2^{ks+k+s})^{\frac{1}{4bb_1}} .$$

Proof. We employ the notation introduced during the proof of theorem 10.8. Our starting point is the inequality (10.18), namely

$$|\mathrm{S}| \leqslant \frac{1}{M} \sum_{y=1}^{M} |\mathrm{S}_N(y)| + M + 1 .$$

As before, we deduce that for any integer $b \geqslant 1$ we have

$$\left\{ \sum_{y=1}^{M} |\mathrm{S}_N(y)| \right\}^{2b} \leqslant M^{2b-1} \sum_{y=1}^{M} |\mathrm{S}_N(y)|^{2b}$$

and consequently it follows that

$$|\mathrm{S}| \leqslant M^{-\frac{1}{2b}} \left\{ \sum_{y=1}^{M} |\mathrm{S}_N(y)|^{2b} \right\}^{\frac{1}{2b}} + M + 1 . \tag{10.25}$$

Exactly as in the proof of inequality (10.22) we obtain

$$\sum_{y=1}^{M} |\mathrm{S}_N(y)|^{2b} \leqslant \sum_{\mathbf{m} \in \Omega_1} \mathcal{N}(\mathfrak{J}, b, k, \mathbf{m}) \left| \sum_{y=1}^{M} e\left\{ \sum_{r=1}^{k} A_r(y)\, m_r \right\} \right|$$

where Ω_1 is the subset of $\mathbb{Z}^{(k)}$ defined by :

$$\Omega_1 = \{ \mathbf{m} \in \mathbb{Z}^{(k)} \text{ with } |m_r| \leqslant b(N^r - 1) \text{ for } 1 \leqslant r \leqslant k \} .$$

We shall now give one of the possible alternative treatments of this last inequality. The first step is to write $A_1(y)\, m_1 + \cdots + A_k(y)\, m_k$ as a polynomial in y, say

$$A_1(y)\, m_1 + \cdots + A_k(y)\, m_k = \beta_0 + \beta_1 y + \cdots + \beta_k y^k ,$$

where the coefficients β_ν are given by

$$\beta_\nu = \frac{\nu + 1}{1} \alpha_{\nu+1} m_1 + \frac{(\nu + 1)(\nu + 2)}{2\,!} \alpha_{\nu+2} m_2 + \cdots + \\ + \frac{(\nu + 1) \ldots (k + 1)}{(k + 1 - \nu)\,!} \alpha_{k+1} m_{k-\nu+1}\,.$$

Let $b_1 \geqslant 1$ be an integer. Raise both sides of the inequality to the power $2\,b_1$ and apply the weighted Hölder inequality to the right hand member. We obtain

$$\left\{ \sum_{\mathbf{m} \in \Omega_1} \mathcal{N}(\mathfrak{J}, b, k, \mathbf{m}) \left| \sum_{y=1}^{M} e\left(\sum_{i=1}^{k} \beta_i y^i \right) \right|^{2b_1} \right\} \left\{ \sum_{\mathbf{m} \in \Omega_1} \mathcal{N}(\mathfrak{J}, b, k, \mathbf{m}) \right\}^{2b_1 - 1} \\ \geqslant \left\{ \sum_{y=1}^{M} | S_N(y) |^{2b} \right\}^{2b_1}.$$

Since there are at most N^{2b} vectors $(x_1, \ldots, x_b, y_1, \ldots, y_b)$ with $\mathbf{x}, \mathbf{y} \in \mathfrak{J}$,

$$\sum_{\mathbf{m} \in \Omega_1} \mathcal{N}(\mathfrak{J}, b, k, \mathbf{m}) \leqslant N^{2b}$$

and so we certainly have

$$\left\{ \sum_{y=1}^{M} | S_M(y) |^{2b} \right\}^{2b_1} \leqslant N^{2b(2b_1 - 1)} \sum_{\mathbf{m} \in \Omega_1} \mathcal{N}(\mathfrak{J}, b, k, \mathbf{m}) \left| \sum_{y=1}^{M} e\left(\sum_{i=1}^{k} \beta_i y^i \right) \right|^{2b_1}. \tag{10.26}$$

We consider next the sum

$$\left| \sum_{y=1}^{M} e(\beta_1 y + \cdots + \beta_k y^k) \right|^{2b_1}$$

In § 5.2 we estimated the sum $| S_M(t) |^{2b}$. The special case of $| S_M(0) |^{2b}$ applies to our present situation and we see that

$$\left| \sum_{y=1}^{M} e(\beta_1 y + \cdots + \beta_k y^k) \right|^{2b_1} = \sum_{\mathbf{l} \in \Omega_2} \mathcal{N}(\mathcal{M}, b_1, k, \mathbf{l})\, e(\beta_1 l_1 + \cdots + \beta_k l_k)$$

where

$$\Omega_2 = \{ \mathbf{l} \in \mathbb{Z}^k \mid | l_r | \leqslant b_1(M^r - 1)\,;\, r = 1, 2, \ldots, k \}\,.$$

Substituting into (10.26) we now have

$$N^{2b(2b_1 - 1)} \sum_{\mathbf{m} \in \Omega_1} \sum_{\mathbf{l} \in \Omega_2} \mathcal{N}(\mathfrak{J}, b, k, \mathbf{m})\, \mathcal{N}(\mathcal{M}, b_1, k, \mathbf{l})\, e(\beta_1 l_1 + \cdots + \beta_k l_k) \tag{10.27}$$

The above double sum can be written as

$$\sum_{\mathbf{l}\in\Omega_2} \mathcal{N}(\mathcal{M}, b_1, k, \mathbf{l}) \sum_{\mathbf{m}\in\Omega_1} \mathcal{N}(\mathfrak{J}, b, k, \mathbf{m})\, e(\boldsymbol{\beta}.\mathbf{l})\,. \tag{10.28}$$

Since the coordinates of $\boldsymbol{\beta}$ are linear functions of $m_1, \ldots, m_k$ we can obviously write $\boldsymbol{\beta}.\mathbf{l} = \mathbf{m}.\boldsymbol{\beta}^*$, where the coordinates of $\boldsymbol{\beta}^*$ are linear functions of $l_1, \ldots, l_k$. We now observe that

$$\begin{aligned}\sum_{\mathbf{m}\in\Omega_1} \mathcal{N}(\mathfrak{J}, b, k, \mathbf{m})\, e(\boldsymbol{\beta}.\mathbf{l}) &= \sum_{\mathbf{m}\in\Omega_1} \mathcal{N}(\mathfrak{J}, b, k, \mathbf{m})\, e(\mathbf{m}.\boldsymbol{\beta}^*)\\ &= \left|\sum_{u=1}^{N} e(u\beta_1^* + \cdots + u^k \beta_k^*)\right|^{2b} \geqslant 0\,.\end{aligned}$$

Because the inner sum in (10.28) is non-negative and since by theorem 10.4(*a*)

$$\mathcal{N}(\mathcal{M}, b_1, k, \mathbf{l}) \leqslant \mathcal{N}(\mathcal{M}, b_1, k)\,,$$

we can majorise (10.28) by

$$\mathcal{N}(\mathcal{M}, b_1, k) \sum_{\mathbf{l}\in\Omega_2} \sum_{m\in\Omega_1} \mathcal{N}(\mathfrak{J}, b, k, \mathbf{m})\, e(\boldsymbol{\beta}.\mathbf{l})\,.$$

and so deduce, as an upper bound for (10.27)

$$N^{2b(2b_1-1)}\, \mathcal{N}(\mathcal{M}, b_1, k) \sum_{\mathbf{m}\in\Omega_1} \mathcal{N}(\mathfrak{J}, b, k, \mathbf{m}) \sum_{\mathbf{l}\in\Omega_2} e(\boldsymbol{\beta}.\mathbf{l})\,. \tag{10.29}$$

Remark. In order to obtain (10.29) we chose to replace $\mathcal{N}(\mathcal{M}, b_1, k, \mathbf{l})$ by its maximum value. However there are other ways of estimating (10.27). For example one could replace $\mathcal{N}(\mathfrak{J}, b, k, \mathbf{m})$ by its maximum value, or one could take absolute values and then use Cauchy's inequality. Of course each of these choices will ultimately lead to different upper bounds for $|S|$. We leave it as an exercise for the reader to find the various upper bounds for $|S|$ which can be obtained in this way.

Returning now to the consideration of (10.29) we see that the summations with respect to $l_1, \ldots, l_k$ can now be carried out. The result, as in chapter 9, § 3 is

$$\left|\sum_{\mathbf{l}\in\Omega_2} e(\beta_1 l_1 + \cdots + \beta_k l_k)\right| \leqslant \prod_{r=1}^{k} \min\left\{2 b_1 M^r, |\sin \pi\beta_r|^{-1}\right\}$$

Our upper bound for (10.29) now becomes

$$N^{2b(2b_1-1)}\, \mathcal{N}(\mathcal{M}, b_1, k) \sum_{\mathbf{m}\in\Omega_1} \mathcal{N}(\mathfrak{J}, b, k, \mathbf{m}) \prod_{r=1}^{k} \min\left\{2 b_1 M^r, |\sin \pi\beta_r|^{-1}\right\}.$$

If s is an integer satisfying $1 \leqslant s < k$, then for r satisfying $1 \leqslant r \leqslant k - s$ we take the estimate

$$\min\,(2 b_1 M^r, |\sin \pi\beta_r|^{-1}) \leqslant 2 b_1 M^r$$

in the above expression. We then obtain the following upper bound for (10.29) :

$$N^{2b(2b_1-1)} \mathcal{N}(\mathcal{M}, b_1, k) \sum_{\mathbf{m} \in \Omega_1} \mathcal{N}(\mathfrak{J}, b, k, \mathbf{m}) (2\, b_1)^{k-s} M^{\frac{1}{2}(k-s)(k-s+1)}$$

$$\prod_{r=k+1-s}^{k} \min \{ 2\, b_1\, M^r, | \sin \pi\beta_r |^{-1} \} .$$

Because $\beta_{k-s+1}, \ldots, \beta_k$ do *not* involve $m_{s+1}, \ldots, m_k$ we can perform the summations with respect to $m_{s+1}, \ldots, m_k$ in the above expression. Furthermore, because of the definition of Ω_1, the pair $\mathbf{x}, \mathbf{y} \in \mathfrak{J}$ is a solution to the system

$$x_1^r + \cdots + x_b^r = y_1^r + \cdots + y_b^r + m_r, \qquad 1 \leqslant r \leqslant s$$

where $-b(N^r - 1) \leqslant m_r \leqslant b(N^r - 1)$, if and only if it is a solution to some system of equations

$$x_1^r + \cdots + x_b^r = y_1^r + \cdots + y_b^r + n_r, \qquad 1 \leqslant r \leqslant k$$

with $\mathbf{n} \in \Omega_1$ and $n_i = m_i$ for $1 \leqslant i \leqslant s$. Thus

$$\sum_{m_1,\ldots,m_s} \sum_{\substack{m_{s+1},\ldots,m_k \\ \mathbf{m} \in \Omega_1}} \mathcal{N}(\mathfrak{J}, b, k, \mathbf{m}) = \sum_{\mathbf{m} \in \Omega_1} \mathcal{N}(\mathfrak{J}, b, k, \mathbf{m})$$

$$= \sum_{\substack{m_1,\ldots,m_s \\ |m_r| \leqslant b(N^r - 1)}} \mathcal{N}(\mathfrak{J}, b, s, \mathbf{m}) .$$

and as an upper bound we obtain :

$$N^{2b(2b_1-1)} (2\, b_1)^{k-s} M^{\frac{1}{2}(k-s)(k-s+1)} \mathcal{N}(\mathcal{M}, b_1, k) \times$$

$$\times \sum_{m_1,\ldots,m_s} \mathcal{N}(\mathfrak{J}, b, s, \mathbf{m}) \prod_{r=k-s+1}^{k} \min (2\, b_1\, M^r, | \sin \pi\beta_r |^{-1}) . \qquad (10.30)$$

As before we appeal to theorem 10.4 to see that

$$\max_{\mathbf{m}} \mathcal{N}(\mathfrak{J}, b, s, \mathbf{m}) = \mathcal{N}(\mathfrak{J}, b, s, \mathbf{0}) = \mathcal{N}(\mathfrak{J}, b, s)$$

and deduce that (10.30) is bounded above by

$$N^{2b(2b_1-1)} (2\, b_1)^{k-s} M^{\frac{1}{2}(k-s)(k-s+1)} \mathcal{N}(\mathcal{M}, b_1, k)\, \mathcal{N}(\mathfrak{J}, b, s) \times$$

$$\times \sum_{m_1,\ldots,m_s} \prod_{r=k-s+1}^{k} \min (2\, b_1\, M^r, | \sin \pi\beta_r |^{-1}) . \qquad (10.31)$$

All that remains now is to consider the sum

$$\sum_{m_1,\ldots,m_{s-1}} \left(\prod_{r=k-s+2}^{k} \min \{ 2\, b_1\, M^r, | \sin \pi\beta_r |^{-1} \} \right) \qquad (10.32)$$

First we recall the definition of the quantities β_v, namely

$$\beta_v = \frac{v+1}{1}\alpha_{v+1} m_1 + \frac{(v+1)(v+2)}{2!}\alpha_{v+2} m_2 + \cdots + \frac{(v+1)\dots(k+1)}{(k+1-v)!}\alpha_{k+1} m_{k-v+1} .$$

Thus each β_v is of the form

$$c_v \alpha_{k+1} m_{k-v+1} + \gamma_v$$

where γ_v is a linear form in $m_1, \dots, m_{k-v}$ and

$$c_v = \frac{(k+1)\dots(v+1)}{(k+1-v)!} \leqslant 2^k .$$

From chapter 9, § 3 we know that generally we have

$$\sum_{x=Q+1}^{Q+T} \min(\mathrm{U}, |\sin(\pi A x \alpha + \beta)|^{-1}) \leqslant 3\left(\frac{AT}{q} + 1\right)(\mathrm{U} + q \operatorname{Log} q),$$

where a/q is a rational approximation to α which satisfies $(a, q) = 1$, $q > 0$ and $|\alpha - a/q| < q^{-2}$. Note that the above upper bound does not depend on β, a fact which is important to the application to the estimation of (10.31).

Splitting off the first term in the product in (10.31) and summing with respect to m_s we obtain the upper bound

$$\sum_{m_1,\dots,m_{s-1}} \left(\prod_{r=k-s+2}^{k} \min\{2 b_1 M^r, |\sin \pi\beta_r|^{-1}\}\right) \times$$
$$\times\, 3\left(2^{k+1} b_1 \frac{N^s}{q} + 1\right)(2 b_1 M^{k-s+1} + q \operatorname{Log} q)$$

Repeating the above process a further $(s - 1)$ times we obtain, as an upper bound for (10.31)

$$3^s \prod_{r=1}^{s} \left(2^{k+1} b_1 \frac{N^r}{q} + 1\right)(2 b_1 M^{k-r+1} + q \operatorname{Log} q)$$

The above expression is majorised by the following slightly neater function

$$(3.2^{k+2} b_1^2)^s M^{\frac{s}{2}(2k-s+1)} \prod_{r=1}^{s} (N^r q^{-1} + 1)(1 + qM^{r-k-1} \operatorname{Log} q) .$$

Upon combining the above upper bound for (10.32) with (10.31) and (10.27) we deduce that an upper bound for

$$\left\{\sum_{y=1}^{M} |\mathrm{S}_N(y)|^{2b}\right\}^{2b_1}$$

is given by

$$c_1(b_1, s, k)\, N^{2b(2b_1-1)}\, M^{\frac{1}{2}k(k+1)}\, \mathcal{N}(\mathcal{M}, b_1, k)\, \mathcal{N}(\mathfrak{J}, b, s)\, \times$$
$$\times \prod_{r=1}^{s} (N^r q^{-1} + 1)(1 + qM^{r-k-1} \log q)$$

where

$$c_1(b_1, s, k) = (2\, b_1)^k\, (3.2^{k+1}\, b^1)^s\,.$$

The theorem now follows from (10.25) and the above upper bound. ∎

5.4 We now complete our discussion of the variations of the Hua-Vinogradov method by giving the details of the third variation which were outlined in § 2. Before stating the theorem we recall the following notation :

M and N are integers satisfying $1 \leqslant M < N$,

f is a real valued function defined on $[N, 2\,N]$ and possessing a derivative of order $(k+1)$ in the interval $(N, 2\,N)$,

the real numbers b and q satisfy $1 \leqslant b$ and $1 \leqslant q < \sqrt{N}$,

m is an integer satisfying $N < m \leqslant \min(N + M, 2\,N - q^2)$,

$$\alpha_r = \frac{f^{(r)}(m)}{r\,!} \quad \text{for} \quad 1 \leqslant r \leqslant k\,,$$

$$\rho_k(xy) = \frac{f^{(k+1)}(m + \theta xy)}{(k+1)\,!}\,(xy)^{k+1} \quad \text{for} \quad x, y \in [0, q)$$

and for $1 \leqslant r \leqslant k$

$$\theta_r(\alpha) = \frac{1}{2\,bq^r} \sum_{|n| < bq^r} \min\,\{\,1, (2\,bq^r\,|\sin \pi n\alpha\,|)^{-1}\,\}\,.$$

Theorem 10.9 *With the above notation, an upper bound for*

$$\left| \frac{1}{N} \sum_{n=N+1}^{N+M} e(f(n)) \right|$$

is given by

$$\max_m \left(\left\{ q^{k(k+1)-4b}\, \mathcal{N}(q, b, k)^2\, (2\,b)^{2k} \prod_{r=1}^{k} \theta_r(\alpha_r) \right\}^{\frac{1}{4b^2}} + \frac{2\,\pi}{q^2} \sum_{x,y} |\,\rho_k(xy)\,| \right) + \frac{2\,q^2}{N}\,.$$

Proof. From our discussion in §§ 2.4-2.7 we see that in order to prove the theorem we have only to obtain an upper bound for

$$q^{-2} \sum_{x=0}^{q-1} \sum_{y=0}^{q-1} e(\alpha_1\, xy + \cdots + \alpha_k\, x^k\, y^k) = q^{-2}\, J_m\,.$$

If we write J_m in the form

$$J_m = \sum_{y=0}^{q-1} R(y)$$

where $R(y)$ denotes the sum

$$R(y) = \sum_{x=0}^{q-1} e(\alpha_1 xy + \cdots + \alpha_k x^k y^k) .$$

then by taking absolute values and applying Hölder's inequality we deduce that

$$| J_m |^{2b} \leqslant q^{2b-1} \sum_{y=0}^{q-1} | R(y) |^{2b} . \tag{10.33}$$

As in § 5.2 we have

$$\begin{aligned} | R(y) |^{2b} &= \{ R(y) \}^b \{ \overline{R(y)} \}^b \\ &= \sum_{\mathbf{m} \in \Omega} \mathcal{N}(q, b, \mathbf{m}) \, e\left(\sum_{r=1}^{k} \alpha_r m_r y^r \right) \end{aligned} \tag{10.34}$$

where the summation is over Ω, the subset of $\mathbb{Z}^{(k)}$, consisting of vectors $\mathbf{m} = (m_1, \ldots, m_k)$ whose coordinates satisfy

$$- b(q-1)^r \leqslant m_r \leqslant b(q-1)^r .$$

If we now denote by $\mathrm{S}(\mathbf{m})$ the sum

$$\mathrm{S}(\mathbf{m}) = \sum_{y=0}^{q-1} e(\alpha_1 m_1 y + \cdots + \alpha_k m_k y^k) .$$

then from (10.33) and (10.34)

$$| J_m |^{2b} \leqslant q^{2b-1} \sum_{\mathbf{m} \in \Omega} \mathcal{N}(q, b, \mathbf{m}) \, | \mathrm{S}(\mathbf{m}) | .$$

By applying Hölder's inequality to the above sum we obtain

$$| J_m |^{4b^2} \leqslant q^{(2b-1)2b} \left\{ \sum_{\mathbf{m} \in \Omega} \mathcal{N}(q, b, \mathbf{m}) \right\}^{2b-1} \left\{ \sum_{\mathbf{m} \in \Omega} \mathcal{N}(q, b, \mathbf{m}) \, | \mathrm{S}(\mathbf{m}) |^{2b} \right\} .$$

Since there are q^{2b} vectors $\mathbf{x}, \mathbf{y} \in \mathbb{Z}^{(b)}$ with coordinates in the interval $[0, q)$ it is trivial that

$$\sum_{m \in \Omega} \mathcal{N}(q, b, \mathbf{m}) \leqslant q^{2b} .$$

and from theorem 10.4(*a*) we also know

$$\mathcal{N}(q, b, \mathbf{m}) \leqslant \mathcal{N}(q, b) .$$

Thus, we certainly have

$$| J_m |^{4b^2} \leqslant q^{4b(2b-1)} \mathcal{N}(q, b) \sum_{m \in \Omega} | \mathrm{S}(\mathbf{m}) |^{2b} .$$

Now we consider the sums $\mathrm{S}(\mathbf{m})$. As in the treatment of $R(y)$ we see that

$$| \mathrm{S}(\mathbf{m}) |^{2b} = \sum_{\mathbf{n} \in \Omega} \mathcal{N}(q, b, \mathbf{n})\, e(\alpha_1 m_1 n_1 + \cdots + \alpha_k m_k n_k)$$

with the consequence that

$$| J_m |^{4b^2} \leqslant q^{4b(2b-1)} \mathcal{N}(q, b) \sum_{\mathbf{m} \in \Omega} \sum_{\mathbf{n} \in \Omega} \mathcal{N}(q, b, \mathbf{n})\, e(\alpha_1 m_1 n_1 + \cdots + \alpha_k m_k n_k)$$
$$\leqslant q^{4b(2b-1)} \mathcal{N}(q, b) \sum_{\mathbf{n} \in \Omega} \mathcal{N}(q, b, \mathbf{n}) \sum_{\mathbf{m} \in \Omega} e(\alpha_1 m_1 n_1 + \cdots + \alpha_k m_k n_k) .$$

Again using the fact that $\mathcal{N}(q, b, \mathbf{n}) \leqslant \mathcal{N}(q, b)$ we have

$$| J_m |^{4b^2} \leqslant q^{4b(2b-1)} \mathcal{N}(q, b)^2 \sum_{\mathbf{n} \in \Omega} \left| \sum_{\mathbf{m} \in \Omega} e(\alpha_1 m_1 n_1 + \cdots + \alpha_k m_k n_k) \right| .$$

It is now convenient to introduce the functions E_r and F_r, defined by

$$E_r(\lambda) = \sum_{|m| \leqslant b(q-1)^r} e(m\lambda)$$

and

$$F_r(\alpha) = \sum_{|n| \leqslant b(q-1)^r} | E_r(\alpha n) | .$$

With this notation our upper bound for $| J_m |^{4b^2}$ becomes :

$$| q^{-2} J_m |^{4b^2} \leqslant q^{-4b} \mathcal{N}(q, b)^2 \sum_{\mathbf{n} \in \Omega} | E_1(\alpha_1 n_1) | \cdots | E_k(\alpha_k n_k) |$$
$$\leqslant q^{-4b} \mathcal{N}(q, b)^2 F_1(\alpha_1) \cdots F_k(\alpha_k) .$$

But, from chapter 9, § 3 we know that

$$| E_r(\lambda) | \leqslant \min \{ 2 bq^r , | \sin \pi\lambda |^{-1} \}$$

and consequently

$$F_r(\alpha) \leqslant \sum_{|n| < bq^r} \min \{ 2 bq^r , | \sin \pi\lambda |^{-1} \}$$
$$\leqslant (2 bq^r)^2 \theta_r(\alpha) ,$$

Thus, we have shown that

$$| q^{-2} T_m |^{4b^2} \leqslant q^{-4b} \mathcal{N}(q, b)^2 (2b)^{2k} q^{k(k+1)} \prod_{j=1}^{k} \theta_j(\alpha_j) .$$

and the theorem follows upon combining the above inequality with (10.13). ■

§ 6 VINOGRADOV'S MEAN VALUE THEOREM

6.1 In this section we shall deduce a more precise upper bound for $\mathcal{N}(q, b, k)$ than is given by theorem 10.6.

Theorem 10.10 *Let s be an integer and suppose that*

$$b \geqslant \frac{1}{4} k(k-3) + sk .$$

If we define σ *and* β *by*

$$\sigma = \frac{1}{2} k^2 \left(1 - \frac{1}{k}\right)^s , \quad \text{and} \quad \beta = \min\left\{\frac{1}{4} k(k+5) - 1, b\right\}$$

then we have

$$\mathcal{N}(q, b, k) \leqslant (2^{4(b-k)} k^{2\beta})^s q^{2b - \frac{1}{2} k(k+1) + \sigma} .$$

The proof of this theorem is rather elaborate. A key step is to obtain a " reduction formula " relating $\mathcal{N}(q_1, b_1, k)$ to $\mathcal{N}(q, b, k)$, where $q_1 < q$ and $b_1 < b$. This is done as theorem 10.13 and then theorem 10.10 will be a trivial consequence. The proof of the reduction formula is " elementary " but rather complicated. We shall need two lemmas.

(i) A " good " upper bound for the number of solutions x_i, which are incongruent modulo p, of the system of congruences

$$x_1^r + \cdots + x_k^r \equiv n_r (\operatorname{mod} p^r) \quad \text{for} \quad 1 \leqslant r \leqslant k .$$

(ii) A rather trivial combinatorial lemma concerned with counting elements in certain finite sets.

These are given as our next two theorems; they will only be used in §§ 6.8 and 6.9.

6.2 Let k and m be fixed positive integers and $p > k$ a fixed prime. If $\boldsymbol{\theta} \in \mathbb{R}^{(k)}$ and $\mathbf{n} \in \mathbb{Z}^{(k)}$, then we denote by $M(\boldsymbol{\theta}, \mathbf{n})$ the number of integral solutions to the following system of congruences and inequalities :

$$\begin{aligned} x_1 + \cdots + x_k &\equiv n_1 (\mathrm{mod}\, p) \\ x_1^2 + \cdots + x_k^2 &\equiv n_2 (\mathrm{mod}\, p^2) \\ &\cdots\cdots\cdots \\ x_1^k + \cdots + x_k^k &\equiv n_k (\mathrm{mod}\, p^k) \end{aligned}$$

$\theta_i \leqslant x_i < mp^k + \theta_i$ for $1 \leqslant i \leqslant k$ and $x_i \not\equiv x_j\, (\mathrm{mod}\, p)$ for $i \neq j$.

We shall later need an estimate for $M(\boldsymbol{\theta}, \mathbf{n})$. The required result is as follows.

Theorem 10.11 *With the above notation we have*

$$\max_{\mathbf{n}} \max_{\boldsymbol{\theta}} M(\boldsymbol{\theta}, \mathbf{n}) \leqslant k\,!\, m^k p^{\frac{1}{2}k(k-1)}.$$

Proof. Denote the left hand side of the above inequality by M. When $k=1$ it is easy to see that we have $M = m$. We now suppose that $k \geqslant 2$. Because the congruence conditions are unaltered if x_i is replaced by $x_i' \equiv x_i\, (\mathrm{mod}\, p^k)$ we may, without loss of generality, suppose that $\boldsymbol{\theta} = \mathbf{0}$ and $m = 1$.

So we are now reduced to counting the number of solutions of the following system of congruences :

$$\begin{aligned} &x_1^r + \cdots + x_k^r \equiv n_r (\mathrm{mod}\, p^r) \quad \text{for} \quad (r = 1, 2, \ldots, k) \\ &x_i \not\equiv x_j (\mathrm{mod}\, p) \quad \text{if} \quad i \neq j\,. \end{aligned} \tag{10.35}$$

In fact we shall prove that if this system does have a solution, then it has precisely $k\,!\, p^{\frac{1}{2}k(k-1)}$ solutions. In order to simplify the linguistics of the argument, which is essentially trivial, we shall consider the following set of systems of congruences, rather than the above system :

For $r = 1, \ldots, k$ denote by $\mathcal{S}_r$ the system :

$$\begin{aligned} x_1^r + \cdots + x_k^r &\equiv n_r (\mathrm{mod}\, p^r) \\ x_1^{r+1} + \cdots + x_k^{r+1} &\equiv n_{r+1} (\mathrm{mod}\, p^r) \\ &\cdots\cdots\cdots \\ x_1^k + \cdots + x_k^k &\equiv n_k (\mathrm{mod}\, p^r) \\ x_i \not\equiv x_j (\mathrm{mod}\, p) \quad &\text{if} \quad i \neq j\,. \end{aligned}$$

Observe that if $\mathbf{y}$ is a solution to the system (10.35), then $\mathbf{y}$ is also a simultaneous solution to the system of congruences $\{ \mathcal{S}_1, \ldots, \mathcal{S}_k \}$ and conversely if $\mathbf{y}$ is a solution to the system of congruences $\{ \mathcal{S}_1, \dots, \mathcal{S}_k \}$, then $\mathbf{y}$ is also a solution of the system (10.35). Thus, we need only count the number of solutions to the system of congruences $\{ \mathcal{S}_1, \dots, \mathcal{S}_k \}$. In fact, we shall prove by induction that the number of solutions of the system $\{ \mathcal{S}_1, \dots, \mathcal{S}_r \}$, with $r \geqslant 2$ is either 0 or $k!\, p^{\frac{1}{2}r(r-1)}$.

If there exists a solution to the system $\mathcal{S}_1$, then by a classical theorem on symmetric functions due to Newton, the x_i are, apart from order, uniquely determined as the roots of a polynomial of degree k in the finite field $\mathbb{F}_p$. Thus, the system $\mathcal{S}_1$ either has 0 or $k!$ solutions. If $\mathcal{S}_1$ has no solutions, then there is nothing move to prove, so we shall suppose that $\mathcal{S}_1$ has $k!$ solutions.

Now we shall count the number of solutions of the system $\{ \mathcal{S}_1, \mathcal{S}_2 \}$. This will be the first step in our induction argument.

Let $y_1, \dots, y_k$ be a solution to $\mathcal{S}_1$. Write $x_i = y_i + pu_i$ for $1 \leqslant i \leqslant k$ and substitute into the system $\mathcal{S}_2$. One then obtains

$$(y_1 + pu_1)^2 + \cdots + (y_k + pu_k)^2 \equiv n_2 (\operatorname{mod} p^2)$$
$$\cdots\cdots\cdots\cdots\cdots\cdots\cdots\cdots\cdots\cdots\cdots\cdots$$
$$(y_1 + pu_1)^k + \cdots + (y_k + pu_k)^k \equiv n_k (\operatorname{mod} p^2)\,,$$

which, after simplification becomes

$$y_1^2 + \cdots + y_k^2 + 2p(y_1 u_1 + \cdots + y_k u_k) \equiv n_2 (\operatorname{mod} p^2)$$
$$\cdots\cdots\cdots\cdots\cdots\cdots\cdots\cdots\cdots\cdots\cdots\cdots\cdots\cdots\cdots$$
$$y_1^k + \cdots + y_k^k + kp(y_1^{k-1} u_1 + \cdots + y_k^{k-1} u_k) \equiv n_k (\operatorname{mod} p^2)\,.$$

By hypothesis we have, for $2 \leqslant r \leqslant k$

$$y_1^r + \cdots + y_k^r = n_r + pt_r$$

and consequently we must have

$$2(y_1 u_1 + \cdots + y_k u_k) \equiv -\, t_2 (\operatorname{mod} p)$$
$$\cdots\cdots\cdots\cdots\cdots\cdots\cdots\cdots\cdots\cdots$$
$$k(y_1^{k-1} u_1 + \cdots + y_k^{k-1} u_k) \equiv -\, t_k (\operatorname{mod} p)\,.$$

Because $y_1, \dots, y_k$ are distinct modulo p at most one can be congruent to zero, so without loss of generality we may suppose that $y_1, \dots, y_{k-1}$ all are different from 0 modulo p. If we assign u_k any value in the range $[0, p)$, then we have a system of $(k - 1)$ linear equations for the $(k - 1)$ unknowns $v_1, \dots, v_{k-1}$. The determinant of this system is

$$\begin{vmatrix} y_1 & \dots & y_k \\ \vdots & & \\ y_1^{k-1} & \dots & y_k^{k-1} \end{vmatrix} = y_1 \cdots y_{k-1} \prod_{i<j} (y_i - y_j)$$

Thus the system of linear equations has, for each choice of u_k, a unique solution for $u_1, ..., u_{k-1}$. Hence each solution of the system $\mathcal{S}_1$ gives rise to p solutions of the system $\mathcal{S}_2$. Consequently the total number of solutions of the system $\{ \mathcal{S}_1, \mathcal{S}_2 \}$ is $k\,!\,p$.

Now assume that the number of solutions of the system $\{ \mathcal{S}_1, ..., \mathcal{S}_r \}$ is $k\,!\,p^{\frac{1}{2}r(r-1)}$. We will compute the number of solutions of the system $\{ \mathcal{S}_1, ..., \mathcal{S}_{r+1} \}$. Let $y_1, ..., y_2$ be a solution of $\{ \mathcal{S}_1, ..., \mathcal{S}_r \}$; write $x_i = y_i + p^r u_i$ for $i = 1, ..., k$ and substitute into the system $\mathcal{S}_{r+1}$. One obtains

$$(y_1 + p^r u_1)^{r+1} + \cdots + (y_k + p^r u_k)^{r+1} \equiv n_{r+1} (\mathrm{mod}\, p^{r+1})$$
$$\cdots\cdots\cdots\cdots\cdots\cdots$$
$$(y_1 + p^r u_1)^k + \cdots + (y_k + p^r u_k)^k \equiv n_k (\mathrm{mod}\, p^{r+1})$$

and after simplification this becomes

$$y_1^{r+1} + \cdots + y_k^{r+1} + (r+1)(y_1^r u_1 + \cdots + y_k^r u_k) p^r \equiv n_{r+1} (\mathrm{mod}\, p^{r+1})$$
$$\cdots\cdots\cdots\cdots\cdots\cdots$$
$$y_1^k + \cdots + y_k^k + k(y_1^{k-1} u_1 + \cdots + y_k^{k-1} u_k) p^r \equiv n_k (\mathrm{mod}\, p^{r+1}) .$$

and by hypothesis we have for m satisfying $r < m \leqslant k$ the equations

$$y_1^m + \cdots + y_k^m = n_m + t_m p^r$$

Thus the above system simplifies to

$$(r+1)(y_1^r u_1 + \cdots + y_k^r u_k) \equiv -\, t_r (\mathrm{mod}\, p)$$
$$\cdots\cdots\cdots\cdots\cdots\cdots \tag{10.36}$$
$$k(y_1^{k-1} u_1 + \cdots + y_k^{k-1} u_k) \equiv -\, t_k (\mathrm{mod}\, p) .$$

As before we can assume that $y_i \equiv 0 \,(\mathrm{mod}\, p)$ for $1 \leqslant i \leqslant k-r$. If we give $u_{k-r+1}, ..., u_k$ any values in the range $[0, p)$, then we have a system of $(k-r)$ linear equations, with non-zero determinant, for the variables $u_1, ..., u_{k-r}$. Thus each choice of $(u_{k-r+1}, ..., u_k)$ uniquely determines $u_1, ..., u_{k-r}$. Hence the system (10.36) has p^r solutions. Thus each solution of $\{ \mathcal{S}_1, ..., \mathcal{S}_r \}$ gives rise to p^r solutions and consequently the total number of solutions of the system $\{ \mathcal{S}_1, ..., \mathcal{S}_{r+1} \}$ is

$$k\,!\,p^{\frac{1}{2}r(r-1)} . p^r = k\,!\,p^{\frac{1}{2}(r+1)r} .$$

This completes the induction step. The theorem now follows as the special case $r = k$. ■

6.3 Suppose that r and $b \geqslant k$ are fixed integers. Recall that for any integer $r > 1$ we defined

$$\mathcal{L}(r) = \{ \mathbf{l} \in \mathbb{Z}^{(b)} \quad \text{with} \quad | \, 0 \leqslant l_i < r \quad \text{for} \quad 1 \leqslant i \leqslant b \} .$$

and consider the subsets $\mathfrak{L}_1$ and $\mathfrak{L}_2$ of $\mathfrak{L}$ given by

$$\mathfrak{L}_1(r) = \{\, \mathbf{l} \in \mathfrak{L}(r) \quad \text{and} \quad \mathbf{l} \quad \text{has a set of } k \text{ } \textit{unequal} \text{ coordinates} \,\}$$

$$\mathfrak{L}_2(r) = \{\, \mathbf{l} \in \mathfrak{L}(r) \quad \text{and} \quad \mathbf{l} \notin \mathfrak{L}_1(r) \,\} .$$

We shall require in theorem 10.13 estimates for the cardinalities of $\mathfrak{L}_1(r)$ and $\mathfrak{L}_2(r)$.

Theorem 10.12 *Set* $\mathfrak{L}(r) = \mathfrak{L}$, $\mathfrak{L}_1(r) = \mathfrak{L}_1$ *and* $\mathfrak{L}_2(r) = \mathfrak{L}_2$. *When then have*

(a) $|\,\mathfrak{L}_1(r)\,| + |\,\mathfrak{L}_2(r)\,| = r^b$

(b) $r(r-1)\dots(r-k+2)\,(k-1)^{b-k+1} \leqslant |\,\mathfrak{L}_2(r)\,| \leqslant \dfrac{(k-1)^b}{(k-1)\,!}\, r^{k-1}$.

Proof. (*a*) This is an utter triviality.

(*b*) To deduce the lower bound in the case $r - k + 2 > 0$ we merely note that if we give $l_1, \dots, l_{k-1}$ any $(k-1)$ distinct values and then give $l_k, \dots, l_b$ any one of these values to obtain $r(r-1)\dots(r-k+2)\,(k-1)^{b-k+1}$ sets $(l_1, \dots, l_b)$ each of which is a vector in $\mathfrak{L}_2$. If $r - k + 2 \leqslant 0$, the left hand side of the inequality is zero.

To deduce the upper bound we consider two cases :

(i) $r > k - 1$ and (ii) $r \leqslant k - 1$.

Case (i). If $\mathbf{l} \in \mathfrak{L}_2$, then its coordinates have at most $(k-1)$ different values. For a given set of $(k-1)$ integers in $[0, r)$ the number of $\mathbf{l}$ with coordinates from this set is $(k-1)^b$. By considering all such sets we conclude that

$$|\,\mathfrak{L}_2(r)\,| \leqslant C_r^{k-1}(k-1)^b \leqslant \frac{(k-1)^b}{(k-1)\,!}\, r^{k-1} .$$

Case (ii). If $r \leqslant k - 1$, then $\mathfrak{L}_2 = \mathfrak{L}$ and so $|\,\mathfrak{L}_1\,| = 0$. By part (*a*)

$$|\,\mathfrak{L}_2(r)\,| = r^{b-k+1}\, r^{k-1} < \frac{(k-1)^b}{(k-1)\,!}\, r^{k-1} .$$

■

6.4 The reduction formula. Suppose that $q > 1$, $b > k$. Define q_1, b_1, H and β by

$$q_1 = q^{1-\frac{1}{k}}, \qquad b_1 = b - k$$

$$\beta = \min\left\{\frac{1}{4}k(k+5) - 1, b\right\} \text{ and } H = 2^{4b}\, k^{2\beta} .$$

The next theorem relates $\mathcal{N}(q, b, k)$ and $\mathcal{N}(q_1, b_1, k)$ and is the heart of the proof of Vinogradov's mean value theorem.

We will omit k from the $\mathcal{N}$ notation whenever there is no possibility of confusion.

Theorem 10.13. *With the above notation we have :*

(a) *If* $b > \frac{1}{4}k(k+5) - 1$, *then*

$$\mathcal{N}(q, b)\, q^{\frac{1}{2}k(k+1) - 2b} \leqslant H\, \mathcal{N}(q_1, b_1)\, q_1^{\frac{1}{2}k(k+1) - 2b_1}.$$

(b) *If* $b \leqslant \frac{1}{4}k(k+5) - 1$, *then*

$$\mathcal{N}(q, b)\, q^{2(1-k^2)} \leqslant H\, \mathcal{N}(q_1, b_1)\, q_1^{2(1-k^2)}.$$

Proof. If $k = 1$ we have $q_1 = 1$, $\beta = \frac{1}{2}$, $H = 2^{4b}$ and we are in case (*a*). By theorem 10.6(*a*)

$$\mathcal{N}(q, b) \leqslant ([q] + 1)^{2b-1},$$

and noticing that $\mathcal{N}(1, b_1) \geqslant 1$ we get

$$\mathcal{N}(q, b) \leqslant ([q] + 1)^{2b-1}\, \mathcal{N}(1, b_1) \leqslant (2\, q)^{2b-1}\, \mathcal{N}(1, b_1)$$

Thus from now on we shall suppose that $k \geqslant 2$ and distinguish two cases :

(i) $q > (2\,k)^k$ and (ii) $1 < q \leqslant (2\,k)^k$.

The first case is difficult while the second is quite easy.

6.5 Case $q > (2\,k)^k$. " Bertrand's postulate " (see exercise 1.8) assures the existence of a prime p in the interval $\left(\frac{1}{2}q^{1/k}, q^{1/k}\right]$. For such a p

$$k < \frac{1}{2}q^{\frac{1}{k}} < p \leqslant q^{\frac{1}{k}}.$$

Let q' be the unique integer which satisfies

$$q_1 \leqslant q' < q_1 + 1.$$

We then have

$$q' \geqslant q^{1-\frac{1}{k}} \geqslant p^{k-1} \quad \text{and} \quad q = q^{\frac{1}{k}}\, q_1 < 2\, pq'.$$

By theorem 10.6(*d*)

$$\mathcal{N}(q, b) \leqslant \mathcal{N}(2\, pq', b) \leqslant 2^{2b}\, \mathcal{N}(pq', b). \tag{10.37}$$

and the problem now is to obtain a " good " upper bound for $\mathcal{N}(pq', b)$ in terms of $\mathcal{N}(q_1, b_1)$, b_1 and q_1.

Recall the definition of the sets $\mathfrak{L}(r)$, $\mathfrak{L}_1(r)$ and $\mathfrak{L}_2(r)$ given in § 6.3. To each $\mathbf{x} \in \mathbb{Z}^{(b)}$ there corresponds a unique " reduced point " $\mathbf{l}(\mathbf{x}) \in \mathfrak{L}(p)$ defined by

$$\mathbf{l}(\mathbf{x}) = (l_1, \dots, l_b) \quad \text{where} \quad l_i \equiv x_i (\operatorname{mod} p) \text{ pour } i = 1, 2, \dots, b .$$

In particular we may define the sets $\mathcal{A}$ and $\mathcal{B}$ by

$$\begin{aligned} \mathcal{A} &= \{ \mathbf{x} \in \mathfrak{L}(pq') \mid \quad \text{with} \quad \mathbf{l}(\mathbf{x}) \in \mathfrak{L}_1(p) \} \\ \mathcal{B} &= \{ \mathbf{x} \in \mathfrak{L}(pq') \mid \quad \text{with} \quad \mathbf{l}(\mathbf{x}) \in \mathfrak{L}_2(p) \} . \end{aligned}$$

As usual, $\mathcal{N}(A, b, k)$ denotes the number of solutions of the system :

$$\begin{aligned} x_1 + \cdots + x_b &= y_1 + \cdots + y_b \\ x_1^2 + \cdots + x_b^2 &= y_1^2 + \cdots + y_b^2 \\ &\vdots \\ x_1^k + \cdots + x_b^k &= y_1^k + \cdots + y_b^k \end{aligned} \tag{10.38}$$

with $\mathbf{x}, \mathbf{y} \in A$, then since $\mathfrak{L}(pq') = \mathcal{A} \cup \mathcal{B}$ we can apply theorem 10.4(b) and deduce that

$$\sqrt{\mathcal{N}(pq', b, k)} \leqslant \sqrt{\mathcal{N}(\mathcal{A}, b, k)} + \sqrt{\mathcal{N}(\mathcal{B}, b, k)} .$$

The integers $\mathcal{N}(\mathcal{A}, b, k)$ and $\mathcal{N}(\mathcal{B}, b, k)$ will now be considered separately.

Since k will be fixed throughout the next few paragraphs we will suppress it.

6.6 The estimation of $\mathcal{N}(\mathcal{A}, b)$. By definition, any $\mathbf{x} \in \mathcal{A}$ has a set of k coordinates which are incongruent modulo p. Let $\mathcal{A}^*$ denote the subset of $\mathcal{A}$ which consists of the vectors whose first k coordinates are all incongruent modulo p.

The set $\mathcal{A}$ is the union of $\binom{b}{k}$ subsets such as $\mathcal{A}^*$ arising from arbitrary selections of k coordinates; by symmetry each such subset has the same number of solutions to (10.38), namely $\mathcal{N}(\mathcal{A}^*, b)$. Hence, by theorem 10.4(b) with $n = \binom{b}{k}$ we have

$$\sqrt{\mathcal{N}(\mathcal{A}, b)} \leqslant C_b^k \sqrt{\mathcal{N}(\mathcal{A}^*, b)}$$

Thus we now have

$$\sqrt{\mathcal{N}(pq', b)} \leqslant C_b^k \sqrt{\mathcal{N}(\mathcal{A}^*, b)} + \sqrt{\mathcal{N}(\mathcal{B}, b)} .$$

6.7 The estimation of $\mathcal{N}(\mathcal{A}^*, b)$. Recall that in (4.2) we defined for any finite subset A of $\mathbb{Z}^{(b)}$

$$\mathbf{T}_A(\boldsymbol{\alpha}) = \sum_{\mathbf{x} \in A} e(\varphi(x_1) + \cdots + \varphi(x_b)) .$$

In this paragraph we consider $\mathbf{T}_{\mathcal{A}^*}(\boldsymbol{\alpha})$ which we write as $\mathbf{T}_{\mathcal{A}}^*(\boldsymbol{\alpha})$. The following notation will be used frequently

$$T_l(\boldsymbol{\alpha}) = \sum_{\substack{0 \leqslant x < pq' \\ x \equiv l (\mathrm{mod}\, p)}} e(\varphi(x)) .$$

If we group the terms of $\mathbf{T}_{\mathcal{A}}^*(\boldsymbol{\alpha})$ into sets such that each $\mathbf{x}$ in the set has the same reduced point $\mathbf{l}$, we obtain

$$\mathbf{T}_A^*(\boldsymbol{\alpha}) = \sum_{l_1, \ldots, l_b}^* T_{l_1}(\boldsymbol{\alpha}) \ldots T_{l_b}(\boldsymbol{\alpha}) ,$$

where each l_i runs over the integers in the interval $[0, p)$ and $\sum^*$ means that the summation is over all sets $(l_1, \ldots, l_b)$ with $l_1, \ldots, l_k$ unequal. From now on, for simplicity, we shall omit $\boldsymbol{\alpha}$ from the notation. It follows that

$$\mathbf{T}_{\mathcal{A}}^* = \sum_{l_1, \ldots, l_k}^* T_{l_1} \ldots T_{l_k} \Big(\sum_l T_l\Big)^{b-k} .$$

Applying Hölder's inequality we deduce that

$$\begin{aligned}
|\mathbf{T}_{\mathcal{A}}^*|^2 &= \Big| \sum_{l_1, \ldots, l_k}^* (T_{l_1} \ldots T_{l_k}) \Big|^2 \cdot \Big| \sum_l T_l \Big|^{2(b-k)} \\
&\leqslant \Big(\sum_{l_1, \ldots, l_k}^* | T_{l_1} \ldots T_{l_k} | \Big)^2 \cdot p^{2(b-k)-1} \sum_l | T_l |^{2(b-k)} \\
&= p^{2(b-k)-1} \sum_l \sum_{l_1 \ldots l_k} | T_{l_1} \ldots T_{l_k} T_l \ldots T_l |^2 \\
&= p^{2(b-k)-1} \sum_l | \mathbf{T}_{\mathcal{A}_l^*} |^2 ,
\end{aligned}$$

where $\mathcal{A}_l^*$ denotes the subset of $\mathcal{A}^*$ which consists of the vectors $\mathbf{x}$ such that $x_i \equiv l$ (mod p) for $k + 1 \leqslant i \leqslant b$. Upon integrating over the unit cube in $\mathbb{R}^{(k)}$,

$$\int_0^1 \cdots \int_0^1 | \mathbf{T}_{\mathcal{A}}^* |^2 \, d\alpha_1 \ldots d\alpha_k \leqslant p^{2(b-k)-1} \sum_l \int_0^1 \cdots \int_0^1 | T_{\mathcal{A}_l^*} |^2 \, d\alpha_1 \ldots d\alpha_k .$$

By theorem 10.5(c) with $\mathbf{n} = \mathbf{0}$ and $A = B = \mathcal{A}^*$ on the left hand side, $A = B = \mathcal{A}_l^*$, on the right hand side, we conclude that

$$\mathcal{N}(\mathcal{A}^*, b) \leqslant p^{2(b-k)-1} \sum_l \mathcal{N}(\mathcal{A}_l^*, b) . \tag{10.39}$$

6.8 The estimation of $\mathcal{N}(\mathcal{A}_l^*, b)$. By theorem 10.4(*c*) we know that $\mathcal{N}(\mathcal{A}_l^*, b)$ is unaltered if $\mathcal{A}_l^*$ is translated by the vector $-\mathbf{l} = (-l, \dots, -l)$. Hence $\mathcal{N}(\mathcal{A}_l^*, b)$ is equal to the number of solutions of the system (10.38) with $\mathbf{x}, \mathbf{y} \in \mathbb{Z}^{(b)}$ and such that

(i) For $1 \leqslant i \leqslant k$ we have $x_i, y_i \in [-l, pq' - l)$ and the x_i and y_i are incongruent modulo p.

(ii) For $k+1 \leqslant i \leqslant b$ we have $x_i \equiv y_i \equiv 0 \pmod p$.

On writing $x_i = pn_i$ and $y_i = pv_i$ for $k+1 \leqslant i \leqslant b$ we see that $\mathcal{N}(\mathcal{A}_l^*, b)$ is equal to the number of integral solutions of the system :

$$x_1^r + \cdots + x_k^r - y_1^r - \cdots - y_k^r = p^r(v_{k+1}^r + \cdots + v_b^r - u_{k+1}^r - \cdots - u_b^r)$$

for $1 \leqslant r \leqslant k$, where $x_i, y_j \in [-l, pq' - l)$, the x_i are incongruent modulo p, the y_j are incongruent modulo p and $u_i, v_j \in [0, q')$.

A crude upper bound for the number of ways we may choose $\{y_j\}$ is $(pq')^k$. For each such choice, the integers $\{x_i\}$ must satisfy :

$$x_1^r + \cdots + x_k^r \equiv y_1^r + \cdots + y_k^r \equiv n_r(y) \pmod{p^r} \quad \text{for} \quad 1 \leqslant r \leqslant k,$$

where $x_i \in [-l, mp^k - l)$, $m = [q'/p^{k-1}] + 1$ and the x_i are mutually incongruent modulo p.

By theorem 10.11 we know that for each choice of the y_j, the number M of possibilities for the x_i satisfies

$$M_m \leqslant k!\, m^k p^{\frac{1}{2}k(k-1)},$$

For each of the possible choices for $\{x_i\}$ and $\{y_j\}$ the integers u_i and v_j must satisfy

$$v_{k+1}^r + \cdots + v_b^r - u_{k+1}^r - \cdots - u_b^r = t_r(\mathbf{x}, \mathbf{y}) \quad \text{for} \quad 1 \leqslant r \leqslant k,$$

with $u_i, v_j \in [0, q')$ and the integers $t_r(\mathbf{x}, \mathbf{y})$ uniquely determined by $\mathbf{x}$ and $\mathbf{y}$. By theorem 10.4(*a*) the number of solutions, $\mathcal{N}(q', b_1, \mathbf{t})$ satisfies :

$$\mathcal{N}(q', b_1, \mathbf{t}) \leqslant \mathcal{N}(q', b_1)$$

and since the greatest integer in both of the intervals $[0, q')$, $[0, q_1)$ is $q' - 1$,

$$\mathcal{N}(\mathcal{A}_l^*, b) \leqslant (pq')^k M \mathcal{N}(q_1, b_1).$$

It now follows from (10.39) that

$$\begin{aligned}\mathcal{N}(\mathcal{A}^*, b) &\leqslant p^{2(b-k)-1} p(pq')^k M_m \mathcal{N}(q_1, b_1) \\ &\leqslant p^{2(b-k)}(pq')^k k!\, \{[q'/p^{k-1}] + 1\}\, p^{\frac{1}{2}k(k-1)} \mathcal{N}(q_1, b_1).\end{aligned}$$

and remarking that

$$q' m \leqslant \frac{2(q')^2}{p^{k-1}}$$

we conclude that

$$\mathcal{N}(\mathcal{A}^*, b) \leqslant 2^k k! \, p^{2b - \frac{1}{2}k(k+1)} (q')^{2k} \mathcal{N}(q_1, b_1),$$

Thus we now have

$$\sqrt{\mathcal{N}(pq', b)} \leqslant C_b^k \left\{ 2^k k! \, p^{2b - \frac{1}{2}k(k+1)} (q')^{2k} \mathcal{N}(q_1, b_1) \right\}^{\frac{1}{2}} + \sqrt{\mathcal{N}(\mathcal{B}, b)}. \tag{10.40}$$

6.9 The estimation of $\mathcal{N}(\mathcal{B}, b)$. We examine the sum $\mathbf{T}_{\mathcal{B}}(\boldsymbol{\alpha})$ which can be written as

$$\mathbf{T}_{\mathcal{B}}(\boldsymbol{\alpha}) = \sum_{\mathbf{l} \in \mathfrak{L}_2(p)} T_{l_1}(\boldsymbol{\alpha}) \dots T_{l_b}(\boldsymbol{\alpha}).$$

For simplicity we now omit $\boldsymbol{\alpha}$ from the notation. Taking absolute values we have

$$| \mathbf{T}_{\mathcal{B}} | \leqslant \sum_{\mathbf{l} \in \mathfrak{L}_2(p)} | T_{l_1} | \dots | T_{l_p} |$$

and using the arithmetic-geometric mean inequality we have, on noting the symmetry, that

$$| \mathbf{T}_{\mathcal{B}} | \leqslant \sum_{\mathbf{l} \in \mathfrak{L}_2(p)} \frac{| T_{l_1} |^b + \cdots + | T_{l_b} |^b}{b} = \sum_{\mathbf{l} \in \mathfrak{L}_2(p)} \frac{| T_{l_j} |^b}{b}$$

An application of Cauchy's inequality gives

$$| \mathbf{T}_{\mathcal{B}} |^2 \leqslant | \mathfrak{L}_2(p) | \sum_{\mathbf{l} \in \mathfrak{L}_2(p)} | \mathbf{T}_{l_1} |^{2b}. \tag{10.41}$$

If we denote by $\mathcal{B}_l$ the subset of $\mathcal{B}$ consisting of vectors $\mathbf{x}$ with $x_i \equiv l \pmod{p}$ for $1 \leqslant i \leqslant b$, then

$$T_l^b = \mathbf{T}_{\mathcal{B}_l}(\boldsymbol{\alpha}).$$

Thus, by integrating (10.41) over the unit cube in $\mathbb{R}^{(k)}$ we obtain

$$\int_0^1 \cdots \int_0^1 | \mathbf{T}_{\mathcal{B}} |^2 \, d\alpha_1 \dots d\alpha_k \leqslant | \mathfrak{L}_2(p) | \sum_{\mathbf{l} \in \mathfrak{L}_2(p)} \int_0^1 \cdots \int_0^1 | \mathbf{T}_{\mathcal{B}_{l_1}} |^2 \, d\alpha_1 \dots d\alpha_k$$

and so by theorem 10.5(c) applied to both sides with $\mathbf{n} = \mathbf{o}$

$$\mathcal{N}(\mathcal{B}, b) \leqslant |\mathfrak{L}_2(p)| \sum_{\mathbf{l} \in \mathfrak{L}_2(p)} \mathcal{N}(\mathcal{B}_{\mathbf{l}_1}, b) \leqslant |\mathfrak{L}_2(p)|^2 \max_{0 \leqslant l < p} \mathcal{N}(\mathcal{B}_l, b) .$$

By theorem 10.4(c) we know that $\mathcal{N}(\mathcal{B}_l, b)$ is unaltered if the set $\mathcal{B}_l$ is translated by the vector $-\mathbf{l} = (-l, \ldots, -l)$. This, $\mathcal{N}(\mathcal{B}_l, b)$ is equal to the number of integral solutions of the system (10.38) which satisfy :

(i) $x_i, y_i \in [-l, pq' - l[$ for $1 \leqslant i \leqslant b$,

(ii) $x_i \equiv y_i \equiv 0 (\mathrm{mod}\, p)$ for $1 \leqslant i \leqslant b$.

Whence, on writing $x_i = pu_i$, $y_i = pv_i$ with $u_i, v_i \in [0, q')$ we conclude that

$$\mathcal{N}(\mathcal{B}_l, b) = \mathcal{N}(q', b) .$$

and so, by theorem 10.4(e), we have

$$\mathcal{N}(q', b) \leqslant (q')^{2k} \mathcal{N}(q, b_1) = (q')^{2k} \mathcal{N}(q_1, b_1) ,$$

Consequently theorem 10.12 gives us

$$\mathcal{N}(\mathcal{B}, b) \leqslant |\mathfrak{L}_2(p)|^2 (q')^{2k} \mathcal{N}(q_1, b_1)$$

$$\mathcal{N}(\mathcal{B}, b) \leqslant \left\{ \frac{(k-1)^b}{(k-1)!} \right\}^2 p^{2k-2} (q')^{2k} \mathcal{N}(q_1, b_1) . \tag{10.42}$$

6.10 The estimation of $\mathcal{N}(q, b)$. From (10.40) and (10.42) we have

$$\sqrt{\mathcal{N}(pq', b)} \leqslant (q')^k \sqrt{\mathcal{N}(q_1, b_1)} \left\{ C_b^k (2^k k! \, p^{2b - \frac{1}{2} k(k+1)})^{\frac{1}{2}} + \frac{(k-1)^b}{(k-1)!} p^{k-1} \right\},$$

By squaring both sides of this inequality we obtain :

$$\mathcal{N}(pq', b) \leqslant \mathcal{N}(q_1, b_1) (q')^{2k} \left\{ C_b^k \sqrt{2^k k!} \, p^{b - \frac{1}{4} k(k+1)} + \frac{(k-1)^b}{(k-1)!} p^{k-1} \right\}^2 .$$

We write the right hand side as $\mathcal{N}(q_1, b_1)(q')^{2k} K^2$ as it now remains to give a " tidy " estimate for K^2.

For convenience we define the quantities δ, μ, β as

$$\delta = \max \left\{ 0, b - \frac{1}{4} k(k+5) + 1 \right\}$$

$$\mu = \max \left\{ k - 1, b - \frac{1}{4} k(k+1) \right\} = k - 1 + \delta$$

$$\beta = \min \left\{ \frac{1}{4} k(k+5) - 1, b \right\} = b - \delta .$$

and we deduce the following facts.

(i) Since $\beta - k - \frac{1}{2} \geqslant 0$ and $k \geqslant 2$,

$$k^{2k-1} \leqslant k^{2\beta-2} \leqslant \frac{1}{4} k^{2\beta}.$$

(ii) Since $k < p$ and $\delta \geqslant 0$,

$$k^b p^{k-1} = k^{\beta+\delta} p^{\mu-\delta} \leqslant k^\beta p^\mu.$$

Furthermore, we observe that

$$\frac{(k-1)^b}{(k-1)!} < k^b \quad \text{and that} \quad 2^k k! < k^{2k-1},$$

then

$$K \leqslant \frac{1}{2}\left\{\binom{b}{k} + 2\right\} k^\beta p^\mu \leqslant 2^{b-1} k^\beta p^\mu.$$

Since in this case we have assumed $q_1 > (2k)^{k-1}$ we now get

$$\left(\frac{q'}{q_1}\right)^{2k} < \left(\frac{q_1+1}{q_1}\right)^{2k} < \left(\frac{q_1+1}{q_1}\right)^{q_1} < e < 4.$$

from which $(q')^{2k} < 4\, q_1^{2k}$. Using this together with the fact that $p \leqslant q^{1/k}$ and (10.37), namely $\mathcal{N}(q, b) \leqslant 2^{2b} \mathcal{N}(pq', b)$, we conclude

$$\frac{\mathcal{N}(q, b)}{\mathcal{N}(q_1, b_1)} \leqslant 2^{2b}.4.q_1^{2k}\, 2^{2b-2}\, k^{2\beta}\, 4\left(q^{\frac{1}{k}}\right)^{2\mu}$$

$$= 2^{4b} k^{2\beta} q_1^{2k}\left(q^{\frac{1}{k}}\right)^{2\mu}.$$

In case (a) we have $\mu = b - k(k+1)$ and recalling that $q^{1/k} = qq_1^{-1}$ we see

$$\frac{\mathcal{N}(q, b)}{\mathcal{N}(q_1, b_1)} \leqslant 2^{4b} k^{2\beta} \frac{q^{2b - \frac{1}{2}k(k+1)}}{q_1^{2b_1 - \frac{1}{2}k(k+1)}}.$$

which is the required result.

In case (b) we have $\mu = k - 1$ which yields

$$\frac{\mathcal{N}(q, b)}{\mathcal{N}(q_1, b_1)} \leqslant 2^{4b} k^{2\beta}\left(\frac{q}{q_1}\right)^{2k^2-2}.$$

6.11 Case (ii), $1 \leqslant q \leqslant (2k)^k$. Let p and q' be integers which satisfy

$$q_1 \leqslant q' < q_1 + 1 \quad \text{and} \quad \frac{1}{2} q^{\frac{1}{k}} < p \leqslant q^{\frac{1}{k}};$$

p is *not necessarily* a prime. From theorem 10.6 parts (*d*) and (*e*) we have

$$\begin{aligned}\mathcal{N}(q, b) &\leqslant \mathcal{N}(2pq', b) \leqslant (2p)^{2b} \mathcal{N}(q', b) \\ &\leqslant (2p)^{2b} (q')^{2k} \mathcal{N}(q', b_1) = (2p)^{2b} (q')^{2k} \mathcal{N}(q_1, b_1).\end{aligned}$$

We now observe that

$$\begin{aligned}(2p)^{2b} q'^{2k} &< \left(2 q^{\frac{1}{k}}\right)^{2b} (2 q_1)^{2k} \\ &< 2^{2b+2k} \left(q^{\frac{1}{k}}\right)^{2(b-\mu)} \left(q^{\frac{1}{k}}\right)^{2\mu} q_1^{2k}\end{aligned}$$

and since $\mu < b$ we certainly have

$$\begin{aligned}2^{2b+2k}\left(q^{\frac{1}{k}}\right)^{2(b-\mu)} &\leqslant 2^{2b+2k}(2k)^{2(b-\mu)} \\ &= 2^{2b+2k}(2k)^{2(\beta-k+1)} \\ &\leqslant 2^{4b+2} k^{2\beta-2} \\ &\leqslant 2^{4b} k^{2\beta}\end{aligned}$$

Hence it now follows that

$$\mathcal{N}(q, b) \leqslant 2^{4b} k^{2\beta} \left(q^{\frac{1}{k}}\right)^{2\mu} q_1^{2k} \mathcal{N}(q_1, b_1). \tag{10.43}$$

In case (*a*) $\mu = b - \frac{1}{4} k(k+1)$ and so (10.43) yields

$$\frac{\mathcal{N}(q, b)}{\mathcal{N}(q_1, b_1)} \leqslant 2^{4b} k^{2\beta} \frac{q^{2b - \frac{1}{2} k(k+1)}}{q_1^{2b_1 - \frac{1}{2} k(k+1)}},$$

which is the required result.

In case (*b*) we have $\mu = k - 1$ and (10.43) yields

$$\frac{\mathcal{N}(q, b)}{\mathcal{N}(q_1, b_1)} \leqslant 2^{4b} . k^{2\beta} \left(\frac{q}{q_1}\right)^{2k^2 - 2},$$

which is the required result.

This completes the discussion of case (ii) and completes the proof of theorem 10.13.

∎

6.12 We are now in a position to prove Vinogradov's mean value theorem. This result was stated as theorem 10.10 at the begining of this section, but for the convenience of the reader we shall restate it now.

Theorem 10.10 *Let s be an integer and suppose that*

$$b \geqslant \frac{1}{4}k(k-3) + sk .$$

If we define β and σ by

$$\beta = \min\left\{\frac{1}{4}k(k+5) - 1, b\right\} \quad \textit{and} \quad \sigma = \frac{1}{2}k^2\left(1 - \frac{1}{k}\right)^s,$$

then we have

$$\mathcal{N}(q, b) \leqslant (2^{4(b-k)}k^{2\beta})^s q^{2b - \frac{1}{2}k(k+1) + \sigma}$$

Proof. *When* $k = 1$ the theorem is a trivial consequence of theorem 10.6(a) so we shall assume from now on that $k \geqslant 2$. Consequently we have

$$b \geqslant sk \geqslant k \quad \text{and} \quad \beta - k \geqslant 0 .$$

Define the function ρ by

$$\rho(q, b) = \mathcal{N}(q, b)\, q^{\frac{1}{2}k(k+1) - 2b}.$$

we shall prove, by induction on s, that

$$\rho(q, b) \leqslant (2^{4(b-k)}k^{2\beta})^s q^{\sigma} ,$$

Case $s = 1$. Let q_0 be the integer defined by

$$q \leqslant q_0 < q + 1 ,$$

then we obviously have

$$\mathcal{N}(q, b) = \mathcal{N}(q_0, b) .$$

and so by theorem 10.6(a) we have

$$\begin{aligned}\rho(q, b) &\leqslant k\,!\, q_0^{2b-k}\, q^{\frac{1}{2}k(k+1) - 2b} \\ &\leqslant k^k\, 2^{2b-k}\, q^{\frac{1}{2}k(k-1)} \\ &\leqslant k^{2\beta - k}\, 2^{2b-k}\, q^{\sigma} \\ &\leqslant 2^{2(b-k)}\, k^{2\beta}\, q^{\sigma} .\end{aligned}$$

since $k^{-k} \leqslant 2^{-k}$.

Case $s > 1$. We now assume the truth of the theorem with s replaced by $s_1 = s - 1$.

Set $q_1 = q^{1-1/k}$ and $b_1 = b - k$ and denote

$$\rho(q_1, b_1), \quad \frac{1}{2}k^2\left(1 - \frac{1}{k}\right)^{s-1}, \quad \min\left(\frac{1}{4}k(k+5) - 1, b_1\right)$$

by ρ_1, s_1 and β_1 respectively.

Since $b \geqslant \frac{1}{4}k(k+5) > k$ and $b_1 > \frac{1}{4}k(k-3) + s_1 k$ we have, by theorem 10.13(a) and the induction hypothesis

$$\rho(q, b) \leqslant 2^{4b} k^{2\beta} \rho_1 \leqslant (2^{4b} k^{2\beta})(2^{4(b_1-k)} k^{2\beta_1})^{s-1} q_1^{\sigma_1}$$
$$\leqslant (2^{4(b-k)} k^{2\beta})^s q^\sigma$$

since $\beta_1 \leqslant \beta$ and $2^{4(b_1-k)} k^{2\beta_1} \leqslant 2^{-4k}(2^{4(b-k)} k^{2\beta})$.

This completes the induction step and the proof of Vinogradov's theorem is complete. ∎

§7 PROOF OF THE PRINCIPAL ESTIMATES

7.1 In this section we shall choose specific values for the parameters in the upper bounds given by theorems 10.7, 10.8 and 10.9. Our objective being the proof of theorems 10.1, 10.2 and 10.3. As the reader will soon see, making such specific choices is rather tedious and messy. We have not made great efforts to extract the best results that are possible by our methods, rather we have chosen to end up with a tidy final result. The reader may find it amusing to make other choices for the parameters and so obtain slightly different estimates. For the readers convenience we shall restate the results which are to be proved.

7.2 The first theorem is as follows.

Theorem 10.1 *Let* $N > 1$ *be a positive integer and* f *the polynomial*

$$f(x) = \alpha_{k+1} x^{k+1} + \cdots + \alpha_1 x,$$

where $k \geqslant 8$. *If* a/q *is a rational approximation to* α_{k+1} *which satisfies* $(a, q) = 1$, $N \leqslant q \leqslant N^{k-1}$ *and* $|\alpha_{k+1} - a/q| < q^{-2}$, *then we have*

$$\left| \sum_{n=1}^{N} e(f(n)) \right| < c(k) N^{1-\frac{1}{\rho}}$$

where $c(k) \leqslant k^{10k}$ *and* $\rho = 4k^2\left(\operatorname{Log} k + \frac{1}{2}\operatorname{Log}\operatorname{Log} x + 3\right)$.

Proof. We are going to employ the upper bound provided by theorem 10.7, namely :

$$N^{1-\frac{1}{4b}}\left\{3(2\,b)^{k-1}\,M^{\frac{1}{2}k(k-1)-4b}\,\mathcal{N}(\mathcal{M},2\,b,k)\left(\frac{N}{q}+1\right)(2\,bM^k+q\,\mathrm{Log}\,q)\right\}^{\frac{1}{4b}}$$

$$+\,M+1\,,$$

where b satisfies $b \geqslant 1$ and $1 \leqslant M \leqslant N$.

By hypothesis we have $N \leqslant q \leqslant N^{k-1}$. For simplicity we shall write $M = N^{1-\eta}$, where η satisfies $0 \leqslant \eta \leqslant 1$ and will be chosen explicitly later. Trivially,

$$Nq^{-1}+1 \leqslant 2 \ \text{ and } \ 2\,bM^k + q\,\mathrm{Log}\,q \leqslant N^k(2\,bN^{-\eta k}+(k-1)\,N^{-1}\,\mathrm{Log}\,N).$$

When we come to choose b we will have $2\,b \geqslant k-1$, in which case the last expression is bounded above by $4\,bN^k$. Then an application of theorem 10.4(c) yields the following upper bound for $|\,S\,|$:

$$|\,S\,| \leqslant N^{1-\frac{1}{4b}}\left\{12(2\,b)^k\,N^k\,M^{\frac{1}{2}k(k-1)-4b}\,\mathcal{N}(M,2\,b,k)\right\}^{\frac{1}{4b}} + N^{1-\eta}+1\,.$$

If b and s satisfy

$$2\,b \geqslant \frac{1}{4}\,k(k-3)+sk\,,$$

then from theorem 10.10 we have

$$\mathcal{N}(M,2\,b,k) \leqslant c_1\,M^{4b-\frac{1}{2}k(k+1)+\sigma}$$

where $\sigma = \dfrac{1}{2}\,k^2\left(1-\dfrac{1}{k}\right)^s$ and $c_1 = \left\{2^{4(2b-k)}\,k^{\frac{1}{2}k(k+5)-2}\right\}^s$. With this assumption, the upper bound for $|\,S\,|$ becomes :

$$|\,S\,| \leqslant c_2\,N^{1-\frac{1}{4b}+\frac{\eta k}{4b}+\frac{\sigma(1-\eta)}{4b}} + N^{1-\eta}+1$$

where $c_2 = (c_1(2\,b)^k\,12)^{\frac{1}{4b}}$.

We choose η so that

$$1-\eta = 1-\frac{1}{4b}+\frac{\eta k}{4b}+\frac{\sigma(1-\eta)}{4\,b}\,,$$

namely, $\eta = \dfrac{1-\sigma}{4\,b+k-\sigma}$. With this choice of η we certainly have

$$|\,S\,| \leqslant c_3\,N^{1-\eta} \tag{10.44}$$

where $c_3 = c_2 + 2$.

Our next step is to choose b and s as specific functions of k so that they satisfy

$$2\,b \geqslant \frac{1}{4}\,k(k-3) + sk \geqslant k - 1\,.$$

First of all we have $\sigma = \dfrac{1}{2}\,k^2\left(1 - \dfrac{1}{k}\right)^s$ and consequently, if

$$s > -\operatorname{Log}\left(\frac{1}{2}\,k^2 \operatorname{Log} k\right)\Big/\operatorname{Log}\left(1 - \frac{1}{k}\right),$$

we have $\sigma < 1/\operatorname{Log} k$. This latter inequality will certainly hold if we take

$$s = \left[k \operatorname{Log}\left(\frac{1}{2}\,k^2 \operatorname{Log} k\right)\right]$$

Since $k \geqslant 8$, we have, for this choice of s, the inequality $\sigma < \dfrac{1}{\operatorname{Log} k} \leqslant \dfrac{1}{2}$. We set

$$b = \left[\frac{1}{8}\,k(k-3) + \frac{1}{2}\,sk + 1\right]$$

and observe that the conditions of theorem 10.10 are satisfied so we do have (10.44).

We shall now compute a relatively simple *lower* bound for η and a simple upper bound for c_3. First of all, we have

$$\begin{aligned}\eta^{-1} &= \frac{4\,b + k - \sigma}{1 - \sigma} \leqslant \left(\frac{1}{2}\,k(k-3) + 2\,ks + 4 + k\right)(1-\sigma)^{-1} \\ &\leqslant \left\{2\,k^2 \operatorname{Log}\left(\frac{1}{2}\,k^2 \operatorname{Log} k\right) + \frac{1}{2}\,k(k-3) + k + 4\right\}(1 + 2\,\sigma)\,,\end{aligned}$$

since $\sigma \leqslant \dfrac{1}{\operatorname{Log} k} \leqslant \dfrac{1}{2}$ upon simplification we see that the above expression is bounded above by

$$4\,k^2\left(\operatorname{Log} k + \frac{1}{2}\operatorname{Log}\operatorname{Log} k + 3\right) = \rho\,.$$

Finally we shall compute a relatively simple upper bound for c_3. No attempt will be made to make the upper bound very tight. We know that

$$\begin{aligned}c_3 &= c_2 + 2 = \{\,12\,c_1(2\,b)^k\,\}^{\frac{1}{4b}} + 2 \\ &= \left\{12(2\,b)^k\left(2^{4(2b-k)}\,k^{\frac{1}{2}k(k+5)-2}\right)^s\right\}^{\frac{1}{4b}} + 2\,.\end{aligned}$$

First of all we observe that

$$\{ 12(2\,b)^k \}^{\frac{1}{4b}} = \exp \frac{k \operatorname{Log} 2\,b + \operatorname{Log} 12}{4\,b} < e^2$$

then we note that $\frac{s}{4\,b} < \frac{1}{2} k$. Consequently we certainly have

$$c_1^{\frac{1}{4} b} \leqslant \left\{ 2^{4(2b-k)} \, k^{\frac{1}{2} k(k+5) - 2} \right\}^{\frac{1}{2} k} \leqslant (2^{8b} \, k^{2k^2})^{\frac{1}{2} k}$$

$$\leqslant 2^{4b/k} \, k^k \leqslant 2^{8k - \operatorname{Log} k} \, k^k, \quad \text{since} \quad 4\,b \leqslant 8\,k^2 \operatorname{Log} k\,.$$

Thus we see that c_3 is certainly bounded above by k^{10k}. This completes the proof of the theorem. ■

7.3 The next theorem is an application of theorem 10.8. It is a result very similar in character to theorem 10.1, but the hypotheses are slightly stronger resulting in a stronger conclusion.

Theorem 10.2 *Let $N > 1$ be an integer and f the polynomial*

$$f(x) = \alpha_{k+1} \, x^{k+1} + \cdots + \alpha_1 \, x\,.$$

If a/q is a rational approximation to α_{k+1} which satisfies : $(a, q) = 1$, $|\,\alpha_{k+1} - a/q\,| < q^{-2}$ and $q = N^\theta$, where $\sqrt{k} \operatorname{Log} k < \theta < k - \sqrt{k} \operatorname{Log} k$.

Then there exist absolute constants A and α such that

$$\left| \sum_{n=1}^{N} e(f(n)) \right| \leqslant A N^{1 - \alpha/k^2 \operatorname{Log} k}\,.$$

Proof. Because we shall not use the theorem in any of our applications we are not going to compute explicit numerical values for A and α. This is left as an exercise for the reader. The theorem is a fairly simple consequence of the upper bound for $|\,S\,|$ given by theorem 10.8, namely

$$c_1 \, N^{1 - \frac{1}{2b_1}} \left\{ M^{-2b_1 + \frac{1}{2} k(k+1)} \, \mathcal{N}(\mathcal{M}, b_1, k) \, \mathcal{N}(\mathfrak{J}, b, s) \prod_{r=1}^{s} \left(\frac{N^r}{q} + 1 \right) \left(1 + \frac{q \operatorname{Log} q}{M^{k-r+1}} \right) \right\}^{\frac{1}{4bb_1}}$$

$$+ M + 1$$

where $c_1 = (b_1^{k+s} \, 3^s . 2^{ks+k+s})^{\frac{1}{4bb_1}}$ and b, b_1, M, s are parameters which satisfy $b \geqslant 1$, $b_1 \geqslant 1$, $1 \leqslant s < k$ and $1 < M \leqslant N$.

From theorem 10.4(*c*), $\mathcal{N}(\mathcal{M}, b_1, k) = \mathcal{N}(M, b_1, k)$ and $\mathcal{N}(\mathfrak{J}, b, s) = \mathcal{N}(N, b, s)$. Also, by theorem 10.10, if

$$b_1 \geqslant \frac{1}{4} k(k - 3) + lk , \tag{10.45}$$

then

$$\mathcal{N}(M, b_1, k) \leqslant c_2 M^{2b_1 - \frac{1}{2} k(k+1) + \sigma_1}$$

where

$$c_2 = \left(2^{4(b_1 - k)} k^{\frac{1}{2} k(k+5) - 2}\right)^l \quad \text{and} \quad \sigma_1 = \frac{1}{2} k^2 \left(1 - \frac{1}{k}\right)^l .$$

We also know that if

$$b \geqslant \frac{1}{4} s(s - 3) + ms \tag{10.46}$$

then

$$\mathcal{N}(N, b, s) \leqslant c_3 N^{2b - \frac{1}{2} s(s+1) + \sigma_2}$$

where

$$c_3 = \left(2^{4(b-k)} s^{\frac{1}{2} s(s+5) - 2}\right)^m \quad \text{and} \quad \sigma_2 = \frac{1}{2} k^2 \left(1 - \frac{1}{k}\right)^m .$$

Assuming that the inequalities (10.45) and (10.46) are satisfied, our upper bound now becomes

$$c_4 N^{1 - \frac{1}{2b_1}} \left\{ M^{\sigma_1} N^{2b - \frac{1}{2} s(s+1) + \sigma_2} \prod_{r=1}^{s} \left(\frac{N^r}{q} + 1\right) (1 + qM^{r-k-1} \operatorname{Log} q) \right\}^{\frac{1}{4bb_1}} + M + 1$$

where c_4 is given by

$$c_4 = \{ c_2 c_3 (b_1^{k+s} 3^s 2^{ks+k+s}) \}^{\frac{1}{4bb_1}} .$$

By hypothesis, $q = N^\theta$ where $\sqrt{k} \operatorname{Log} k < \theta < k - \sqrt{k} \operatorname{Log} k$. Thus if we now fix s such that $s < \theta < k - s$, then $N^r q^{-1} + 1 \leqslant 2$ for $1 \leqslant r \leqslant s$ and

$$\prod_{r=1}^{s} (N^r q^{-1} + 1) \leqslant 2^s .$$

Furthermore, writing $M = N^{1-\eta}$ where $0 \leqslant \eta < 1$ will be chosen explicitly later,

$$1 + \frac{q \operatorname{Log} q}{M^{k+1-r}} \leqslant 1 + \frac{N^{k-s}(k-s) \operatorname{Log} N}{N^{(1-\eta)(k+1-r)}}$$
$$\leqslant 2(k-s)\, N^{\eta(k-r+1)} .$$

Hence it follows that the product

$$\prod_{r=1}^{s} (N^r q^{-1} + 1)(1 + qM^{r-k-1} \operatorname{Log} q)$$

is bounded above by

$$\left\{ 4(k-s)\, N^{\frac{1}{2}\eta(2k-s+1)} \right\}^s$$

The upper bound for $| \mathrm{S} |$ now becomes

$$| \mathrm{S} | \leqslant c_5\, N^{1-\tau} + N^{1-\eta} + 1$$

where

$$\tau = \frac{1}{4\, bb_1} \left\{ \frac{1}{2} s(s+1) - \sigma_2 - \sigma_1(1-\eta) - \frac{1}{2} \eta s(2\,k - s + 1) \right\}$$

$$c_5 = c_4 \left\{ 4^s (k-s)^s \right\}^{\frac{1}{4bb_1}} .$$

We now choose η so that $\tau = \eta$, namely

$$\eta = \frac{\frac{1}{2} s(s+1) - \sigma_1 - \sigma_2}{4\, bb_1 - \sigma_1 + \frac{1}{2} s(2\,k - s + 1)} .$$

Finally we choose the parameters b, b_1, s, l and m. For simplicity we take

$$b = A_1\, k^2 \operatorname{Log} k, \quad b_1 = A_2\, s^2, \quad s = [\sqrt{k} \operatorname{Log} k]$$

where A_1 and A_2 are sufficiently large absolute constants. Then we take l and m to be $B_1\, k \operatorname{Log} k$ and $B_2\, k \operatorname{Log} k$ for suitable B_1 and B_2 such that (10.45) and (10.46) are satisfied. With this choice of l and m we certainly have $\sigma_1 < \frac{1}{2}$ and $\sigma_2 < \frac{1}{2}$. Also, we obtain a lower bound for η of the form

$$\eta > \frac{\alpha}{k^2 \operatorname{Log} k} .$$

where α is an absolute constant.

Thus we now have an upper bound of the form

$$c_6\, N^{1-\alpha/k^2\, \mathrm{Log}\, k}$$

where $c_6 = c_5 + 2$. All that remains now is to show that c_6 is bounded above by an absolute constant. A trivial inspection shows that with our choice of b, b_1 and s that this is so. ∎

7.4 Our final estimation is the following theorem, which as we remarked earlier, will be needed in chapter 11. For this reason we shall be quite explicit in our choice of constants. This will mean that the details of the proof are rather tedious and messy; unfortunately there seems to be no way of avoiding this and we apologize in advance !

Theorem 10.3 *Let N, M be integers satisfying $1 < N$ and $1 \leqslant M \leqslant N$. Suppose that f is a real, infinitely differentiable function defined on $[N, 2N]$. If f has the property that for some fixed $\lambda \geqslant \frac{1}{2}$ we have*

$$\frac{N^{\lambda-r}}{4^r} \leqslant \frac{|f^{(r)}(x)|}{r!} \leqslant N^{\lambda-r},$$

for all $u \in (N, 2N)$ and all $r \geqslant 1$, then

$$|S| = \left| \sum_{n=N+1}^{N+M} e(f(n)) \right| \leqslant AN^{1-\alpha/\lambda^2}.$$

where A and α are absolute constants. A possible choice for A and α being $A = 3$, $\alpha = (49\,152)^{-1}$.

Proof. The theorem is a fairly straightforward consequence of theorem 10.9, but the details will be a little taxing on the reader's concentration. We recall the notation of § 5.4 : M and N are integers satisfying $1 \leqslant M < N$, the real numbers b and q satisfy $1 \leqslant b$ and $1 \leqslant q < \sqrt{N}$, m is an integer satisfying

$$N < m \leqslant \min(N + M, 2N - q^2),$$

$$\alpha_r = \frac{f^{(r)}(m)}{r!} \quad \text{for} \quad 1 \leqslant r \leqslant k,$$

$$\rho_k(xy) = \frac{f^{(k+1)}(m + \theta xy)}{(k+1)!}(xy)^{k+1} \quad \text{where} \quad 0 \leqslant x, y \leqslant q - 1;$$

$$\theta_r(\alpha) = \frac{1}{2bq^r} \sum_{|n|<bq^r} \min\{1, (2bq^r |\sin \pi n\alpha|)^{-1}\} \quad \text{for} \quad r = 1, 2, \ldots, k.$$

With our later application of the mean value theorem in mind we write $b = lk^2$. Furthermore, since $1 \leqslant q < \sqrt{N}$, it is convenient to take $q = N^{\frac{1}{2} - \eta}$. The new parameters l and η will be chosen later.

A division of the proof into two cases, namely when N is " large " and when N is " small " is necessary, but the precise division is fairly arbitrary. We have chosen a division which will simplify some of the numerical estimations which are going to occur later. Unfortunately we cannot give any more concrete motivation for our particular choice of cases, which is :

$$\text{(i) } N^{\eta} \geqslant 2^{2lk} \text{ and (ii) } N^{\eta} < 2^{2lk}.$$

Naturally case (i) is the more difficult. It will involve a term by term analysis of the upper bound for $|S|N^{-1}$ provided by theorem 10.9. For case (ii) we shall simply make use of the trivial bound $|S| \leqslant N$.

Case (i); $N^{\eta} \geqslant 2^{2lk}$. From theorem 10.9, an upper bound for $|S|N^{-1}$ is given by

$$\max_m \left\{ q^{k(k+1)-4b} \mathcal{N}(q, b, k)^2 (2b)^{2k} \prod_{r=1}^{k} \theta_r(\alpha_r) \right\}^{\frac{1}{4b^2}} + \frac{2q^2}{N} + \max_m \left\{ \frac{2\pi}{q^2} \sum_{x,y} |\rho_k(xy)| \right\}.$$

7.5 The sum $\sum |\rho_k(xy)|$. Our first step is to show that the hypothesis on f ensures that the sum $\sum |\rho_k(xy)|$ is " small ". For all $m \in [N, 2N - 1]$ we have, since $x, y \in [0, q)$,

$$\rho_k(xy) = \frac{|f^{(k+1)}(m + \theta xy)|}{(k+1)!} (xy)^{k+1}$$
$$\leqslant N^{\lambda-k-1} q^{2(k+1)}$$

Consequently it follows that

$$2\pi q^{-2} \sum_{x,y} |\rho_k(xy)| \leqslant 2\pi N^{\lambda} \left(\frac{q^2}{N}\right)^{k+1}.$$

Thus we now have, as upper bound for $|S|N^{-1}$,

$$\frac{|S|}{N} \leqslant \max_m \left\{ q^{k(k+1)-4b} \mathcal{N}(q, b, k)^2 (2b)^{2k} \prod_{r=1}^{k} \theta_r(\alpha_r) \right\}^{\frac{1}{4b^2}} + \frac{2q^2}{N} + 2\pi N^{\lambda}\left(\frac{q^2}{N}\right)^{k+1}. \quad (10.47)$$

7.6 The term $\mathcal{N}(q, b, k)^2$. We are going to use theorem 10.10 to infer an upper bound for $\mathcal{N}(q, b, k)$. Recall that $b = lk^2$ and set $s = \left[\left(l - \frac{1}{4}\right)k + \frac{3}{4}\right]$. The hypotheses of theorem 10.10 are then satisfied since

$$sk \leqslant \left(l - \frac{1}{4}\right)k^2 + \frac{3}{4}k \leqslant b - \frac{1}{4}k(k-3).$$

Hence we conclude that

$$\mathcal{N}(q, b, k) \leqslant (2^{4(b-k)}\, k^{2\beta})^s\, q^{2b - \frac{1}{2}k(k+1) + \sigma},$$

where

$$\sigma = \frac{1}{2}k^2\left(1 - \frac{1}{k}\right)^s \quad \text{et} \quad \beta = \min\left\{b, \frac{1}{4}k(k+5) - 1\right\}.$$

and (10.47) is bounded above by

$$\frac{|S|}{N} \leqslant \max_m \left\{(2b)^{2k}\,(2^{4(b-k)}\,k^{2\beta})^{2s}\, q^{2\sigma} \prod_{r=1}^{k} \theta_r(\alpha_r)\right\}^{\frac{1}{4b^2}} + \frac{2q^2}{N} + 2\pi N^{\lambda}\left(\frac{q^2}{N}\right)^{k+1}.$$

7.7 The product $\prod \theta_r(\alpha_r)$. The hypothesis that $N^{\eta} \geqslant 2^{2lk}$ together with the hypotheses on f imply the following inequalities :

$$\begin{aligned} 0 < 2\,bq^r\,|\,\alpha_r\,| &\leqslant 2\,lk^2\,N^{-r\eta}\,N^{\lambda - \frac{1}{2}r} \leqslant 2\,lk^2\,N^{-\eta}\,N^{\lambda - \frac{1}{2}r} \\ &\leqslant (2\,lk)^2\,N^{-\eta}\,N^{\lambda - \frac{1}{2}r} \leqslant 2^{2lk}\,N^{-\eta}\,N^{\lambda - \frac{1}{2}r} \leqslant N^{\lambda - \frac{1}{2}r}. \end{aligned}$$

We now set $h = [2\lambda]$ and choose the parameter k to be $k = 2h$. Then, for r satisfying $2\lambda \leqslant h + 1 \leqslant r \leqslant 2h = k$, we have

$$2\,bq^r\,|\,\alpha_r\,| \leqslant 1.$$

It follows that for all integers $n \in (-bq^r, bq^r)$ where $h + 1 \leqslant r \leqslant 2h$,

$$0 < |\,n\alpha_r\,| \leqslant \frac{1}{2}$$

Consequently for such n and r we have

$$|\sin(\pi n\alpha_r)| \geqslant 2\,|\,n\alpha_r\,|.$$

and from the definition of θ_r we see that for $1 + h \leqslant r \leqslant 2h = k$

$$(2\,bq^r)^2\,\theta_r(\alpha_r) \leqslant 2\,bq^r + 2\sum_{n=1}^{bq^r}(2\,n\,|\,\alpha_r\,|)^{-1}.$$

It will be sufficient for later applications to have a comparatively crude upper bound for the last expression, namely

$$| \alpha_r |^{-1} + | \alpha_r |^{-1} \sum_{n=1}^{bq^r} n^{\gamma-1} < | \alpha_r |^{-1} + | \alpha_r |^{-1} \int_2^{bq^r+1} (x-1)^{\gamma-1}\, dx + | \alpha_r |^{-1}$$

$$\leqslant \frac{2}{| \alpha_r |} \frac{(bq^r)^\gamma}{\gamma}, \tag{10.48}$$

where γ satisfies $0 < \gamma < 1$ and will be chosen explicitly later. Applying the hypotheses of the theorem and recalling that $q = N^{\frac{1}{2}-\eta}$, we see that

$$\begin{aligned} q^{2r} | \alpha_r | &\geqslant \left(N^{\frac{1}{2}-\eta} \right)^{2r} 2^{-2r} N^{\lambda-r} \\ &> \left(\frac{1}{2} N^{\frac{1}{2}-\eta} \right)^{2r} 2^{-2r} N^{\lambda-r} \\ &\geqslant 2^{-4r} N^{\lambda-2r\eta} . \end{aligned}$$

From (10.48) we now conclude that for $r \in [h+1, 2h]$,

$$\begin{aligned} \theta_r(\alpha_r) &\leqslant \frac{2}{4 b^2 q^{2r}} q^{2r} 2^{4r} N^{2r\eta-\lambda} \frac{(bq^r)^\gamma}{\gamma} \\ &< 2^{4r} N^{2r\gamma-\lambda} \frac{q^{r\gamma}}{\gamma} . \end{aligned}$$

Quite trivially, for all r,

$$0 < \theta_r(\alpha) \leqslant 1 ,$$

and so we obtain the following inequalities :

$$\prod_{r=1}^{k} \theta_r(\alpha_r) \leqslant \prod_{r=h+1}^{k} \theta_r(\alpha_r) \leqslant \prod_{r=h+1}^{k} \left(2^{4r} N^{2r\eta-\lambda} \frac{q^{r\gamma}}{\gamma} \right) .$$

The product $\prod_{r=h+1}^{k} q^{r\gamma} \gamma^{-1}$ is majorised by $(q^{k\gamma} \gamma^{-1})^k$ and so

$$\prod_{r=1}^{k} \theta_r(\alpha_r) \leqslant q^{\gamma k^2} \gamma^{-k} \prod_{r=h+1}^{k} 2^{4r} N^{2r\eta-\lambda} = q^{\gamma k^2} \gamma^{-k} 2^{2h(3h+1)} N^{\eta h(3h+1)-\lambda h} .$$

Thus the upper bound for $| S | N^{-1}$ is now

$$\left\{ (2b)^{2k} (2^{4(b-k)} k^{2\beta})^{2s} q^{2\sigma+\gamma k^2} \gamma^{-k} (4 N^\eta)^{h(3h+1)} N^{-\lambda h} \right\}^{\frac{1}{4b^2}} + \frac{2 q^2}{N}$$
$$+ 2 \pi N^\lambda \left(\frac{q^2}{N} \right)^{k+1} .$$

7.8 Our next step is to choose γ as a function of l and k and so obtain an upper bound for the exponent $2\,\sigma + \gamma k^2$. With our choice of $S = \left[\left(l - \frac{1}{4}\right) k + \frac{3}{4}\right]$, we have

$$4\,s \geqslant (4\,l - 1)\,k$$

and consequently it follows that

$$2\,\sigma = k^2\left(1 - \frac{1}{k}\right)^s \leqslant k^2\, e^{-\frac{s}{k}} \leqslant k^2\, e^{\frac{1}{4} - l}.$$

If we now choose γ as

$$\gamma = \frac{1}{21}\, e^{-l},$$

then we have

$$2\,\sigma + \gamma k^2 \leqslant k^2\left(e^{\frac{1}{4} - l} + \frac{1}{21}\, e^{-l}\right) \leqslant \frac{4}{3} k^2\, e^{-l}.$$

In order to simplify the typography we shall define the quantities δ and ε by

$$\delta = (3\, l^2\, k^2\, e^l)^{-1} \quad \text{and} \quad \varepsilon = (4\, l^2\, k^4)^{-1} = \frac{1}{4\, b^2}.$$

It follows that

$$\left\{ (2\,b)^{2k}\, (2^{4(b-k)}\, k^{2\beta})^{2s}\, q^{4b^2\delta} (21\, e^l)^k\, ((4\, N^{\eta})^{h(3h+1)}\, N^{-\lambda h})^{4b^2\varepsilon} \right\}^{1/4b^2} +$$

$$+ \frac{2\, q^2}{N} + 2\,\pi N^{\lambda}\left(\frac{q^2}{N}\right)^{k+1}.$$

7.9 Now we are going to obtain a fairly simple upper bound for the expression

$$D = (21\, e^l)^k\, (2\, b)^{2k}\, (2^{4(b-k)}\, k^{2\beta})^{2s}.$$

Our aim is to show that

$$D \leqslant (4\, k)^{\frac{2b^2}{k}}.$$

Recall that $b = lk^2$ and observe that

$$s \leqslant k.l, \quad \beta \leqslant b \quad \text{and} \quad 21\, e^l = 21\, e.e^{l-1} < 64.4^{l-1} = 16.4^l < 16.4^{kl}.$$

So we certainly have

$$D^{\frac{1}{2}k} \leqslant 2\,b.4.2^l(2^{4(b-k)}\,k^{2b})^l$$
$$= 4.2^l.2\,lk^2.2^{-4kl}(2^{4b}\,k^{2b})^l$$
$$\frac{8\,lk^2\,2^l}{2^{4kl}} \leqslant \frac{8\,lk^2}{2^{3kl}} \leqslant \frac{8\,lk^2}{(2\,kl)^3} < 1\,,$$

which implies that $D \leqslant (4\,k)^{4blk} = (4\,k)^{4b^2/k}$.

Consequently our upper bound is now

$$\frac{|\,\mathrm{S}\,|}{N} \leqslant (4\,k)^{\frac{1}{k}}\,q^\delta\,\{\,(4\,N^\eta)^{h(3h+1)}\,N^{-\lambda h}\,\}^\varepsilon + \frac{2\,q^2}{N} + 2\,\pi N^\lambda\left(\frac{q^2}{N}\right)^{k+1}.$$

7.10 The final stage of the argument consists in the explicit choice of the parameters η and l. We choose η to be

$$\eta = \frac{3\,\lambda}{4(3\,h+1)}$$

and make the following observations for use later. Since $h = [2\,\lambda] \leqslant 2\,\lambda$, $h > 2\,\lambda - 1$ and $\lambda > \frac{1}{2}$,

(i) $\frac{3}{32} < \eta < \frac{1}{8}$ car $h = [2\,\lambda]$ et $\lambda \geqslant \frac{1}{2}$;

(ii) $\lambda - (3\,h+1)\,\eta = \frac{1}{4}\lambda \geqslant \frac{1}{8}h$;

(iii) $\lambda - 2\,\eta(2\,h+1) = -\frac{2}{3}\eta$.

From our definition of η and (i), (ii) above it follows that

$$\left(\frac{1}{2} - \eta\right)\delta - \varepsilon h\,\{\,\lambda - (3\,h+1)\,\eta\,\} < \frac{13}{32}\delta - \frac{h^2\,\varepsilon}{8}$$
$$= \frac{13}{32}\,\frac{e^{-l}}{3\,l^2\,k^2} - \frac{1}{32\,l^2\,k^4}\cdot\frac{k^2}{4}$$
$$= \frac{52\,e^{-l} - 3}{384(lk)^2}\,.$$

We shall denote this last expression by $-\,\rho$. Recalling that $q = N^{\frac{1}{2}-\eta}$ we now have

$$\frac{|\,\mathrm{S}\,|}{N} < (4\,k)^{1/k}\,2^{2h(3h+1)\varepsilon}\,N^{-\rho} + 2\,N^{-2\eta} + 2\,\pi N^{-\frac{2}{3}\eta}\,. \tag{10.49}$$

It is easily seen that

$$\rho = \frac{1}{128(lk)^2} - \frac{13\,e^{-l}}{96(lk)^2} < \frac{1}{128(lk)^2} < \frac{\eta}{12(lk)^2} < \frac{\eta}{6}$$

thus quite trivially

$$\rho - 2\,\eta < -\frac{1}{2}\eta \text{ and } \rho - \frac{2}{3}\eta < -\frac{1}{2}\eta\,.$$

Now write (10.49) as

$$\frac{|\,S\,|}{N} < N^{-\rho}\left\{(4\,k)^{1/k}\,2^{2h(3h+1)\varepsilon} + 2\,N^{\rho-2\eta} + 2\,\pi N^{\rho-\frac{2}{3}\eta}\right\}$$
$$< N^{-\rho}\left\{(4\,k)^{1/k}\,2^{8h^2\varepsilon} + 2\,N^{-\eta/2} + 2\,\pi N^{-\eta/2}\right\},$$

and since $N^\eta \geqslant 2^{2lk}$ we conclude that

$$\frac{|\,S\,|}{N} \leqslant \left\{(4\,k)^{1/k}\,2^{8h^2\varepsilon} + (2 + 2\,\pi)\,2^{-lk}\right\}N^{-\rho} = BN^{-\rho}\,. \tag{10.50}$$

This completes the discussion of case (i).

7.11 Case (ii). We now suppose that $N^\eta < 2^{2lk}$ and we shall deduce that (10.50) is trivially satisfied. From the trivial estimate for $|\,S\,|$ we have

$$|\,S\,| \leqslant N = N^\rho . N^{1-\rho}\,,$$

All that remains is to prove that $N^\rho \leqslant B$. This is easy, for we have

$$N^\rho < N^{\frac{1}{128(lk)^2}} < (N^\eta)^{\frac{1}{12(lk)^2}} < 2^{\frac{1}{6lk}} < 2^{\frac{1}{6k}} < (4\,k)^{\frac{1}{k}}\,,$$

Thus

$$|\,S\,| < BN^{1-\rho}\,.$$

7.12 To complete the proof we shall choose numerical values for the parameters l, ε, ρ etc. We choose $l = 4$, then $e^l > 52$ and so $\rho > 0$. Also, since $k \leqslant 4\,\lambda$ we have

$$\rho^{\lambda^2} = \frac{\lambda^2}{3.2^7\,k^2}\cdot\frac{3 - 52\,e^{-4}}{4^2} > \frac{1}{3.2^7.4^2}\,\frac{3 - 1}{4^2} = \frac{1}{3.2^{14}} = (49\,152)^{-1}$$

that is, $\rho > (49\,152)^{-1}\,\lambda^{-2} = \alpha\lambda^{-2}$.

Now we are going to obtain an upper bound for $8h^2\varepsilon$. We recall that $\varepsilon = (4l^2k^4)^{-1}$ and $k = 2h \geqslant 2$, thus with $l = 4$ we certainly have

$$8h^2\varepsilon \leqslant 2^{-7} = (128)^{-1}.$$

Finally we are in a position to compute an upper bound for B. We have

$$\begin{aligned} B &= (4k)^{\frac{1}{k}}\, 2^{8h^2\varepsilon} + (2 + 2\pi)\, 2^{-lk} \\ &< (4.2)^{\frac{1}{2}}.2^{\frac{1}{128}} + 10.2^{-8} < 3. \end{aligned}$$

and our final conclusion is that

$$|S| < 3N^{1-\alpha/\lambda^2}$$

where $\alpha = (49\,152)^{-1}$. ■

Notes to chapter 10

VINOGRADOV'S MEAN VALUE THEOREM

We have seen the importance of estimating the number of integral solutions of the system of diophantine equations

$$x_1^r + \cdots + x_t^r = y_1^r + \cdots + y_t^r \qquad (1 \leqslant r \leqslant k)$$

which satisfy $1 \leqslant x_i, y_i \leqslant \mathrm{P}$. Let us denote the number of such solutions by $\mathcal{N}(\mathrm{P}, t, k)$. One can use the Hardy-Littlewood-Vinogradov analytic method to obtain an asymptotic formula for $\mathcal{N}(\mathrm{P}, t, k)$, provided that t is sufficiently large compared to k. (For a readable introduction to this technique see Davenport [3] and Birch [1].) For the problem in hand Hua [3] carried out the details and we shall now describe his conclusions.

Suppose that $k \geqslant 2$ and define t_0, as a function of k, by the following table :

k	2	3	4	5	6	7	8	9	10
t_0	3	8	23	62	156	380	889	2 034	4 595

and for $k \geqslant 10$

$$t_0(k) = [k^2(3 \operatorname{Log} k + \operatorname{Log}\operatorname{Log} k + 4)] \,.$$

If $t > t_0(k)$, then as P tends to infinity

$$\mathcal{N}(\mathrm{P}, t, k) = \mathfrak{J}(k, t)\, \mathfrak{S}(k, t)\, \mathrm{P}^{2t - \frac{1}{2}k(k+1)} + o\left(\mathrm{P}^{2t - \frac{1}{2}k(k+1)}\right)$$

where $\mathfrak{J}(k, t)$ and $\mathfrak{S}(k, t)$ are defined by

$$\mathfrak{J}(k, t) = \int_{-\infty}^{+\infty} \cdots \int_{-\infty}^{+\infty} \left| \int_0^1 e(\beta_1 x + \beta_2 x^2 + \cdots + \beta_k x^k)\, dx \right|^{2t} d\beta_1 \ldots d\beta_k$$

and

$$\mathfrak{S}(k, t) =$$

$$\sum_{q_1=1}^{\infty} \cdots \sum_{q_k=1}^{\infty} \sum_{\substack{h_1 < q_1 \\ (h_1\, q_1) = 1}} \cdots \sum_{\substack{h_k < q_k \\ (h_k\, q_k) = 1}} \left| \frac{1}{q_1 \cdots q_k} \sum_{x=1}^{q_1 \cdots q_k} e\left(\frac{h_1}{q_1} x + \cdots + \frac{h_k}{q_k} x^k\right) \right|^{2t} .$$

The integral $\mathfrak{J}(k, t)$ is convergent if $t > \frac{1}{4}k(k+2)$ and the series $\mathfrak{S}(k, t)$ is convergent if $t > \frac{1}{4}k(k+1)+1$ and divergent if $t = \frac{1}{4}k(k+1)+1$. Hua was forced to impose the more stringent condition $t > t_0(k)$ in order to be certain that the " error " term in the asymptotic formula for $\mathcal{N}(\mathrm{P}, t, k)$ is smaller than the principal term.

In the opposite direction Hua [1] had earlier investigated the size of the *least* value of t such that the system

$$x_1^r + \cdots + x_t^r = y_1^r + \cdots + y_t^r \qquad (1 \leqslant r \leqslant k)$$
$$x_1^{k+1} + \cdots + x_t^{k+1} \neq y_1^{k+1} + \cdots + y_t^{k+1}$$

has an integral solution. Denoting this least integer by $M(k)$, then Hua proved that

$$M(k) \geqslant (k+1)\left\{\left[\frac{\operatorname{Log}\frac{1}{2}(k+2)}{\operatorname{Log}\left(1+\frac{1}{k}\right)}\right]+1\right\}.$$

It is clear that this lower bound for $M(k)$ is asymptotic to $k^2 \operatorname{Log} k$ as k tends to infinity.

Exercises to chapter 10

10.1 (i) Let $\{ \lambda_r \}$ be a sequence of distinct non-zero complex numbers. Show that

$$2 \sum_{r=1}^{n} \lambda_r = \lambda_1 \left(1 + \frac{\lambda_1 + \lambda_2}{\lambda_1 - \lambda_2}\right) + \lambda_n \left(1 - \frac{\lambda_{n-1} + \lambda_n}{\lambda_{n-1} - \lambda_n}\right) + \sum_{r=2}^{n-1} \lambda_r \left(\frac{\lambda_r + \lambda_{r+1}}{\lambda_r - \lambda_{r-1}} - \frac{\lambda_{r-1} + \lambda_r}{\lambda_{r-1} - \lambda_r}\right).$$

If $\{ a_r \}$ is a sequence of real numbers, distinct modulo π, show that

$$2 \sum_{r=1}^{n} e^{2ia_r} = e^{2ia_1} \{ 1 + i \operatorname{cotg} (a_2 - a_1) \} + e^{2ia_n} \{ 1 - i \operatorname{cotg} (a_n - a_{n-1}) \}$$
$$+ \sum_{r=2}^{n-1} i\, e^{2ia_r} \{ \operatorname{cotg} (a_{r+1} - a_r) - \operatorname{cotg} (a_r - a_{r-1}) \} .$$

(ii) Suppose that the sequence $\{ a_n \}$ satisfies

$$0 < \theta \leqslant a_2 - a_1 \leqslant a_3 - a_2 \leqslant \cdots \leqslant a_n - a_{n-1} \leqslant \varphi < \pi .$$

prove that

$$\left| \sum_{r=1}^{n} e^{2ia_r} \right| \leqslant \frac{1}{2} \operatorname{cotg} \frac{\theta}{2} + \operatorname{tg} \frac{\varphi}{2} .$$

(For further results of this type see Mordell [1].)

10.2 Let f be a real valued function defined on the interval $[A, B]$. Suppose that for $A \leqslant x \leqslant B$.

(i) $| f'(x) | \leqslant \frac{1}{2}$ and (ii) $f''(x) \geqslant 0$.
Show that

$$S = \sum_{A \leqslant n \leqslant B} e(f(n)) - \int_A^B e(f(x))\, dx + O(1) .$$

where the implied " O " constant is absolute.
[Hint : Use the Abel summation formula and integration by parts to deduce that

$$S = \int_{[A]+1}^{[B]} e(f(x))\, dx + \int_{[A]+1}^{[B]} \left\{ x - [x] - \frac{1}{2} \right\} 2\, \pi i f'(x)\, e(f(x))\, dx + O(1) ,$$

then expand $x - [x] - \frac{1}{2}$ as a Fourier series and show that the second of the above integrals is bounded by an absolute constant.]

10.3 Let f be a real differentiable function in the interval (A, B). Suppose that f' is monotone and satisfies $f'(x) \geqslant m > 0$ in (A, B). Prove that

$$\int_A^B e(f(x))\,dx = O\left(\frac{1}{m}\right).$$

10.4 Suppose that the real valued function f is twice differentiable in the interval (A, B) and satisfies $f''(x) \geqslant r > 0$, prove that

$$\int_A^B e(f(x))\,dx = O\left(\frac{1}{\sqrt{r}}\right).$$

CHAPTER 11

The error term in the Prime Number Theorem (II)

We return to the problem of estimating the error term in the Prime Number Theorem. In chapter 4 we showed how an error term could be derived from the knowledge of a zero free region for $\zeta(s)$ and, in turn, how one could derive a zero free region from information about the size of $|\zeta(s)|$ in the critical strip. In this chapter we use theorem 10.3 to give estimates for $|\zeta(s)|$ which are better than those found in theorem 2.3 and so lead to a better error term.

The structure of the chapter is extremely simple. As theorem 11.1 we shall deduce a " good " upper bound for $|\zeta(s)|$ in the critical strip. This theorem is an improvement upon theorem 2.3. Theorem 11.1 is then used in conjunction with theorem 4.4 to deduce a new zero free region for $\zeta(s)$ to the left of the line $\mathcal{R}(s)=1$. Finally, we shall use this zero free region together with theorem 4.6 to deduce as theorem 11.3 a bound for the magnitude of the error term in the Prime Number Theorem.

Theorem 11.1 *There exist absolute constants K and a with the following property. If* $0 \leqslant \xi \leqslant \frac{1}{2}$, $\sigma \geqslant 1 - \xi$ *and* $t \geqslant 3$, *then*

$$|\zeta(\sigma + it)| \leqslant Kt^{a\xi^{3/2}}(\operatorname{Log} t)^{2/3} .$$

Possible values for K and a are $K = 2\,100$ *and* $a = 86$.

Proof. For $\sigma \geqslant 1 - \xi$, $t \geqslant 3$ and $X \geqslant 1$ formula (2.7) gives

$$\zeta(s) = \sum_{n \leqslant x} n^{-s} + \frac{\{x\}}{x^s} + \frac{x^{1-s}}{s-1} - s\int_x^\infty \frac{\{u\}}{u^{s+1}}\,du .$$

Let us now define the functions S and R by

$$S(x) = \sum_{3 \leqslant n \leqslant x} n^{-s} \qquad (x \geqslant 3),$$

$$R(x) = 1 + \frac{1}{2^s} + \frac{\{x\}}{x^s} + \frac{x^{1-s}}{s-1} - s \int_x^\infty \frac{\{u\}}{u^{s+1}} du$$

then we have

$$\zeta(s) = S(x) + R(x)$$

which implies

$$|\zeta(s)| \leqslant |S(x)| + |R(x)|.$$

Quite trivially we see that

$$|R(x)| \leqslant 1 + \frac{1}{2^\sigma} + \frac{x^{1/2}}{t} + \frac{1}{x^{1/2}} + \frac{\sigma + t}{\sigma x^{1/2}}. \tag{11.1}$$

If we now take $t \geqslant 4\pi$ and $X = (t/2\pi)^2 \geqslant 4$, then

$$|R(x)| \leqslant 1 + \frac{1}{\sqrt{2}} + \frac{1}{2\pi} + \frac{1}{2} + \frac{1}{2} + 4\pi < 16.$$

Thus, with $X = (t/2\pi)^2$ we have

$$|\zeta(s)| \leqslant \left| \sum_{2 \leqslant n \leqslant X} n^{-s} \right| + 16.$$

We next estimate the sum S(X) when $X = (t/2\pi)^2$. First define the integer $M \geqslant 2$ by

$$2^M < \left(\frac{t}{2\pi}\right)^2 \leqslant 2^{M+1}.$$

The sum S(X) can now be written as

$$S(x) = \sum_{m=1}^{M} Z_m$$

where the quantities Z_m are defined in the following way :

$$Z_m = \sum_{n=2^m+1}^{2^{m+1}} n^{-s} \qquad \text{for} \quad 1 \leqslant m < M$$

and

$$Z_M = \sum_{n=2^M+1}^{[x]} n^{-s}.$$

Consequently we certainly have

$$| \zeta(s) | \leqslant \sum_{m=1}^{M} | Z_m | + 16 .$$

Theorem 10.3 will be employed to estimate the sums $| Z_m |$. If Q, Q' are any integers which satisfy $1 \leqslant Q < Q' = \min (X, 2Q - 1)$, then by Abel lemma (theorem 1.8),

$$\left| \sum_{n=Q+1}^{Q'} n^{-\sigma - it} \right| \leqslant 2 Q^{-\sigma} \max_{Q \leqslant r \leqslant Q'} \left| \sum_{n=Q}^{r} e^{-it \operatorname{Log} n} \right| . \tag{11.2}$$

so we must estimate the sum

$$\max_{Q \leqslant r \leqslant Q'} \left| \sum_{n=Q}^{r} e^{-it \operatorname{Log} n} \right| .$$

Let λ be the real number defined by

$$\frac{t}{2\pi} = Q^{\lambda} .$$

Since $Q < X = (t/2\pi)^2$ we see that $\lambda > \frac{1}{2}$ and quite trivially we have

$$\lambda < (\operatorname{Log} t)/(\operatorname{Log} Q) \,..$$

If we define the real function f to be

$$f(u) = - \frac{t}{2\pi} \operatorname{Log} u .$$

then for $r = 1, 2, \ldots$ we have

$$\frac{| f^{(r)}(u) |}{r\,!} = \frac{t}{2\pi} \cdot \frac{1}{ru^r} = \frac{Q^{\lambda}}{r . u^r} ,$$

Because $1 \leqslant r < 2^r$ we have, for all $u \in [Q, 2Q]$, the following inequalities

$$\frac{Q^{\lambda - r}}{4^r} \leqslant \frac{Q^{\lambda}}{ru^r} = \frac{| f^{(r)}(u) |}{r\,!} \leqslant \frac{Q^{\lambda}}{u^r} \leqslant Q^{\lambda - r} .$$

Thus all the hypotheses of theorem 10.3 are satisfied and

$$\max_{Q \leqslant r \leqslant Q'} \left| \sum_{n=Q}^{r} n^{-it} \right| \leqslant A Q^{1 - \alpha/\lambda^2} \leqslant AQ \exp \left\{ - \alpha \frac{(\operatorname{Log} Q)^3}{(\operatorname{Log} t)^2} \right\} .$$

Hence upon taking $Q = 2^m$ for $1 \leqslant m < M$ we deduce from (11.2) that

$$| Z_m | \leqslant 2\, A\, 2^{m(1-\sigma)} \exp\left\{ - \alpha \frac{(m \operatorname{Log} 2)^3}{(\operatorname{Log} t)^2} \right\}.$$

Since $1 - \sigma \leqslant \xi$,

$$| S(x) | \leqslant 2\, A \sum_{m=1}^{M} \exp\left\{ m\xi \operatorname{Log} 2 - \alpha \frac{(m \operatorname{Log} 2)^3}{(\operatorname{Log} t)^2} \right\}.$$

In order to majorise the above sum we shall write $\alpha = \beta + \gamma$, where β and γ are positive numbers to be chosen explicitly later. We then trivially have

$$| S(x) | \leqslant 2\, A \max_{u \geqslant 0} \left\{ \exp\left(u\xi - \frac{\beta u^3}{(\operatorname{Log} t)^2} \right) \right\} \cdot \int_0^\infty \exp\left\{ - \frac{\gamma u^3}{(\operatorname{Log} t)^2} \right\} \frac{du}{\operatorname{Log} 2}.$$

The maximum in the first expression occurs when $3\, \beta u^2 = \xi (\operatorname{Log} t)^2$ and is equal to

$$\exp\left\{ \frac{2}{3\sqrt{3\,\beta}} \cdot \xi^{\frac{3}{2}} \operatorname{Log} t \right\}.$$

The integral can be written as

$$\frac{\gamma^{-\frac{1}{3}}}{\operatorname{Log} 2} (\operatorname{Log} t)^{\frac{2}{3}} \int_0^\infty \exp(-\, v^3)\, dv$$

and this last integral is trivially majorised by

$$\int_0^1 dv + \int_1^\infty v^{-3}\, dv = \frac{3}{2}.$$

Thus we now have

$$| S(x) | \leqslant 2\, A \frac{3\, \gamma^{-1/3}}{2 \operatorname{Log} 2} \cdot t^{a\xi^{3/2}} (\operatorname{Log} t)^{2/3}$$

where $a = \dfrac{2}{3}\sqrt{3\,\beta}$. Hence for $\sigma \geqslant 1 - \xi$ and $t \geqslant 4\,\pi$,

$$| \zeta(s) | < \left(\frac{3\, A\gamma^{-1/3}}{\operatorname{Log} 2} + 16 \right) t^{a\xi^{3/2}} (\operatorname{Log} t)^{2/3}.$$

By a choice of $\beta > 0$ we can take a to be any real number greater than $2/3\sqrt{3\,\alpha}$, where α is the constant in theorem 10.3. For example we can take $\alpha = (49\,152)^{-1}$ which means that $a > 85$. Thus we can take $a = 86$ which means that $\beta = (49\,923)^{-1}$

and $\gamma > (4.10^6)^{-1}$. Thus we can take $K = 2\,100$. That is, we have shown that for $\sigma \geqslant 1 - \xi$ and $t \geqslant 4\pi$ we have

$$|\zeta(s)| < 2\,100\, t^{86\xi^{3/2}} (\text{Log } t)^{2/3}$$

The above inequality also holds for $3 \leqslant t \leqslant 4\pi$, because in this case we have, from (11.1) with $X = 4$,

$$\begin{aligned} |\zeta(s)| &\leqslant |S(4)| + |R(4)| \\ &< \left(\frac{1}{3^{1/2}} + \frac{1}{4^{1/2}}\right) + \left(1 + \frac{1}{2^{1/2}} + \frac{2}{3} + \frac{1}{2} + \frac{\frac{1}{2} + 4\pi}{\frac{1}{2}\cdot 2}\right) \\ &< 2\,100\,. \end{aligned}$$

This completes the proof of the theorem. ■

We will now use the upper bound given by the above theorem together with theorem 4.4 to deduce the existence of a zero free region for $\zeta(s)$ which is " larger " than the zero free region given in theorem 4.5. Before we state the next theorem we shall introduce some notation.

Let λ be the function defined for all $t > e$ by

$$\lambda(t) = (\text{Log } |t|)^{2/3} (\text{Log Log } |t|)^{1/3}\,.$$

We shall denote by $R_c(l)$ the region of the complex plane defined by

$$R_c(l) = \{ s = \sigma + it \mid \sigma \geqslant 1 - c\lambda^{-1}(t)\,, |t| \geqslant l \}\,.$$

Theorem 11.2 *There exist absolute constants c and t_0 such that in the region $R_c(t_0)$ we have*

(a) $\zeta(s) \neq 0$,

(b) $Z(s) = O(\lambda(|t|))$ *uniformly in σ as $|t|$ tends to infinity.*

A possible value for c is $(8\,757)^{-1}$. The " O " constant and t_0 can be determined effectively.

Proof. The theorem is a simple consequence of theorems 4.4 and 11.1. We choose ξ in theorem 11.1 to be

$$\xi = \left(\frac{b}{a}\,\frac{\text{Log Log } t}{\text{Log } t}\right)^{2/3}$$

where a is the constant in theorem 11.1 and b is a constant to be chosen explicitly later. Because

$$t^{a\xi^{3/2}} = \exp(a\xi^{3/2} \log t) = \exp(b \log\log t) = (\log t)^b\,.$$

we conclude that for $\sigma \geqslant 1 - \xi$ and $|t| \geqslant 3$,

$$|\zeta(s)| \leqslant K(\operatorname{Log} t)^{b+\frac{2}{3}} .$$

In particular, for any $T \geqslant 3$ an upper bound for $|\zeta(s)|$ in the two regions

$$\{ s = \sigma + it : \sigma \geqslant 1 - \xi, |t - T| \leqslant 3\xi \}$$

and

$$\{ s = \sigma + it : \sigma \geqslant 1 - \xi, |t - 2T| \leqslant 3\xi \}$$

is given by

$$M = K_1(\operatorname{Log} T)^{b+\frac{2}{3}}$$

where K_1 is a suitable absolute constant. With the notation of theorem 4.4 we define η by

$$\frac{1}{\eta} = \frac{A}{\xi} \operatorname{Log} \frac{KM}{\xi} .$$

Thus

$$\frac{1}{\eta} = A\left(\frac{a}{b}\right)^{\frac{2}{3}} \left(\frac{\operatorname{Log} T}{\operatorname{Log}\operatorname{Log} T}\right)^{\frac{2}{3}} \operatorname{Log} \left\{ \frac{(\operatorname{Log} T)^{b+\frac{4}{3}}}{(\operatorname{Log}\operatorname{Log} T)^{\frac{2}{3}}} KK_1\left(\frac{a}{b}\right)^{\frac{2}{3}} \right\}$$

If we now choose t_0 to be so large as to satisfy $(\operatorname{Log}\operatorname{Log} t_0)^{\frac{2}{3}} > KK_1(a/b)^{\frac{2}{3}}$, then for $T > t_0$,

$$\frac{1}{\eta} < A\left(\frac{a}{b}\right)^{\frac{2}{3}} \left(b + \frac{4}{3}\right) (\operatorname{Log} T)^{\frac{2}{3}} (\operatorname{Log}\operatorname{Log} T)^{\frac{1}{3}} . \tag{11.3}$$

We choose b to minimise (11.3), namely $b = \frac{8}{3}$, and obtain

$$\frac{1}{\eta} < A(3a)^{2/3} (\operatorname{Log} T)^{2/3} (\operatorname{Log}\operatorname{Log} T)^{1/3}$$

A trivial computation shows that

$$c^{-1} = A(3a)^{2/3} = 216(258)^{2/3} \leqslant 8\,757 .$$

The theorem now follows from theorem 4.4. ■

Our final theorem gives an estimate for the error term in the Prime Number Theorem. The result is a consequence of theorems 11.2 and 4.6.

Theorem 11.3 *There exists an absolute constant α with the property that as x tends to infinity,*

$$\psi(x) = x + \mathrm{O}\left\{ x \exp\left(- \frac{\alpha (\mathrm{Log}\, x)^{\frac{3}{5}}}{(\mathrm{Log}\,\mathrm{Log}\, x)^{\frac{1}{5}}} \right) \right\}.$$

The implied " O " constant can be determined effectively and a possible choice for α is $(105)^{-1}$. The corresponding asymptotic formulae hold for the functions $M(x)$, $\Pi(x)$, $\theta(x)$ and $\pi(x)$.

Proof. We shall use the notation of chapter 4, § 4. Let c and t_0 be the constants of theorem 11.2 and recall that

$$\lambda(t) = (\mathrm{Log}\,|\,t\,|)^{2/3}\,(\mathrm{Log}\,\mathrm{Log}\,|\,t\,|)^{1/3}\,.$$

Choose $t_1 \geqslant t_0$ so that $\zeta(s)$ has no zeros in the rectangle

$$\{\, s = \sigma + it \mid 1 - c\lambda^{-1}(t_1) \leqslant \sigma \leqslant 1,\ |\,t\,| \leqslant t_1 \}$$

and $c \left| \dfrac{\lambda'(t_1)}{\lambda^2(t_1)} \right| < 1$ (to satisfy the condition on $\eta'(t)$). Then define the function η by :

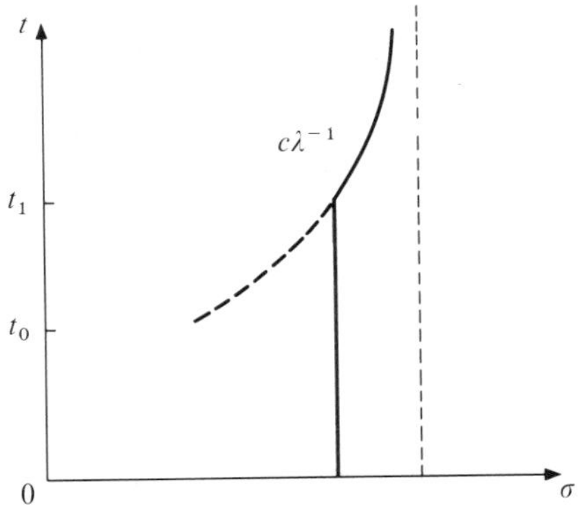

Fig. 15

$$\eta(t) = \begin{cases} c\lambda^{-1}(t) & \text{if} \quad |\,t\,| \geqslant t_1 \\ c\lambda^{-1}(t_1) & \text{if} \quad |\,t\,| \leqslant t_1\,. \end{cases}$$

With this choice of η, theorem 11.2 tells us that all the conditions of theorem 4.6 are satisfied. Thus it remains to compute a " nice " lower bound for

$$H_1(x) = \inf_{t \geqslant 1} \{\, \eta(t)\,\mathrm{Log}\, x + \mathrm{Log}\, t \,\}\,.$$

Let δ be a fixed real number satisfying $0 < \delta \leqslant 1$ and which will be chosen explicitly later. We shall suppose that x is so large that

$$(\mathrm{Log}\, x)^{\delta} > (\mathrm{Log}\, t_1)\,.$$

The following three inequalities are trivial to see :

(i) For t satisfying $\mathrm{Log}\, t_1 \leqslant \mathrm{Log}\, t \leqslant (\mathrm{Log}\, x)^{\delta}$ we have

$$H_1(x) \geqslant \frac{c\,\mathrm{Log}\, x}{\delta^{1/3} (\mathrm{Log}\, t)^{2/3}\,(\mathrm{Log}\,\mathrm{Log}\, x)^{1/3}} + \mathrm{Log}\, t\,. \tag{11.4}$$

(ii) For t satisfying $\operatorname{Log} t \geqslant (\operatorname{Log} x)^{\delta}$ we have

$$H_1(x) \geqslant (\operatorname{Log} x)^{\delta} . \tag{11.5}$$

(iii) For t satisfying $\operatorname{Log} t \leqslant \operatorname{Log} t_1$ we have

$$H_1(x) \geqslant \eta(t_1) \operatorname{Log} x . \tag{11.6}$$

By the arithmetic-geometric mean inequality we have for all $u > 0$, $v > 0$

$$\frac{u+v}{5} \geqslant \left(\frac{u}{3}\right)^{3/5} \left(\frac{v}{2}\right)^{2/5}$$

Consequently the right-hand side of (11.4) is at least equal to

$$5\left(\frac{c}{3\,\delta^{1/3}}\right)^{3/5} \left\{\frac{\operatorname{Log} x}{(\operatorname{Log} t)^{2/3} (\operatorname{Log} \operatorname{Log} x)^{1/3}}\right\}^{3/5} \left\{\frac{\operatorname{Log} t}{2}\right\}^{2/5}$$

$$= 5\left(\frac{c^3}{324\,\delta}\right)^{1/5} \frac{(\operatorname{Log} x)^{3/5}}{(\operatorname{Log} \operatorname{Log} x)^{1/5}} . \tag{11.7}$$

If we now choose $\delta = \frac{3}{5}$, then the right-hand sides of (11.5) and (11.6) are both greater than (11.7) for all sufficiently large values of x. Hence we certainly have, for all sufficiently large x,

$$H_1(x) \geqslant 5\left(\frac{c^3}{324\,\delta}\right)^{1/5} \frac{(\operatorname{Log} x)^{3/5}}{(\operatorname{Log} \operatorname{Log} x)^{1/5}} .$$

Upon taking $c = (8\,757)^{-1}$ we have the conclusion of the theorem by appealing to theorem 4.6 and taking α to be any number less than $9.53.10^{-3}$. In particular we can take $\alpha = (105)^{-1}$. This completes the proof of the theorem. ■

Bibliography

[1] D. Allison, *On obtaining zero free regions for the zeta function from estimates of $M(x)$*, Proc. Cambridge Philos. Soc. 67 (1970), 333-337. **MR.** 40 4211.

[1] D. R. Anderson, J. J. Stiffler, *Lower bounds for the maximum moduli of certain classes of trigonometric sums*, Duke Math. J. 30 (1963), 171-176. **MR.** 26 2404.

[1] J. V. Armitage, *Zeta functions with a zero at $s = 1/2$*, Inventiones Math. 15 (1972), 199-205. **MR.**

[1] P. Bachman, J. Hadamard, E. Maillet, *Propositions transcendantes de la théorie des nombres*, Encyclopédie des sciences mathématiques I 17 (1910), 215-387.

[1] C. L. Baker, F. J. Gruenberger, *The first six million prime numbers*, The RAND Corporation, Santa Monica, published by the Microcard Foundation, Madison Wisconsin (1959).

[1] P. T. Bateman, *Proof of a conjecture of Grosswald*, Duke Math. J. 25 (1957), 67-72. **MR.** 19 1040.

[2] P. T. Bateman, *Lower bounds for $\sum h(m)/m$ for arithmetic functions similar to real residue characters*, J. Math. Anal. Appl. 15 (1966), 2-20. **MR.** 33 7313.

[3] P. T. Bateman, H. Diamond, *Asymptotic distribution of Beurling's generalised prime numbers*, studies in Number Theory (ed. by W. J. Levecque), vol. 6 M.A.A. (1969). **MR.** 39 4105.

[4] P. T. Bateman, E. Grosswald, *On Epstein's zeta function*, Acta Arith. 9 (1964), 365-373. **MR.** 31 3392.

[1] A. Beurling, *Analyse de la loi asymptotique de la distribution des nombres premiers généralisés*, Acta Math. 68 (1937), 255-291.

[1] B. J. Birch, *Forms in many variables*, Proc. Royal Soc. (London) ser. A 265 (1962), 245-263. **MR.** 12 345.

[1] H. Bohr, H. Cramér, *Die neuere entwicklung der analytischen Zahlentheorie*, Enzyklopädie der mat. Wiss. II 8 (1922), 722-849.

[1] E. Bombieri, *Sulle formule di A. Selberg generalizzate per classi di funzioni aritmetiche e le applicazioni al problema del resto nel " Primzahlsaty "*, Riv. Mat. Univ. Parma 3 (1962), 393-440. **MR.** 27 4804.

[2] E. Bombieri, H. Davenport, *Small differences between prime numbers*. Proc. Roy. Soc. ser. A 293 (1966), 1-18. **MR.** 33 7314.

[1] R. Brauer, *On the zeta function of algebraic number fields*, Amer. J. Math. 69 (1947), 243-250. **MR.** 8 567.

[2] R. Brauer, *On Artin's L series with general group characters*, Ann. of Math. 48 (1947), 502-514. **MR.** 8 503.

[1] R. Breusch, *Zur Verallgemeinerung des Bertrandschen Postulats, dass zwischen x und 2 x stets Primzahlen liegen*, Math. Zeit. 34 (1932), 505-526.

[2] R. Breusch, *Another proof of the prime number theorem*, Duke Math. J. 21 (1954), 49-53. **MR.** 16 904.

[3] R. Breusch, *An elementary proof of the Prime Number Theorem with remainder term*, Pacific J. Math. 10 (1960), 487-497. **MR.** 22 4685.

[1] E. D. Cashwell, C. J. Everett, *The ring of number-theoretic functions*, Pacific J. Math. 9 (1959), 975-985. **MR.** 21 7226.

[2] J. W. S. Cassels, *Footnote to a note of Davenport and Heilbronn*, J. London Math. Soc. 36 (1961), 177-184, **MR.** 26 3881.

[4] J. W. S. Cassels, A. Fröhlich, *Algebraic Number Theory*, Academic press (1967).

[1] A. M. Cohen, M. J. E. Mayhew, *On the difference* $\pi(x) - \mathrm{Li}\,(x)$, Proc. London Math. Soc. 18 (1968), 691-713. **MR.** 38 2102.

[1] I. G. van der Corput, *Sur l'hypothèse de Goldbach pour presque tous les nombres pairs*, Acta Arith. 2 (1937), 266-290.

[2] J. G. van der Corput, *Sur le reste dans la démonstration élémentaire du théorème des nombres premiers*, Colloque sur la Théorie des Nombres, Bruxelles (1955), 163-182. **MR.** 18 112.

[1] H. Cramér, *Ein Mittelwertsatz in der Primzahltheorie*, Math. Zeit. 12 (1922), 147-53.

[2] H. Cramér, *On the order of magnitude of the difference between consecutive prime numbers*, Acta Arith. 2 (1936), 396-403.

[1] N. G. Čudakov, *On certain trigonometric sums containing prime numbers*, Doklady Akad. Nauk. SSSR (N.S.) 58 (1947), 1291-1294. **MR.** 9 333.

[2] N. G. Čudakov, Yu. V. Linnik, *On a class of completely multiplicative functions*, Doklady Akad. Nauk SSSR (N.S.) 74 (1950), 193-196. **MR.** 12 393.

[3] N. G. Čudakov, K. A. Rodorski, *On generalised characters*, Doklady Akad. Nauk SSSR (N.S.) 73 (1950), 1137-1139. **MR.** 12 393.

[1] H. Davenport, *On the series for* $L(1)$, J. London Math. Soc. 24 (1949), 224-233. **MR.** 11 162.

[2] H. Davenport, *Eine Bermerkung über Dirichelt's L Funktionens*, Nach. Akad, Wiss. Göttingen, Math. Phys. Kl. 11 (1966), 203-212. **MR.** 36 120.

[3] H. Davenport, *Analytic methods for diophantine equations and diophantine inequalities*, Campus Publishers (Ann. Arbor) 1962.

[4] H. Davenport, *Multiplicative number theory*, Markham, Chicago (1967).

[5] H. Davenport, H. Heilbronn, *On the zeros of certain Dirichlet series I, II*, J. London Math. Soc. 11 (1936), 181-185 and 307-312.

[1] H. Delange, *Théorèmes Taubériens*, Publications du Sém. de Math. d'Orsay 1961/2. Sec. de Math. d'Orsay (1963).

[1] J.-M. Deshouillers, *Thèse*, Univ. Paris VI (1972), Centre de Doc. CNRS A.O. 7 364.

[2] J.-M. Deshouillers, *Sur le problème de Goldbach*, Acta Arith. (to appear).

[1] H. Diamond, *The Prime Number Theorem for Beurling's generalised numbers*, J. No. Theory 1 (1969), 200-207. **MR.** 39 4106.

[2] H. Diamond, *A set of generalised integers showing Beurling's theorem to be sharp*, Illinois J. Math. 14 (1970), 29-34. **MR.** 40 5556.

[3] H. Diamond, J. Steinig, *An elementary proof of the Prime Number Theorem with a remainder term*, Invent. Math. 11 (1970), 199-258. **MR.** 43 6169.

[1] P. G. L. Dirichlet, *Beweis des Satzes, dass jede unbegrenzte arithmetische Progression, deren erstes Glied und Differenz granze Zahlen ohne gemeinschaftlichen Faktor sind, unendlich viele Primzahlen enthält*, Abh. Akad. Berlin (1837), 45-71.

[1] P. D. T. A. Elliot, *On the size of* $L(1, \chi)$, J. reine angew. Math. 2.36 (1969), 26-36. **MR.** 40 2619.

[1] P. Erdös, *On a new method in elementary number theory which leads to an elementary proof of the prime number theorem*, Proc. Nat. Acad. Sci. (Washington) 35 (1949), 374-384. **MR.** 10 595.

[2] P. Erdös, A. E. Ingham, *Arithmetical Tauberian theorems*, Acta Arith. 9 (1964), 341-356. **MR.** 31 1228.

[1] T. Estermann, *On Goldbach's problem : Proof that almost all even positive integers are sums of two primes*, Proc. London Math. Soc. 44 (1938), 307-314.

[1] W. Feit, *Characters of finite groups*, W. A. Benjamin INC New York (1968).

[1] P. Finsler, *Über die Primzahlen zwischen n und 2 n*, Festschrift zum 60 Geburtstag von Prof. Dr. A. Speiser. Zürich, Orell-Fussli (1954), 188-122. **MR.** 7 243.

[1] P. X. Gallagher, *Primes in progressions to prime power modulus*, Invent. Math. 16 (1972), 191-201. **MR.** 12 345.

[1] C. F. Gauss, *Werke*, Bd 2, 444-447, 520.

[1] E. Grosswald, *Oscillation theorems*, Conf. on the theory of arithmetic functions, Springer Lecture Notes 251 (1972). **MR.** 12 345.

[1] J. Hadamard, *Sur la distribution des zéros de la fonction* $\zeta(s)$ *et ses conséquences arithmétiques*, Bull. de la Soc. Math. de France 24 (1896), 199-220.

[1] H. Halberstam, *Progress towards hypothesis H*, Journées arithmétiques Françaises, mai (1971), Université de Provence, Marseille.

[2] H. Halberstam, K. F. Roth, *Sequences*, Clarendon Press, Oxford (1966).

[3] H. Halberstam, H.-E. Richert, *Sieve methods*, Markham press (to appear).

[1] R. S. Hall, *Theorems about Beurling's generalised primes and the associated zeta function*, Ph. D. thesis, Univ. of Illinois (1967).

[1] D. Hanson, *On the products of primes*, Canad. Math. Bull. 15 (1972), 33-37. **MR.** 12 345.

[1] G. H. Hardy, *Ramanujan*, Cambridge Univ. Press (1940).

[2] G. H. Hardy, *Divergent series*, Oxford (1949).

[3] G. H. Hardy, J. E. Littlewood, *Contributions to the theory of the Riemann zeta function and the distribution of primes*, Acta Math. 41 (1918), 119-196.

[4] G. H. Hardy, E. M. Wright, *An introduction to the theory of numbers*, 4th edition (Oxford) Clarendon Press (1964).

[1] C. B. Haselgrove, *A disproof of a conjecture of Polya*, Mathematika 5 (1958), 141-145. **MR.** 21 2291.

[2] C. B. Haselgrove and J. C. P. Miller, *Tables of the Rieman zeta function*, Royal Society Tables, vol. 6 (Cambridge) (1960).

[1] H. Heilbronn, *On real characters*, Acta Arith. 2 (1937), 212-213.

[1] E. Hecke, *Mathematische Werke*, Göttingen (1959).

[1] H. Helson, *Convergence of Dirichlet Series* (to appear).

[1] D. Hensley, I. Richards, *Primes in intervals* (to appear).

[1] E. M. Horadam, *An unsolved problem in number theory*, Bull. Soc. Math. Grèce (N.S.) 9 (1968), 143-147. **MR.** 40 4198.

[1] L. K. Hua, *On Tarry's problem*, Quart. J. Math. (Oxford) 9 (1928), 315-320.

[2] L. K. Hua, *An improvement of Vinogradov's mean value theorem and several applications*, Quart. J. Math. (Oxford) 20 (1949), 48-61. **MR.** 10 597.

[3] L. K. Hua, *On the number of solutions of Tarry's problem*, Acta Sci. Sinica 1 (1952), 1-76. **MR.** 16 337.

[4] L. K. Hua, *Die Abschätzung von Exponentialsummen und ihre Anwendung in der Zahlentheorie*, Enzyklopädie der mathematischen Wissenschaften I. 2 Heft 13, Teil 1, Teubner, Leipzig (1959).

[5] L. K. Hua, *Additive prime number theory*, Amer. Math. Soc. translations (1965).

[1] M. Huxley, *On the difference between consecutive primes*, Invent. Math. 15 (1972), 164-170. **MR.** 45 1856.

[1] A. E. Ingham, *The distribution of prime numbers*, Cambridge Tracts in Mathematics and Mathematical Physics, N° 30 (1932), reprinted by Hafner, New York (1964).

[2] A. E. Ingham, *On two conjectures in the theory of numbers*, Amer. J. Math. 64 (1942), 313-319. **MR.** 12 345.

[3] A. E. Ingham, *Some Tauberian theorems connected with the prime number theorem*, G. London Math. Soc. 20 (1945), 171-180. **MR.** 8 147.

[1] P. T. Joshi, *The size of $L(1, \chi)$ for real non-principal residue characters χ with prime modulus*, G. No. Theory 2 (1970), 58-73. **MR.** 12 345.

[1] M. Kalecki, *A simple elementary proof of $M(x) = o(x)$*, Acta Arith. 13 (1967), 1-7. **MR.** 36 2574.

[2] M. Kalecki, *A short elementary proof of the prime number theorem*, Prace Mat. 13 (1969), 51-55. **MR.** 41 6772.

[2] J. Karamata, *Sur les inversions asymptotiques des produits de convolutions*, Bull. Akad. Serbe Sci. (N.S.) 20 Cl. Sci. Math.-Nat. Sci. Mat. 3 (1957), 11-32. **MR.** 21 1961.

[1] J. Karamata, *Sur les inversions asymptotiques des produits de convolutions*, Glas Srpske Akad. Nauka 228 Od. Prirod-Mat. Nauka (N.S.) 13 (1957), 23-57. **MR.** 21 1960.

[1] J. Katai, *On oscillations of number theoretic functions*, Acta Arith. 13 (1967), 107-122. **MR.** 36 2577.

[2] J. Katai, *On investigations in the comparative prime number theory*, Acta Math. Acad. Sci. Hung. 18 (1967), 379-391. **MR.** 36 1405.

[3] J. Katai, *On sets characterising number theoretic functions*, Acta Arith. 13 (1968), 315-320. **MR.** 40 123.

[4] J. Katai, *On the distribution of arithmetical functions on the set of primes plus one*, Composito Math. 19 (1972), 278-289.

[1] R. Kershner, *Determination of a van der Corput-Landau absolute constant*, Amer. J. Math. 57 (1935), 840-846.

[1] Š. Knapowski, *On sign changes of the difference* $\pi(x) - \text{Li}(x)$, Acta Arith. 7 (1961), 107-119. **MR.** 24 A3142.

[2] S. Knapowski, *On new " explicit formulas " in prime number theory I*, Acta Arith. 5 (1958), 1-14. **MR.** 12 345.

[3] S. Knapowski, *On new " explicit formulas " in prime number theory II*, Acta Arith. 6 (1960), 23-35. **MR.** 12 345.

[4] S. Knapowski, *On Siegel's theorem*, Acta Arith. 14 (1968), 417-424. **MR.** 37 4034.

[5] S. Knapowski, P. Turán, *Comparative prime number theory I, ..., VIII*, Acta Math. Hung. 13 (1962), 299-364 ; 14 (1963), 31-78 and 241-268. **MR.** 26 3682/3.

[6] S. Knapowski, P. Turán, *Further developments in comparative prime number theory I*, Acta Arith. 9 (1964), 23-41. **MR.** 29 75.

[7] S. Knapowski, P. Turán, *Further developments in comparative prime number theory II*, Acta Arith. 10 (1964), 293-313. **MR.** 30 4739.

[8] S. Knapowski, P. Turán, *Further developments in comparative prime number theory III, IV, V*, Acta Arith. 11 (1965), 115-127, 147-161, 193-202. **MR.** 31 4773, 32 99.

[9] S. Knapowski, P. Turán, *Further developments in comparative prime number theory VI*, Acta Arith. 12 (1966), 85-96. **MR.** 12 345.

[10] S. Knapowski, P. Turán, *Further developments in comparative prime number theory VII*, Acta Arith. 21 (1972), 193-201. **MR.** 12 345.

[1] G. A. Kolesnik, *Distribution of prime numbers of the form* $[n^c]$, Math. Zametki (1967), 117-128. **MR.** 12 345.

[1] G. Kreisel, *Mathematical significance of consistency proofs*, J. Symbolic Logic 23 (1958), 155-182. **MR.** 12 345.

[1] M. P. Kuhn, *Eine Verbesserung des Restglieds beim elementaren Beweis des Primzahlsatzes*, Math. Scand. 3 (1955), 75-99. **MR.** 17 587.

[1] E. Landau, *Handbuch der Lehre von der Verteilung der Primzahlen*, Teubner, Leipzig (1909) (Reprinted by Chelsea Publishing Co., New York (1953)).

[2] E. Landau, *Über einige ältere Vermutungen und Behauptungen in der Primzahltheorie*, Math. Zeit. 1 (1918), 1-24 and 213-219.

[3] E. Landau, *Vorlesungen über Zahlentheorie*, Hirgel, Leipzig (1927).

[1] J. Leech, *Note on the distribution of prime numbers*, J. London Math. Soc. 32 (1957), 56-58. **MR.** 18 642.

[2] R. S. Lehman, *On the difference* $\pi(x) - \text{Li}(x)$, Acta Arith. 11 (1966), 397-410. **MR.** 34 2546.

[1] D. H. Lehmer, *On the exact number of primes less than a given limit*, Illinois J. Math. 3 (1959), 381-388. **MR.** 21 5613.

[2] D. H. Lehmer, S. Selberg, *A sum involving the function of Möbius*, Acta Arith. 6 (1960), 111-114. **MR.** 22 6762.

[1] D. N. Lehmer, *List of prime numbers from* 1 *to* 10,006,721, Carnegie Inst. of Washington No. 18 (Washington) 1914.

[1] B. Levin, *On the distribution of primes in an arithmetic progression*, Izv. Akad. Nauk. Uz SSR Fiz.-Mat. Nauk (1961), 15-28. **MR.** 25 2048.

[1] N. Levinson, *A motivated account of an elementary proof of the prime number theorem*, Amer. Math. Monthly 76 (1969), 225-245. **MR.** 39 2712.

[1] J. H. v. Lint, H-E. Richert, *Über die summe* $\sum \mu^2(n)/\varphi(n)$, Indag. Math. 26 (1964), 582-587. **MR.** 30 123.

[1] J. E. Littlewood, *The quickest proof of the prime number theorem*, Acta Arith. 18 (1969), 83-86. **MR.** 19 123.

[1] M. E. Low, *Real zeros of the Dedekind zeta function of an imaginary quadratic field*, Acta Arith. 14 (1968), 117-140. **MR.** 11 234.

[1] F. Mertens, *Über eine zahlentheoretische Funktion*, Sitzungsber. Akad. Wien 106 Abt. 2a (1897), 761-830.

[2] F. Mertens, *Über die numerische Prüfung dieser Vermutung*, Sitzungsber. Akad. Wien 110 Abt. 2a (1901), 1053-1102.

[1] R. J. Miech, *A number theoretic constant*, Acta Arith. 15 (1969), 119-137. **MR.** 39 142.

[1] T. Mitsui, *On the prime ideal theorem*, J. Math. Soc. Japan 20 (1968), 233-247. **MR.** 36 6362.

[1] H. L. Montgomery, *Topics in multiplicative number theory*, Lecture Notes in Math. No. 227, Springer, Berlin (1971).

[2] H. L. Montgomery, *Corrélations dans l'ensemble des zéros de la fonction zêta*, Pub. Math. Univ. de Bordeaux 1 (1972), 1-9.

[1] L. J. Mordell, *On the Kusmin-Landau inequality for trigonometric sums*, Acta Arith. 4 (1958), 3-9. **MR.** 20 123.

[2] L. J. Mordell, *On a cyclotomic resolvent*, Arch. Math. 13 (1962), 486-487. **MR.** 26 123.

[1] G. Neubauer, *Eine empirische Untersuchung zur Mertenschen Funktionen*, Numer. Math. 5 (1963), 1-13. **MR.** 27 5721.

[1] A. Ȯgg, *Modular forms and Dirichlet series*, Benjamin (1969).

[1] R. E. A. C. Paley, *A theorem on characters*, J. London Math. Soc. 7 (1932), 28-32.

[1] H. R. Pitt, *Tauberian Theorems*, Tata Institute monographs No. 2 (1958).

[1] G. Polya, *Verschiedene Bermerkungen zur Zahlentheorie*, Jahrber. deutsch. Math.-Ver. 28 (1919), 31-40.

[1] J. Popken, *A measure for the differential-transcendence of the zeta function of Riemann*, Number theory and Analysis (Papers in honor of E. Landau) 245-255. Plenum N.Y. (1969). **MR.** 41 3415.

[1] K. Prachar, *Primzahlverteilung*, Die Grundlehren der Math. Wiss. kd. XCL, Springer, Berlin (1957).

[1] I. Pyateckiĭ-Šapiro, *On the distribution of prime numbers of the form* $[f(n)]$, Mat. Sbornik 33 (1953), 559-566. **MR.** 15 854.

[1] G. Rhin, *Sur la répartition modulo* 1 *de la suite* $f(p)$, Acta Arith. 23 (1973), 15-123. **MR.** 46 123.

[1] G. Ricci, *Recherches sur l'allure de la suite* $(p_{n+1} - p_n)(\text{Log}\, p_n)^{-1}$, Colloque sur la théorie des nombres, Bruxelles (1955), 93-106.

[1] B. Riemann, *Ueber die Anzahl der Primzahlen unter eine gegebenen Grösse*, Monat. Preuss. Akad. Wiss. (Berlin) (1859), 671-680 (Oeuvres math. (1898), 165-176.)

[1] J. B. Rosser, *Real roots of Dirichlet L series*, Bull. Amer. Math. Soc. 55 (1949), 906-913. **MR.** 11 332.

[2] J. B. Rosser, *Real roots of real Dirichlet L series*, J. Research Nat. Bur. Standards 45 (1950), 505-514. **MR.** 12 804.

[3] J. B. Rosser, L. Schoenfeld, *Approximate formulas for some functions of prime numbers*, Illinois J. Math. 6 (1962), 64-94. **MR.** 25 1139.

[4] J. B. Rosser, L. Schoenfeld, J. M. Yohe, *Rigorous computation and the zeros of the Riemann zeta function*, Information Processing 68 (Proc. IFIP Congress, Edinburgh 1968), vol. 1, Mathematics-Software, 70-76. **MR.** 41 2892.

[1] L. A. Rubel, *An abelian theorem for number theoretic sums*, Acta Arith. 6 (1960), 175-177 and 523. **MR.** 22 10947.

[1] B. Saffari, *Sur la fausseté de la conjecture de Mertens*, C.R. Acad. Sci. (Paris) A-B 271 (1970), A 1097-A 1101. **MR.** 43 6167.

[1] A. Schinzel, *Sur une conséquence de l'hypothèse de Goldbach*, Bulgar. Akad. Nauk. Izv. Math. Inst. 4 (1959), 35-38. **MR.** 25 123.

[1] L. Schoenfeld, *An improved estimate for the summatory function of the Möbius function*, Acta Arith. 15 (1968), 221-233. **MR.** 39 2716.

[1] I. Schur, *Einige Sätze über Primzahlen mit anwendungen auf Irreduzibilitätsfragen, Sitzungsbericht* d. Preuss. Akad. de Wiss. (Berlin) (1929), 125-136 and 370-391.

[1] S. L. Segal, *A tauberian theorem for Dirichlet convolutions*, Illinois J. Math. 13 (1969), 316-320. **MR.** 39 143.

[2] S. L. Segal, *A general Tauberian theorem of Landau-Ingham type*, Math. Zeit. 111 (1969), 159-167. **MR.** 40 2624.

[2] A. Selberg, *An elementary proof of the prime number theorem*, Ann. of Math. 50 (1949), 305-313. **MR.** 10 595.

[3] A. Selberg, *An elementary proof of the prime number theorem for arithmetic progressions*, Canadian J. Math. 2 (1950), 66-78. **MR.** 11 420.

[1] S. Selberg, *Über die Verteilung einiger Klassen quadratfreier Zahlen, die aus einer gegeben Anzahl von Primfaktoren zusammengesetzt sind*, skr. Norske Vid. Akad. Oslo 1 No. 5 (1942). **MR.** 6 57.

[1] D. Shanko, *Quadratic residues and the distribution of primes*, Math. Tables Aids to Comp. 13 (1959), 272-284. **MR.** 21 7186.

[1] H. N. Shapiro, *On the number of primes less than or equal to x*, Proc. Amer. Math. Soc., 1 (1950), 346-348. **MR.** 12 80.

[2] H. N. Shapiro, *Tauberian theorems and elementary prime number theory*, Comm. Pure Appl. Math. 12 (1959), 579-610. **MR.** 20 123.

[3] H. N. Shapiro, *On the convolution ring of arithmetic functions*, Comm. Pure Appl. Math. 25 (1972), 287-336. **MR.** 46 123.

[1] C. L. Siegel, *Gesammelte Abhandlungen*, Springer-Verlag (1966).

[1] W. Sierpiński, *Sur l'équivalence de deux hypothèses concernant les nombres premiers*, Bulgar. Akad. Nauk Izv. Mat. Inst. 4 (1959), 3-6. **MR.** 25 5027.

[1] S. Skewes, *On differences* $\pi(x) - \mathrm{Li}(x)$ *II*, Proc. London Math. Soc. 5 (1955), 48-70. **MR.** 16 676.

[1] R. Spira, *Calculation of Dirichlet L functions*, Math. Comp. 23 (1969), 489-498. **MR.** 37 123.

[1] H. Stark, *On the zeros of Epstein's zeta function*, Mathematika 14 (1967), 47-55. **MR.** 35 6633.

[2] H. Stark, *L functions and character sums for quadratic forms I, II*, Acta Arith. 14 (1968), 35-50 and 15 (1969), 307-317. **MR.** 40 123.

[3] H. Stark, *A problem in comparative prime number theory*, Acta Arith. 18 (1971), 311-320. **MR.** 45 1234.

[1] W. Stás, *Über eine abschätzung des Restglieds im Primzahlsatz*, Acta Arith. 5 (1959), 427-434. **MR.** 22 1555.

[2] W. Stás, *Über die Umkehrung eines satzes von Ingham*, Acta Arith. 6 (1961), 435-446. **MR.** 26 3679.

[1] J. Steinig, *The sign changes of certain arithmetic functions*, Comm. Math. Helv. 44 (1969), 385-400. **MR.** 41 1658.

[1] J. J. Sylvester, *On Tchebycheff's theory of the totality of prime numbers comprised within given limits*, Amer. J. Math. 4 (1881), 230-247.

[2] J. J. Sylvester, *On arithmetic series*, Messenger of Maths. 21 (1892), 1-19 and 87-120.

[1] T. Tatuzawa, *On a theorem of Siegel*, Jap. J. Math. 21 (1951), 163-178. **MR.** 14 452.

[2] T. Tatuzawa, *On the number of primes in an arithmetic progression*, Jap. J. Math. 21 (1951), 313-333, **MR.** 15 202.

[1] P. L. Tchebycheff, *Sur la fonction qui détermine la totalité des nombres premiers inférieurs à une limite donnée*, J. de Math. 17 (1852), 341-365.

[2] P. L. Tchebycheff, *Mémoire sur les nombres premiers*, J. de Math. 17 (1852), 366-390.

[3] P. L. Tchebycheff, *Lettre de M. le professeur Tchébychev à M. Fuss, sur un nouveau théorème relatif aux nombres premiers dans les formes* $4n + 1$ *et* $4n + 3$, Bull. cl. phys.-math. Acad. St. Petersburg 11 (1853), 208.

[1] R. Teuffel, *Beweise für zwei Sätze von H. F. Scherk über Primzahlen*, Jber. Deutsch. Math. Ver. 58 (1955), 43-44. **MR.** 17 587.

[1] E. C. Titchmarsh, *The Fourier Integral*, Clarendon press, Oxford (1948).

[2] E. C. Titchmarsh, *Theory of functions*, Clarendon press, Oxford (1949).

[3] E. C. Titchmarsh, *The theory of the Riemann zeta function*, Clarendon press, Oxford (1951)

[1] P. Turán, *On some approximative Dirichlet polynomials in the theory of the zeta function of Riemann*, Danske Vid. Selske. Mat. Fys. Medd. 24 N° 17 (1948). **MR.** 5 123.

[2] P. Turán, *On the remainder term in the prime number formula I*, Acta Math. Acad. Sci. Hungar. 23 (1951), 48-63. **MR.** 10 123.

[3] P. Turán, *On the remainder term in the prime number formula II*, Acta Math. Acad. Sci. Hungar. 23 (1951), 155-165. **MR.** 10 123.

[4] P. Turán, *Real zeros of Dirichlet's L functions*, Acta Arith. 5 (1959), 309-314. **MR.** 23 A127.

[5] P. Turán, *On the twin prime problem III*, Acta Arith. 14 (1968), 399-407. **MR.** 15 123.

[1] G. Valiron, *Théorie des Fonctions*, Masson.

[1] C.-J. de la Vallée Poussin, *Recherches analytiques sur la théorie des nombres ; Première partie : La fonction* $\zeta(s)$ *de Riemann et les nombres premiers en général*, Annales de la Soc. Sci. de Bruxelles 20 (1896), 183-256.

[2] C.-J. de la Vallée Poussin, *Sur la fonction* $\zeta(s)$ *de Riemann et le nombre des nombres premiers inférieurs à une limite donnée*, Mem. de l'Acad. roy. de Belgique LIX (1898).

[1] R. C. Vaughan, *On Goldbach's problem*, Acta Arith. 22 (1972), 21-48. **MR.** 45 123.

[1] I. M. Vinogradov, *The method of trigonometric sums in the theory of numbers*, Interscience (1954).

[1] A. Walfisz, *Über die Wirksamkeit einiger Abschätzungen trigonometrischer Summen*, Acta Arith. 4 (1958), 108-180. **MR.** 15 123.

[2] A. Walfisz, *Weylsche Exponentialsummen in der neueren Zahlentheorie*, VEB Deutscher Verlag der Wissenschaften, Berlin (1963).

[1] A. Weil, *Jacobi sums as " Grossencharaktere "*, Trans. Amer. Math. Soc. 73 (1952), 487-495. **MR.** 11 123.

[1] H. Weyl, *Über die Gleichverteilung von Zahlen mod. Eins*, Math. Ann. 77 (1916), 313-52.

[1] N. Wiener, *Tauberian theorems*, Annals of Math. 2 (1932), 1-100.

[1] K. Wiertelak, *On the application of Turán's method to the theory of Dirichlet L-functions*, Acta Arith. 19 (1971), 249-259. **MR.** 45 123.

[1] E. Wirsing, *Elementare Beweise des Primzahlsatzes mit Restgleid II*, J. Reine Angew. Math. 214/15 (1964), 1-18. **MR.** 29 3457.

Author index

Index of terms